AF816940

Sustainable Communities through Digital Transformation

This book brings together cutting-edge exploratory research findings to show how a vision for sustainable communities can be enabled by digital transformation. It attempts to apply existing knowledge about digital transformation and sustainable communities and compare, interpret, diagnose, and evaluate a variety of digital solutions to ascertain their suitability for the delivery of a more sustainable built environment. Chapters examine a breadth of issues, including how digital transformations could:

- Provide digital/physical working/living environments that anticipate emerging lifestyles
- Blend digital engagements into the physical engagements within the built environment
- Support business and social activity in physical and online venues
- Use advanced information and community-oriented technologies for efficient management of urban services
- Promote sustainability
- Express narratives that celebrate the experience of place and community
- Leverage transformation of educational systems
- Foster linkages between universities, and between universities and businesses
- Facilitate working relationships among small and large companies
- Foster new processes and arrangements for innovation in the built environment

By comparing the key principles of digital transformation with those of sustainable communities, the contributors seek to justify or discount the applicability of digital change for achieving more sustainable communities. The research presented in this book is essential reading for architecture, urban planning, quantity surveying, building surveying, real estate, and construction management professionals and academics.

Yusuf Arayici is Professor of Built Environment at Northumbria University.

Niraj Thurairajah is Associate Professor of Built Environment at Northumbria University, where he leads the Sustainable Construction Futures research group.

Bimal Kumar is currently Professor of Innovative Digital Construction Technologies at the University of Strathclyde in Glasgow, UK.

Spon Research

Publishes a stream of advanced books for built environment researchers and professionals from one of the world's leading publishers. The ISSN for the Spon Research programme is ISSN 1940-7653 and the ISSN for the Spon Research E-book programme is ISSN 1940-8005

Life-Cycle Greenhouse Gas Emissions of Commercial Buildings
An Analysis for Green-Building Implementation Using A Green Star Rating System
Cuong N. N. Tran, Vivian W. Y. Tam and Khoa N. Le

Data-driven BIM for Energy Efficient Building Design
Saeed Banihashemi, Hamed Golizadeh and Farzad Pour Rahimian

Successful Development of Green Building Projects
Tayyab Ahmad

BIM and Construction Health and Safety
Uncovering, Adoption and Implementation
Hamed Golizadeh, Saeed Banihashemi, Carol Hon and Robin Drogemuller

Digitalisation in Construction
Recent Trends and Advances
Edited by Chansik Park, Farzad Pour Rahimian, Nashwan Dawood, Akeem Pedro, Dongmin Lee, Rahat Hussain and Mehrtash Soltani

Managing Projects with Smart Technologies
Developing Technological Competency for Project Managers
Bon-Gang Hwang, Jasmine Ngo and Hanjing Zhu

Innovations, Disruptions and Future Trends in the Global Construction Industry
Edited by Temitope Omotayo, Temitope Egbelakin, Olabode Ogunmakinde, and Adebayo Sojobi

Power in Megaproject Decision-making
A Governmentality Approach
Jessica Pooi Sun Siva and Thayaparan Gajendran

Occupational Diseases in the Construction Industry
Imriyas Kamardeen and Abid Hasan

Sustainable Communities through Digital Transformation

Edited by Yusuf Arayici, Niraj Thurairajah, and Bimal Kumar

Routledge
Taylor & Francis Group
LONDON AND NEW YORK

First published 2025
by Routledge
4 Park Square, Milton Park, Abingdon, Oxon OX14 4RN

and by Routledge
605 Third Avenue, New York, NY 10158

Routledge is an imprint of the Taylor & Francis Group, an informa business

© 2025 selection and editorial matter, Yusuf Arayici, Niraj Thurairajah, and Bimal Kumar; individual chapters, the contributors

The right of Yusuf Arayici, Niraj Thurairajah, and Bimal Kumar to be identified as the authors of the editorial material, and of the authors for their individual chapters, has been asserted in accordance with sections 77 and 78 of the Copyright, Designs and Patents Act 1988.

All rights reserved. No part of this book may be reprinted or reproduced or utilised in any form or by any electronic, mechanical, or other means, now known or hereafter invented, including photocopying and recording, or in any information storage or retrieval system, without permission in writing from the publishers.

Trademark notice: Product or corporate names may be trademarks or registered trademarks, and are used only for identification and explanation without intent to infringe.

British Library Cataloguing-in-Publication Data
A catalogue record for this book is available from the British Library

ISBN: 9781032449036 (hbk)
ISBN: 9781032462042 (pbk)
ISBN: 9781003380559 (ebk)

DOI: 10.1201/9781003380559

Typeset in Times New Roman
by Newgen Publishing UK

Contents

Editor Biographies

Yusuf Arayici is Professor of Built Environment at Northumbria University. Since 2000, his research has ranged from building information modelling and process improvement to energy management and sustainability. He has led substantial research groups over a prolonged period through continuous cycles of funded research projects with a sum of £10M, graduated many PhD students, and published more than 100 papers and five books about Building Information Modelling (BIM). Currently, he is researching on deep learning and cognitive digital twins for healthcare FM and AI-supported heritage BIM.

Niraj Thurairajah is Associate Professor of Built Environment at Northumbria University, where he leads the Sustainable Construction Futures research group. His research delves into the complexities of the built environment, focusing on human and process-related aspects. This includes how organisations and individuals navigate technological innovation, apply new economic theories within built environment contexts, and implement decarbonisation strategies to enhance climate resilience. He has authored over 100 peer-reviewed publications covering areas such as the digital built environment, construction economics and procurement, decarbonisation, and education. He played leading roles in research and industry engagement projects such as Green Skills, ORBEE, ACBEE, ExcelDL, ISLAND I & II, CKE, and H2B innovation grants. His work integrates academic rigour with practical industry insights, significantly contributing to the field of sustainable construction.

Bimal Kumar is currently Professor of Innovative Digital Construction Technologies at the University of Strathclyde in Glasgow, UK. He leads the Digital Construction, Procurement and Law research group. He is a Fellow of the Institution of Civil Engineers and a Chartered Engineer. He is also a Fellow of the British Computer Society, a Member of the American Society of Civil Engineers and the Institution of Engineers of India, and a Senior Fellow of the UK Higher Education Academy. He is well known for his work on the application of artificial intelligence/knowledge-based systems to structural design – this has resulted in three books and numerous publications. He is also the founding Secretary/Treasurer of the European Group for Structural Engineering Applications of Artificial Intelligence (EG-SEA-AI), now called European Group of Intelligent Computing in Engineering (EG-ICE). His main research interests are the design and development of systems and advanced IT tools to facilitate the integration of the design and construction processes, intelligent buildings, modelling building regulation and codes representation and processing, BIM, and knowledge management in engineering.

Contributors

Samuel Adeniyi Adekunle, University of Johannesburg, South Africa

Onaopepo Adeniyi, Northumbria University, UK

Douglas Aghimien, De Montfort University, UK

Clinton Aigbavboa, University of Johannesburg, South Africa

Opeoluwa Akinradewo, University of Johannesburg, South Africa

John Aliu, Federal University of Technology Minna, Nigeria

Seraphim Alvanides, Northumbria University, UK

Merve Anaç, Hasan Kalyoncu University, Türkiye

Yusuf Arayici, Northumbria University, UK

John Oluwaseun Atolagbe, Federal University of Technology Akure, Nigeria

Nadia Bertolino, Università di Pavia, Italy

Gökhan Demirdöğen, Yıldız Technical University, Türkiye

Dat Tien Doan, Auckland University of Technology, New Zealand

Andrew Ebekozien, University of Johannesburg, South Africa

Obuks Ejohwomu, University of Manchester, UK

Ali Ghaffarianhoseini, Auckland University of Technology, New Zealand

Aleksander Gil, Northumbria University, UK

Paul Greenhalgh, Northumbria University, UK

Bülent Haznedar, Gaziantep University, Türkiye

Oluwadamilare Olamide Ilesanmi, Federal University of Technology Akure, Nigeria

Bahriye Ilhan Jones, Northumbria University, UK

Zeynep Işık, Yıldız Technical University, Türkiye

Jiayi Jin, Northumbria University, UK

Georgios Kapogiannis, University of Aberdeen, Qatar

Mark Kirk, Northumbria University, UK

Bimal Kumar, University of Strathclyde, UK

David Laurance, Northumbria University, UK

Yan Ma, Auckland University of Technology, New Zealand

David Manase, Oryx Universal College/Liverpool John Moore's University Qatar

Leo Moreton, Northumbria University, UK

Kevin Muldoon-Smith, Northumbria University, UK

Ayodeji Emmanuel Oke, Federal University of Technology Akure, Nigeria

Oluwapelumi Ruth Oluwaleye, Federal University of Technology Akure, Nigeria

Jonathan Pearson, Northumbria University, UK

Steve Philips, University of Westminster, UK

Pavithra Rathnasiri, Northumbria University, UK

Shashwat Shashwat, Northumbria University, UK

Valentah Siamuzwe-Manase, Bond University, Australia

Jane Loxley, Northumbria University, UK

Niraj Thurairajah, Northumbria University, UK

Ayse Ozbil Torun, Northumbria University, UK

Pedro Miguel Valadas Vicente, Northumbria University, UK

Jennifer Wiles, Northumbria University, UK

Mingyu Zhu, Newcastle University, UK

Kishor Zingre, Northumbria University, UK

1 Introduction

Yusuf Arayici

Sustainable communities are human- or society-centred interpretations of sustainable development comprised of four key pillars: environmental sustainability, physical sustainability, economic sustainability, and social sustainability. Of those pillars, social sustainability is the most challenging due to its qualitative and intangible aspects, such as quality of life and well-being, that must be ensured for the long term, encapsulating the needs of today as well as for future generations. A sustainable community's vision has eight key characteristics. These are:

- **Active, inclusive, and safe** – *fair, tolerant, and cohesive with a strong local culture and other shared community activities*
- **Well run** – *with effective and inclusive participation, representation, and leadership*
- **Environmentally sensitive** – *providing places for people to live that are considerate of the environment*
- **Well designed and built** – *featuring quality built and natural environment*
- **Well connected** – *with good transport services and communication linking people to jobs, schools, health care, and other services*
- **Thriving** – *with a flourishing and diverse local economy*
- **Well served** – *with public, private, community, and voluntary services that are appropriate to people's needs and accessible to all*
- **Fair for everyone** – *including those in other communities, now and in the future*

However, meeting the key dimensions of sustainable communities can be challenging due to their complex and interrelated features, such as socio-economic, environmental, and physical requirements, which all correlate with each other.

The complex relations between social, environmental, and economic issues overlap with and emerge in the concept of sustainability, which aims at improving and maintaining the quality of human life within the carrying capacity of supporting eco-systems. Such eco-systems are driven by inter-organisational and cross-industry collaboration, open systems for research and development (R&D), and workers who have the aptitudes for and skills required by digital transformation.

DOI: 10.1201/9781003380559-1

Digital transformation is seen as an enabler for creating sustainable communities in which information and knowledge are the primary assets and resources. Societal and communal activities depend on a huge volume of cross-disciplinary knowledge and information. The rise in digital transformation is driven by Information and Communication Technologies (ICT) and the increasing pace of technological change. Since the marginal cost of manipulating, storing, and transmitting information and knowledge is virtually zero, the application of knowledge to all aspects of communities and societies is greatly facilitated and the digital intensity of economic, societal, and environmental activities greatly increased. Increasing the volume of digitally intensive activities is essential if societies are to be sustainable.

Digital transformation can increase productivity growth and prosperity through innovation. It is important to highlight that digital transformation is not just about science and technology: it is about highly skilled individuals adding value to all sectors as well as about sustainable communities. Therefore, there is a mutual relationship between digital transformation and sustainable communities. It is assumed that increasing digital transformation to meet society's aspirations will enable for a better environment and communities. The aim of digital transformation is to create sustained, competitive, and dynamic societies, communities, and economies, capable of sustainable growth with more and better jobs and greater social inclusion.

Digital transformation is also an enabler for societies to transform from the industrial age into the knowledge age. In this regard, knowledge-intensive communities are a vision towards which to make progress and ultimately to achieve transformation towards sustainable communities and societies.

Hence, this book aims to discuss the digital transformation context in relation to sustainable communities to highlight the level of match and conformance between the two to justify the suitability of the former for achieving the vision of the latter. The social facet of sustainability focuses on the interaction between "real estate development" and the surrounding local community, an interaction that is mainly linked to the topic of sustainable communities.

In the context of a sustainable built environment, digital transformation is the process of using digital technologies to create or modify existing practices in the built environment, and to meet people's needs on a sustainable community basis, including environmental sustainability, social sustainability, quality of life, and well-being. Its goal is to convert manual and labour-based professional practices into information-based digital practices with various levels of automation and data processing, converting raw data to information and knowledge that addresses sustainable communities. Thus, it is deemed as enabling societies and nations to transform, make progress, and ultimately achieve a transformation towards knowledge societies that have the features and characteristics of sustainable communities.

Significant match and conformance can be seen between the key themes of sustainable communities and digital transformation. Therefore, based on this conformance, it can be hypothetically said that sustainable communities are possible through true strategy and the implementation of digital transformation in the built environment. This hypothesis will be further advocated through the evidential studies and practices of digital design and construction contained in this book. Accordingly, using this analysis of digital transformation in the built environment, the hypothesis can then be validated, and relevant recommendations are made for digital transformation in the built environment, paving the way for sustainable communities.

This book binds together cutting-edge exploratory research findings regarding how a vision for sustainable communities can be enabled by digital transformation. It attempts to apply existing knowledge about digital transformation and sustainable communities and to conduct comparative interpretation and diagnosis with regard to digital transformation's suitability for sustainable communities. There is a need to develop a strong vision with strong evidence about digital transformation and its contribution to social sustainability, including quality of life and well-being.

A sustainable future is an existential challenge, and the existing digital arsenal must transform to address this challenge. Therefore, this book assesses digital transformation enabling sustainable communities as a way forward for the built environment. For that purpose, it compares key principles of digital transformation with the features of sustainable communities to justify or discount the applicability of the former for the latter.

This book explores business perspectives in Chapter 2 in terms of the implementation of digital technologies for improved and sustainable business development. Taking into consideration various facets of digital transformation in the built environment business, Chapter 2 explains the business processes analysis and strategic management context. Chapter 3 provides a closer look at digital transformation with the example of building information modelling (BIM) implementation in the New Zealand construction industry context and discusses the pros and cons of such digital transformation with insights and reflections from BIM experts. It is asserted that motivation for achieving sustainability goals and efficiencies with BIM adoption is limited by people, process, and policy-related challenges.

Chapter 4 investigates the adoption of robotics technologies in the context of the Nigerian construction industry by looking at local and international companies practicing in the construction industry in Nigeria. This chapter elaborates the benefits of the implementation of robotic technologies: decreased reliance on direct labour, increased productivity, increased occupational safety, improved quality, and increased control over the process's outcome. It also interrogates the stakeholder's level of awareness about robotic technologies.

Chapter 5 provides a critical example of sustainable facilities management for healthcare facilities. Facilities management (FM) is one of the most critical parts of healthcare facilities since the effectiveness of FM directly affects patients' lives. However, lean healthcare FM necessitates the right information flow and monitoring of healthcare key performance indicators (KPIs). Therefore, this chapter provides a healthcare framework consisting of big data analytics and BIM to enable FM information queries and monitoring of FM KPIs by considering lean management philosophy.

Chapter 6 focuses on BIM as an enabler for circularity and proposes a holistic circular building life cycle approach that includes the main steps and principles for design, construction, use, and end-of-life phases along with the indicators. For successful circularity adoption, key stakeholders should embrace integrated and inclusive methods and systems of design, construction, use, and deconstruction. Since most design decisions affect the rest of the building life cycle phases, the design phase requires particular attention, consideration, planning, and programming at both product and process levels. Based on the accurate design data and information flow for the succeeding processes, the construction, operation and maintenance, and end-of-life phases – applying the phase-specific principle and generating the necessary data for the other phases and relevant stakeholders – can complete the loop.

Chapter 7 introduces a new perspective to construction industry challenges such as productivity, labour shortage, and unattractiveness to prospective workers due to the accident rate. This chapter explores human augmentation technology and its adoption in the construction industry to improve productivity. Considering the low productivity, loss of workforce, and unattractive image of the construction industry due to its high accident rate, the study presents a viable solution through the adoption of human augmentation to increase industry productivity and provide a lifeline in the event of an accident on site. Observing the identified technologies critically, it is evident that they emanate from the medical field. Thus, they are being incorporated into the construction industry to solve its problems. The construction industry also needs to be ready if another global pandemic strikes in the future.

On the other hand, Chapter 8 provides an architectural perspective to the social-spatial implications of digital transformation with various ICT tools and technologies such as the Internet of Things (IoTs) and digital twins. Rapid spread of new devices such as cameras, microphones, sensors, computing, and network access in non-computer products is paving the way for the transition from the realm of artifacts into the realm of architecture. It becomes the responsibility of an architect (or spatial designer) to know how to deal with this apparatus and the associated challenges. However, adaptation of these artifacts or systems is predominantly focused on optimising efficiency, cost, and sustainability, thereby ignoring socio-spatial matters in terms of interactions between environments, buildings, and their inhabitants. Hence, this chapter explores how digital transformation and its artifacts have transcended the overall realm of architecture, enabling a social-spatial understanding of dynamic activities in spaces and places with digital–spatial systems.

Chapter 9 elaborates on the correlation between digital technologies and eco-communities. It draws an architectural reflection on the role played by digital technologies in fostering collective imagination within a particular category of sustainable communities as eco-communities. The digital tools and communication channels can pave the way for architectural counternarrative, where real and virtual spheres co-exist. It concentrates on three case studies across England: Old Hall, Hockerton, and Landmatters eco-communities. A critical assessment of potential and challenges imposed by digital technologies and architectural design is carried out in the discussion, leading to context-specific reflections on digital technologies and the local eco-communities.

On the other hand, Chapter 10 focuses on the contribution of spatial modelling for active school travel – a key component of walkable and sustainable environments. It elaborates on a data-led methodology that explores the effect of street network design and school social characteristics on walking to primary schools in urban and rural locations in England. School environments are spatially modelled using a wide range of street connectivity measures, including those from space syntax. This chapter argues that street network layout design should be considered in planning and urban design, particularly in areas surrounding primary schools or when locating new schools, to address some of our most important health (e.g., childhood obesity), environmental (e.g., carbon emissions), and economic (e.g., public health costs) concerns.

Chapter 11 provides environmental analysis on the rise of urban air temperature in relation to urban growth. Use of facade materials to enhance outdoor conditions is limited. This chapter gives a computational model to analyse the effectiveness of building materials in urban microclimates. It uses a numerical simulation tool to model and simulate an urban environment with building materials, roads, and pathways. Full-force weather conditions (air temperature, relative humidity, wind velocity, and direction) are taken into consideration in the model and simulation.

Chapter 12 provides a critical diagnosis about enhancing flood resilience in the built environment. Data-driven decision-making is introduced as a transformational approach for planning, investment analysis, risk assessment, and general decision-making from the perspective of the built environment. To achieve a sustainable future, this chapter emphasises the use of the built assets' digital data to make them resilient against climate-induced catastrophes.

Chapter 13 evaluates the performance of a green building called "Heritage Place," located in Ikoyi, Lagos, Nigeria. The objectives include assessing the building's overall performance, gauging user satisfaction, establishing a correlation between user satisfaction and building performance over time, identifying any defects exhibited by the building, and providing suggestions for its improvement. This study utilises a survey research approach and administers questionnaires to the building's occupants. This chapter contributes to the existing understanding of green building performance assessment and provides valuable insights into enhancing building sustainability and user satisfaction.

Chapter 14 is a heritage BIM research project from Türkiye that argues historical structures, as repositories of cultural significance, epitomise the aesthetic and material attributes of their respective epochs, underscoring the need for their preservation and transmission to subsequent generations. Contemporary archiving systems exhibit considerable diversity, shaped by technological progress and regional variations. Digital imaging, 3D scanning, written sources, video and audio recordings, databases, and mobile applications constitute the array of technologies employed in archiving historical buildings. The BIM system, widely effective in contemporary construction, incorporates physical, functional, and technological data, fostering collaboration across multiple disciplines. This chapter endeavours to establish a heritage building information management (HBIM) supported by an archive model, facilitating interventions in historic buildings. This archival model aims to centralise diverse information within a single platform, offering a three-dimensional perspective and aiding researchers in comparing material and technical properties across different periods. The proposed model in this chapter were exemplified within the context of historical civil architecture buildings in Gaziantep, Türkiye.

Chapter 15 is another case study reflection from Australia on 3D printing, which is becoming a game changer for the construction industry. This chapter examines the applicability of 3D printing in construction housing projects in Australia and provides a comprehension of the best value-for-money utilisation of this technology in the industry. It articulates 3D printing with comparative case study analysis with facts and figures as a viable alternative to the traditional construction methods used in housing projects.

Chapter 16 introduces a framework of commercial building adaptation and reuse. The situation of commercial properties is a critical measure of the effectiveness of the real estate market, economic performance, and urban resilience. Buildings that are vacant provide a window into how well local economies are doing and how building space is evolving in these diverse places. This chapter outlines a conceptual framework that describes commercial property vacancy across the commercial real estate building life cycle through initial construction to final demolition and redevelopment, as well as implementation of this conceptual framework within a market context, comparing six major commercial property locations. It provides an original perspective on the changing nature of demand and the broader socioeconomic context.

Chapter 17 diagnoses the level of BIM implementation in the United Kingdom at regional level, compares the level of advancement of BIM adoption and use in the construction firms located in different regions across the United Kingdom. This chapter provides a greater understanding of BIM integration within UK construction companies. Comparison across every UK region, analysing companies of varying sizes and different construction sectors, aims to uncover the benefits and hindrances found with BIM adoption at a national level.

Chapter 18 presents an original experimental method for revealing fossil fuel related stranded assets in the built environment, based on a novel adaptation of existing data associated with environmental performance certificates and associated fossil fuel use baseline datasets in England. The experimental research in this chapter, based on a case study investigation in Newcastle-upon-Tyne, uncovers fossil fuel exposure throughout the building energy labelling criteria, not just those with the worst ratings.

Chapter 19 concentrates on heritage building information modelling (HBIM) and the reality capture survey using tools such as terrestrial laser scanning, drone-mounted LiDAR sensors, and photogrammetry. This HBIM modelling process is commonly criticised as complex and time-consuming. Hence, this chapter explores the semantic segmentation of point cloud data with artificial intelligence in academic and industrial research. It explores applications of the supervised machine learning, deep learning, and algorithmic modelling workflows that continually strive for semi-automation in HBIM generation.

2 Digital Transformation as a Business Differentiation Strategy for Sustainable Built Environment

Valentah Siamuzwe-Manase, David Manase, and Georgios Kapogiannis

Introduction

The world is changing. The physical, biological, environmental, and social world is under increasing pressure. There is growing scientific evidence that indicates that man's activities on the earth are having a life-negating impact on the world around him (Pitts, 2004). One crucial way this is demonstrated is climate change. Scientific research says that the main cause of this change is the increased number of emissions of greenhouse gases into the atmosphere, as a consequence of man's activities, so much more than the earth can absorb (Pitts, 2004). As well known that the main human influence on global climate is emissions of the key greenhouse gases – carbon dioxide (CO2), methane, and nitrous oxide. The accumulation of these gases in the atmosphere strengthens the greenhouse effect. The concentrations of these gases in the atmosphere have now reached levels unprecedented for tens of thousands of years. Various commentators have stated that climate change has been perceived as the single greatest global threat to humans (UN, 2023a).

This realisation has driven governments at various climate change events to decide to work together and draw targets that have to be reached globally if life on earth is to continue sustainably (UN, 2023a; UN, 2023b). Buildings in the built environment are major consumers of energy. Identification as a major contributor to emissions leading to climate change means that the industry should be a key player in providing innovative solutions to climate change and sustainability problems. Generally, the built environment is not well known for innovation capabilities that could go towards resolving climate change and sustainability problems.

Other industry sectors, such as the manufacturing sector, have outperformed the built environment sector in terms of innovation. In recent years, digital transformation of the built environment has been embraced as holding the keys or solutions to the malaise of climate change and the inability to achieve sustainable targets (WBCSD, 2021). This is in part because of successes of digital transformation in other sectors. There are now many factors encouraging organisations in the built environment to adopt advances in digital transformation practices and for these factors to be promoted more effectively from organisation planning phases through to strategy formulation and implementation. Therefore, digital transformation

DOI: 10.1201/9781003380559-2

should be viewed not only from the perspective of technology alone, but also as part of the built environment organisation strategy to achieve that competitive advantage over competitors.

Increasingly, businesses in the built environment are beginning to seek ways in which digital platforms and frameworks can start to address climate change and sustainability-generated problems and develop strategies to suit. Against this background, a need arises to explore how built environment organisations can utilise digital transformation proficiencies and development as a business strategy that helps them achieve competitive advantage by doing so.

The discussion in this chapter aims to explore and outline how built environment organisations can utilise digital transformation to achieve a competitive advantage in the marketplace. To achieve this, the chapter reviews the body of knowledge on digital transformation in the built environment sector. The study also gives an outline of the organisations' appraisal to understand its business environment. The key drivers of change and how the organisation then goes about crafting and executing strategy is outlined. The chapter further outlines digital transformation as a concept that can enable competitive advantage to organisations that embrace it.

Digital Transformation Development

Digital transformation development refers to the process of planning, implementing, and managing digital transformation initiatives within an organisation or business. It involves adopting and integrating digital technologies, processes, and strategies to drive organisational change and achieve specific business objectives. Key stages involved in digital transformation development include the following:

- **Planning and Strategy**

  This stage involves assessing the current state of the organisation, identifying digital transformation goals, and developing a strategic plan. It includes conducting a comprehensive analysis of the organisation's current operations, identifying areas that can benefit from digital transformation, setting clear goals and objectives, and developing a roadmap for implementation.

- **Technology Adoption**

  Once the strategic plan is in place, the next stage is to identify and adopt the appropriate digital technologies that align with the organisation's goals and objectives. This may include technologies such as cloud computing, data analytics, Internet of Things (IoT) devices, artificial intelligence (AI), robotics, and other emerging technologies.

- **Implementation**

  Implementation involves the actual deployment of digital technologies and processes across the organisation. This may include setting up new systems, integrating existing systems, training employees on how to use the new technologies, and implementing new processes and workflows to leverage the digital tools effectively.

- **Change Management**

  Digital transformation involves significant changes to how the organisation operates, and change management is crucial to ensure successful adoption and integration of new technologies and processes. This may include providing training and support to employees, addressing any resistance to change, and communicating the benefits and impact of digital transformation to stakeholders.

- **Data Governance and Security**

  Digital transformation often involves the collection, storage, and analysis of large amounts of data, and data governance and security are critical considerations. Organisations need to establish robust data governance policies, data management processes, and security measures to protect sensitive information, comply with relevant regulations, and ensure data privacy and security.

- **Continuous Improvement**

  Digital transformation is an ongoing process, and organisations need to continuously assess and refine their digital initiatives to ensure they are delivering the desired outcomes. This may involve monitoring key performance indicators (KPIs), analysing data, identifying areas for improvement, and making necessary adjustments to optimise digital transformation initiatives.

- **Collaboration and Stakeholder Engagement**

  Digital transformation often involves collaboration with various stakeholders, including employees, customers, partners, and suppliers. Effective collaboration and stakeholder engagement are essential for the success of digital transformation initiatives. This may include fostering a culture of innovation, encouraging feedback and input from stakeholders, and involving them in the decision-making process.

Digital transformation development is a complex and ongoing process that requires careful planning, execution, and management. It involves adopting and integrating digital technologies, processes, and strategies to drive organisational change, achieve business objectives, and stay competitive in today's digital-driven business landscape.

Digital Transformation in the Built Environment

As mentioned before, digital transformation refers to the integration of digital technologies into all aspects of an organisation's operations, processes, and strategies to fundamentally change the way it operates and delivers value. Thus, digital transformation in the built environment sector refers to the adoption and integration of digital technologies, processes, and strategies to enhance the efficiency, productivity, and sustainability of construction projects and operations. Digital transformation is changing the built environment by revolutionising traditional practices and introducing new ways of designing, planning, executing, and managing construction projects (WBCSD, 2021). Some key areas where digital transformation is making an impact in built environment are mentioned by several commentators, including Sacks et al. (2018), Behzadi (2016), and Balaguer and Abderrahim (2008):

- **Building Information Modelling (BIM)**

  BIM is a digital representation of a building's physical and functional characteristics that enables collaboration and coordination among different stakeholders throughout the project lifecycle. BIM can be used to create virtual representations of buildings, allowing architects, engineers, and construction teams to collaborate and optimise design decisions. BIM allows for improved design visualisation, clash detection, and project coordination, leading to reduced rework, increased efficiency, and better project outcomes. This can lead to more sustainable building designs that incorporate energy-efficient features, use of eco-friendly materials, and waste reduction strategies, resulting in buildings with lower environmental footprints and higher performance.

- **IoT**

  IoT devices, such as sensors, wearables, and drones, are being used in built environments to collect real-time data on equipment, materials, and workers. This data can be analysed to optimise construction processes, monitor safety, and improve decision-making. IoT and connected devices can be used to create smart buildings that optimise resource utilisation, improve occupant comfort, and enhance overall operational efficiency. Examples include smart lighting systems, automated HVAC controls, and occupancy sensors, which can optimise energy usage, reduce waste, and create more sustainable built environments.

- **Augmented Reality (AR) and Virtual Reality (VR)**

  AR and VR technologies are used in construction for visualisation, design review, and simulation of construction processes. AR and VR can improve communication, enhance design accuracy, and allow for virtual walkthroughs, reducing errors and rework.

- **Robotics and Automation**

  Robotics and automation technologies, such as 3D printing, autonomous equipment, and robotic process automation (RPA), are being used in the built environment to automate repetitive tasks, improve accuracy, and enhance productivity.

- **Cloud Computing and Data Analytics**

  Cloud computing allows for centralised data storage, collaboration, and access to information from anywhere, while data analytics enables data-driven decision-making. Cloud computing and data analytics are used in construction for project management, cost estimation, scheduling, and performance tracking.

- **AI**

  AI is used in the built environment for various applications, such as predictive analytics, risk management, and predictive maintenance. AI can analyse large amounts of data, identify patterns, and provide insights for better decision-making and improved project outcomes.

- **Mobile Technologies**

  Mobile technologies, such as smartphones and tablets, are used in the built environment for communication, collaboration, and real-time access to project information. Mobile technologies enable on-site data collection, reporting, and tracking, improving communication and decision-making.

- **Energy Management and Optimisation**

  Digital solutions can be employed to monitor and manage energy consumption in buildings, enabling real-time data collection, analysis, and optimisation of energy usage. This can result in more efficient energy management practices, such as demand response, energy benchmarking, and predictive maintenance, leading to reduced energy consumption, lower operating costs, and improved sustainability performance.

- **Lifecycle Asset Management**

  Digital tools can be utilised to manage the entire lifecycle of a building, including maintenance, repair, and replacement of assets. Predictive maintenance and asset tracking technologies can optimise asset utilisation, reduce downtime, and extend asset lifespan, resulting in improved sustainability performance and cost savings.

- **Data-Driven Decision-Making**

  Digital transformation can enable organisations to collect and analyse large volumes of data related to building performance, resource utilisation, and occupant behaviour. This data can be leveraged to make data-driven decisions, identify areas for improvement, and implement sustainability initiatives that align with the organisation's goals and values.

- **Enhanced Customer Experience**

  Digital technologies can also enhance the customer experience in the sustainable built environment. For example, building occupants can use smart building apps to monitor and control their energy usage, provide feedback on comfort levels, and participate in sustainability programmes, leading to improved occupant satisfaction and loyalty.

Overall, digital transformation in the built environment is reshaping the industry by improving collaboration, increasing productivity, reducing errors and rework, enhancing safety, and optimising project outcomes. It is transforming traditional construction practices and providing new opportunities for built environment organisations to innovate and stay competitive in the rapidly changing construction landscape. In the built environment, digital transformation can play a significant role in achieving business differentiation by enabling organisations to adopt innovative approaches to design, construction, operation, and maintenance that promote sustainability, reduce environmental impact, and enhance overall performance.

Business Strategy, Competitive Advantage, and Digital Transformation

Strategic management includes understanding the strategic position of an organisation's strategic choices for the future and turning strategy into action (Whittington et al., 2023). It involves environmental scanning (both external and internal), strategy formulation and strategic implementation, and evaluation and control. Thus, it can be said that, for any organisation that is formulating a strategic path for its business, a study of its internal and external business environment in arriving at an effective strategy is key. To that effect, the following models/tools are generally utilised in the evaluation of the business environment, strategy formulation, implementation, evaluation, and control:

- PESTEL Analysis
- Porter's 5 Framework
- SWOT Analysis

Strategic Analysis Tools

PESTEL Analysis

Businesses operate in a complex business environment that is constantly changing. Whittington et al., (2023) have defined the external business environment of an organisation in three layers. Businesses need to know what is happening in these environments and understand the impacts, if any, of the activities in these environments and tailor their strategies to deal with these with regard to the commercial advantage of the business. The farthest layer of the business environment is the macro-environment. This consists of the broad environmental factors that impact business and other players in its industry to a greater or lesser extent (Whittington et al., 2023).

To help organisations understand this business layer, organisations can use the PESTEL Analysis framework. PESTEL Analysis is a useful tool for understanding market growth or decline, and as such the position, potential, and direction for a business based on an understanding of the external factors in the macro-environment impacting on a business's industry (Whittington et al., 2023).

PESTEL Analysis is a business measurement tool and an acronym for political, economic, social, technological, legal, and environmental factors. These headings are a framework for reviewing a strategy or position, direction of a company, a marketing proposition, or idea. The framework analyses what macro-environment factors are setting future trends that might impinge on organisations (Whittington et al., 2023). What factors have an impact and to what degree will differ from country to country and from industry to industry. Even though these factors are interrelated, their impact will not all be equally important to all businesses. As such, organisations need to identify which key factors affect them more than others. These factors are called key drivers of change, which Whittington et al. (2023) define as forces that are likely to affect the structure of an industry, sector, or market.

Porter's Five Forces Framework

From a strategic management point of view, it is useful for managers in any organisation to understand the competitive forces acting on and between organisations in the same industry or sector since this will determine the attractiveness of the industry and the way in which individual organisations might choose to compete. This understanding may inform important decisions about product/market strategy and whether to leave or enter industries (Whittington et al., 2023). Frost (2021) says that the industry environment lies at the core of the macro-environment and the impact that the macro-environment has on a firm in an industry is through its effects on the industry environment.

Grant (2021) states that the basic premise that underlies industry analysis is the level of profitability, which is influenced by the structure of that industry, whether it be a monopoly industry or an industry in perfect competition as defined by industrial organisation (IO) economics. He suggests that by examining the structural features in an industry and how these interact, it is possible to predict the competitive behaviour in an industry and, consequently, the resulting profitability levels. Porter (1998) formulated Porter's Five Forces Competition Framework to facilitate the analysis of industry structures and their impact on industry profitability. The framework suggests that there are three sources of 'horizontal' competition: competition from substitutes, new entrants, and established rivals; and two sources of 'vertical' competition: the bargaining power of suppliers and buyers (Figure 2.1) (Grant, 2021). The strength of each of these competitive forces is determined by several key structural variables, as outlined in Figure 2.1.

The structural determinants of the five forces of competition include:

- **Threat of Substitutes**

  As in Figure 2.1, there are factors that enhance consumer substitution. In the construction industry, for example, substitution threats come from mobile

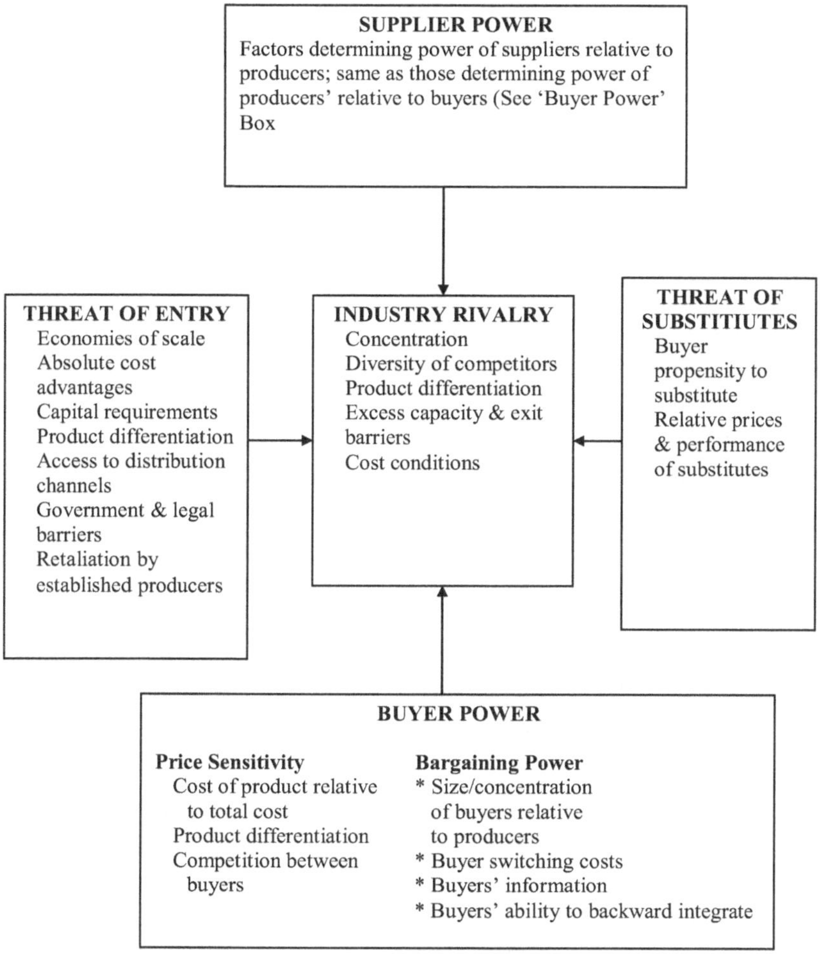

Figure 2.1 The Structural Determinants of the Five Forces of Competition.
Source: Grant (2006).

accommodation provision such as caravans and boats. To some extent, prefabricated accommodation such as the flat-pack house packages sold by firms such as IKEA may be a threat to traditional construction as we know it. However, the performance of such alternative accommodation does not meet most of society's requirements and, therefore, is not a significant threat to construction of brick and mortar with foundations in the ground.

- **Threat of New Entrants**

In construction, threats on new entrants are most significant at lower levels where small construction companies compete for market share. This is because, for instancc, to become a large contractor from a medium-sized contractor, there is a need for high capital investment. This may be a barrier to new entrants. However, the threat of new entrants may be more in the form of consultancy services. As an example, as new forms of procurement/contracts such as private finance initiatives were being introduced into the industry; many other professionals not strictly belonging to the construction industry, such as financial and investment consultants, were entering the construction market and acting as advisors to different parties, clients, and contractors alike.

This can seen as a threat to traditional consultants such as consultant quantity surveyors, whose commercial role is being overtaken. Furthermore, some of the new forms of contracts included pre-contract involvement of contractors as consultants and/or post-construction involvement in the operation and maintenance of a building within the construction contract. This means that the industry boundaries and, therefore, the competitors are becoming blurred as businesses are integrating backwards and forward.

- **Bargaining Power of Buyers**

  The construction industry's largest client in the United Kingdom is the government in various forms of public sector institutions (e.g., education, health, custodial). Because it is such a large client, the government has a lot of power over what happens in the industry. Government expenditure plays a big part in industry activity. As such, government legislation, targets, budget, and procurement processes have a large impact in determining the industry's landscape. Government targets and guidelines on digital transformation, for example, are driving the direction of the industry in terms of the extent to which the industry is addressing the subject. Over and above government influence, large private companies also have an impact on the industry. With reference to digital transformation, there is a recognition of the importance of clients as a driver of change in the construction industry. Private clients are being driven to embrace digital transformation to create and achieve value and achieve competitive advantage.

- **Bargaining Power of Suppliers**

  The construction industry's supply chain is fragmented in non-specialist products like aggregate, cement, bricks, and blocks. Furthermore, there are a lot of small-size trade contractors that make that supply chain fragmented as well. However, there are few large-sized construction industries to which both material and trade suppliers seek to provide a service. In that respect, the suppliers have little power while their larger buyers, e.g., contractors, have high bargaining power. However, in specialist supplies such as furniture and equipment, there are few suppliers. Consequently, such suppliers have greater bargaining power. Regarding digital transformation, there are a few consultancies that have developed their knowledge in this field. To that end, such consultancies have high bargaining power in terms of their fees as well as in determining what is built because of their expertise.

- **Industry Rivalry**

  Competition levels are different depending on which level a contractor is operating. For smaller to medium-sized firms where there are a lot of players with fewer bids and less client budget scope, competition is intense, and margins are low, sometimes even as low as 2% as a result of competition. However, for larger firms or higher-level operations where there are few large players, though competition is healthy, it is less intense because there are fewer competitors competing for bids at that level and client spend is larger.

  The combination of these two factors allows for healthier margins and greater chance of success in winning tenders. At this level, firms tend to target a win in at least one in every three bids that they tender for. Despite this, larger firms also tend to put in measures with the intention of staying ahead of their competitors. Taking a business-wide focus, for example, on digital transformation implementation, firms would gain credentials and stay ahead of their competitors. Overall, competition is relatively high across all levels, but particularly among larger firms/contractors, to enter this market and gain competitive advantage.

SWOT Analysis and the Resource-Based View of Strategy

The IO model suggests that the profitability of an organisation is largely determined by factors from its external environment. It assumes that (Frost, 2021):

- The external environment determines strategies
- Firms control similar resources
- Resources are mobile across firms
- Rational managers work in all firms
- Logical analysis determines industry attractiveness

However, the resource-based view (RBV) suggests that an organisation's profitability is determined by characteristics inside the organisation. It focuses on developing/obtaining valuable resources and capabilities that are difficult or impossible for rivals to imitate (Grant, 2021; Frost, 2021). Contrary to the IO model, it assumes that (Frost, 2021):

- The internal environment is more influential than the external
- Firms acquire different resources which they utilise differently.
- Resources are not highly mobile.
- Combination/integration of resources and capabilities form competitive advantage.
- Resource integration develops capabilities

Grant (2021), however, says that for an organisation's strategy to be successful, it needs to tailor its resources and capabilities in such a way that they are addressing the

Figure 2.2 The Basic Strategy Framework/ Janus the God of the Gates (Grant, 2021).
Source: Grant, R.M. (2021), *Contemporary Strategy Analysis*, 11th Edition, Wiley.

opportunities and threats that the external environment is presenting (Figure 2.2). Managers need to be aware of what is in their internal as well as external environment. Therefore, rather than focus on the external environment and the threats and opportunities that the organisation presents to define its activities, it must focus on defining its resources and capabilities because these are more stable than the external environment.

These resources and the capabilities offer the organisation a more stable basis on which to define its identity and in turn what strategic tools it has to tackle the changing variables in its business environment. Knowledge and understanding of an organisation's capabilities and resources will then determine what strategy it should follow given the circumstances in its environment.

This is a more realistic and practical approach because it is no use if an organisation formulating a strategy does not have the capability and resources to execute, no matter how suitable the strategy is to address environmental factors (Figure 2.3). Based on this view, which marries the external and internal environment, it is therefore necessary to recognise the external environment factors impacting an organisation as well as its own internal resources and capabilities and to analyse how well these relate to assess the suitability of an organisation's strategy to the market. Whittington et al. (2023) say that the strategic capability of the organisation is made up of resources and competencies.

Whittington et al. (2023) define strategy as the direction and scope of an organisation over the long term, which achieves advantage in the changing environment through the configuration of resources and competences with the aim of fulfilling stakeholder expectations. Grant (2021) says that strategy is a link between the firm and its external business environment (Figure 2.2). Strategy formulation needs to incorporate a full understanding of the business environment in which the organisation is operating to create a strategy that addresses the market condition.

The right strategy formulation is so crucial because strategy determines the makeup and structure of a business in its entirety as it is long-term and determines the resources and competencies of the organisation. Key to all this is the need, ultimately, to achieve competitive advantage or survival in the case of some public

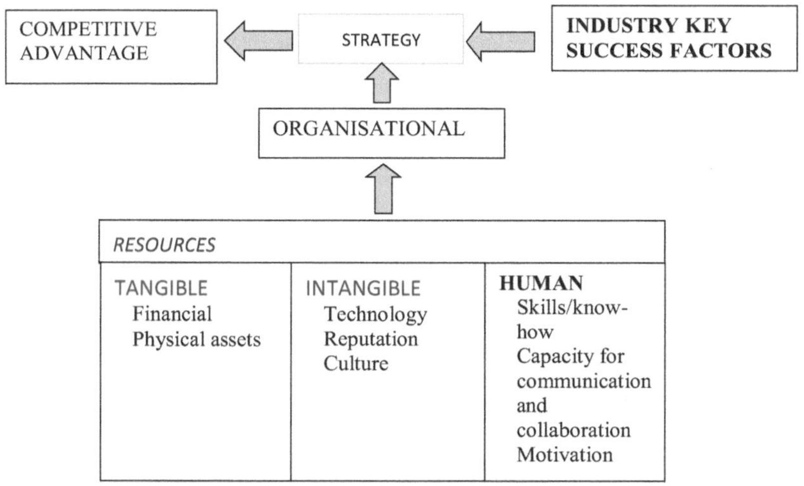

Figure 2.3 Links among Resources, Capabilities & Competitive Advantage (Grant, 2021).
Source: Grant, R.M. (2021), *Contemporary Strategy Analysis*, 11th Edition, Wiley.

sector organisations. It is therefore key that the strategy that a business formulates is so different and effective that it will give it a sustainable/long-term advantage over its competitors. Thompson et al. (2022) say that a company can achieve competitive advantage when an attractive segment of the market prefers its products/services over those of its competitors.

In the built environment sector, the various capabilities inherent in digital transformation encourage built environment organisations to adopt practices through their planning systems that enable them to have competitive advantage over their competitors. Built environment organisations embarking on and accommodating digital transformation methods to enable sustainable solutions may find themselves at a competitive advantage because they will:

- Be meeting environmentally conscious client requirements: These cover many large construction clients, in both the private and public sector. As such, they will be the favoured choice.
- Be ahead of the race: Quick-thinking companies can begin to formulate strategies that enable them to implement sustainable construction. By doing so, they will gain a competitive advantage over rivals who are yet to change their operations to meet market demands and will have the learning/experience curve advantage over slower contenders.

To be able to take advantage of this change in the environment and digital transformation, it cannot be over-stressed that businesses in the built environment need to formulate strategies that can differentiate them from what others are doing in the

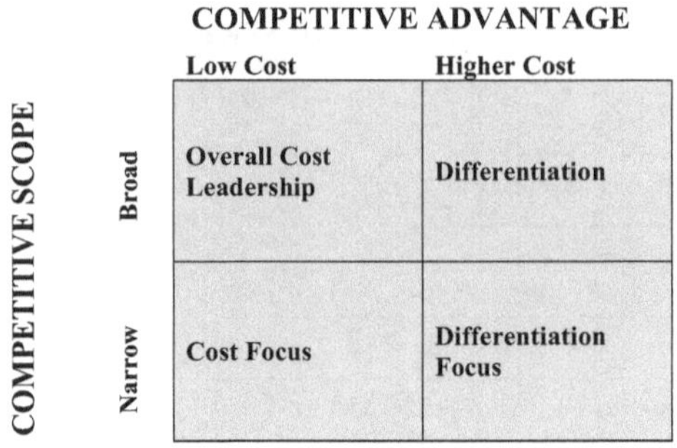

Figure 2.4 Porter's 3 Generic Strategies.
Source: Adopted from Porter (1998).

long term (Thompson et al., 2022). Porter (2004) defined three generic strategies for achieving competitive advantages that can be adopted about sustainable development (Figure 2.4):

- **Overall Cost Leadership**

  Emphasises efficiency. By producing high volumes of standardised products, a firm hopes to take advantage of economies of scale and experience curve effects. 'Cost advantage' is the focus. This does not imply low selling costs but rather low operating costs that may then lend itself to low selling costs because of low operating costs.

- **Differentiation**

  Involves creating products that are perceived as unique *industry wide*. Differentiation can take many forms: design or brand image, technology, features, customer service (Porter, 1998). The unique features or benefits should provide superior value for the customer if this strategy is to be successful to a point of eliminating price sensitivity. This can provide considerable insulation from competition.

- **Focus**

  The firm concentrates on a select few target markets with the aim of providing superior service to the selected niche (Porter, 1998). Competitive advantage is generated specifically for the niche.

These principles are still generally applicable even in this day of global markets. However, it is important to realise that while these strategies are the core strategy, they are not exclusive. Even though a product is differentiated, it still needs to be priced competitively, and even though low-cost leadership is sought as the core strategy, at times market demands dictate that there is more to market demand for a product than just its price. Thus, the manufacturer still must make his product appealing to the market.

Core Competency Analysis

Whichever strategy the business formulates to meet market conditions, the organisation must have the strategic capability to meet market demand at the level at which it is required to succeed (Whittington et al., 2023). Analysing the strategic ability of a business involves assessing the organisation's resources and competencies and seeing if they fit into the environment that the business is operating in; where they do not, strategic development then looks at how strategic capability can be changed to fit the environment (Whittington et al., 2023).

One of the most important aspects of assessing a business's strategic capability is the assessment of the business's competence level in performing key aspects of its business. A business's competencies can range from being merely competent in performing an activity that is basic to operating in the business it is in to being a core or distinct competence (Thompson et al., 2022).

Core competencies are those competencies that critically underpin the organisation's competitive advantage (Whittington et al., 2023). They are activities that a business performs better than its competitors, are central to the business's strategy, and does better than other internal activities (Thompson et al., 2022). Most often, a core competence is knowledge based and resides more in the people and less in physical assets (Whittington et al., 2023). Whittington et al. (2023) say that, to achieve long-term competitive advantage, it is important for businesses to keep on improving their core competencies to keep ahead of the competition. This is because core competencies change over time.

Thus, digital transformation requires that an organisation adopting it should have the relevant resources and competencies necessary to implement it. Better still, for digital transformation to be a business differentiator that gives a business a competitive advantage, a business needs, in line with what Whittington et al. (2023) say, to have core competencies in this area. The technical aspect of digital transformation development indicates a need for built environment organisations to develop the skills of its workforce so that it is technically capable of understanding the needed design/construction measures that are necessary in achieving sustainable development.

With time, not only do they need to have these competences, but they also need to develop them so much so that they become core competencies that will give them the required edge to achieve a competitive advantage. While the built environment workforce needs digital skills, they also need a change in behaviour. It can be surmised that at a lower level, this is an individual's behaviour, but at a higher level, there needs to be a change in the organisation's behaviour, in the way it addresses digital transformation in the built environment. Consequently, management in built environment organisations needs to begin to look at how it will manage this change to ensure that the consequences are positive and advantageous to achieving the goal.

Change Management

In the fast-moving/changing business environment, organisational success has become directly attributable to organisations' ability to handle and sustain strategic change (Wilson, 1992). Peters and Waterman (2004) described change as the master key to corporate success and competitive advantage. Scholars such as Thompson et al. (2022) state that for an organisation to stay ahead of its competitors, it must continually improve its competencies that give it competitive advantage, thus in some ways alluding to what Peters and Waterman state. Consequently, Thorne (1991) states that managers should be masters of change. The role of management in such situations is to establish some rationality or predictability out of the chaos that characterises the change process (Wilson, 1992).

Burnes (2017) identifies three models of change:

1. **Incremental**

   Pettigrew et al (1992) described incremental change as change that is successive, limited, and in negotiated shifts.

2. **Punctuated Equilibrium**

   This model suggests that organisations evolve through relatively stable periods that are punctuated by relatively short bursts of fundamental change (revolutionary periods), which periods determine the need/basis for change (Romanelli and Tushman, 1994)

3. **Continuous Transformation**

   This model views change as a continuous process in which organisations must continuously improve themselves in a fundamental manner to maintain competitiveness and survive (Burnes, 2017).

Central to the concept of change is the fact that organisations exist in a business environment that is changing, whether continuously or otherwise. Organisations must therefore adapt to these changes by formulating strategies that address these changes. Herein lies the link between strategy and change in that strategy seeks to address change, whether reactively and proactively, in the business environment of an organisation to ensure profitable existence of an organisation. Strategy will reallocate resources and develop capabilities internally that meet the changes of the environment, even to the extent of changing organisational structure and culture. It is this internal change due to of digital transformation that is under examination in this discussion. It is this link that supports the argument that the ability of the organisation to manage digital transformation change successfully is key in achieving competitive advantage. The Janus model of inward and outward facing management stresses the importance of management having both an outward appreciation/understanding of the changes in the organisation's environment and an inward understanding of what the organisation's resources and capabilities are and how these meet demands of the external environment and the changes thereof.

There are three schools of thought that address how change can be managed in an organisation. The first is the individual perspective school, which suggests that change can be achieved by influencing an individual's behaviour through the environment that an individual is exposed to (Burnes, 2017). The second is the group dynamics school, which says that an individual's behaviour is influenced by a group's values, norms, and roles. Thus, managing change, and in this discussion digital transformation change, should be exercised at a group level because, if the group's values/norms/values are modified to achieve the required change, then the individual's behaviour will follow suit (Cummings and Huse, 2018). The third school of thought, which is the school that this research adopts, is the Open Systems school of thought. This school of thought suggests that change and, in this case, digital transformation change should be addressed with reference to the organisation as an entity. It says that an organisation consists of several interacting subsystems and that a change in one subsystem affects the other and in turn the performance. This system is based on a method of describing and evaluating these subsystems to determine how they need to be changed to improve the overall functioning of the organisation (Burnes, 2017). Moreover, it sees organisations as 'open' systems in two respects – to the external environment and internally in that they interact with the other organisational subsystems.

Thus, internal changes in one area affect other areas, and these have an impact on the external environment and vice versa (Buckley and Casson, 2010). The objective of the Open System is to structure the functions of business in such a manner that overall business objectives are collectively pursued through clearly defined lines of coordination and interdependence (Mullins and Rees, 2023). Mullins argues that there are four principal organisational subsystems the change can be managed through: organisational goals and values subsystems; technical subsystems; psychosocial subsystems; and the managerial subsystem.

The Open System is concerned with understanding organisations in their entirety and therefore attempts to take a holistic rather than particularistic perspective, which is reflected in its approach to change (Burnes, 2017). As such, change must be addressed at each of the identified subsystems as change in one affects the other. Organisational behaviour has emphasised the roles, to name but a few, of training, interpersonal skills, organisational development, communication, and effective groups in achieving and sustaining organisational change (Wilson, 1992). However, these cannot be done in isolation but must interconnect across the organisation.

An organisation in the built environment, therefore, seeking to adopt digital transformation as part of its strategy, needs to address the resources and capabilities that it has and assess their strategic fit. If these do not meet the requirements, there will be a need for resource reallocation, capability development, and even a whole cultural change, which will entail that its employees always must incorporate sustainability in their practices. This is in line with Mullins and Rees (2023) illustrated in the arguments discussed above which implicitly suggests that digital transformation practices should be incorporated into the design of the building right from inception. Digital transformation in the built environment may also require, as Mullins says, a change in the technical.

Digital Transformation as a Strategy for Business Differentiation and Competitive Advantage

Digital transformation can help built environment organisations gain competitive advantage. However, it must be recognised that, to do this, some organisations would need to be doing something significantly different from what the rest of the industry is doing. Digital transformation could be used/implemented to provide the competitive advantage for built environment organisations such as:

- Ensuring that digital transformation runs right through all its processes and procedure.
- Management leading the way.
- Knowledge is the most important asset: knowledge of not only the subject but also of the best practice solutions that work.
- The need for tangible evidence from projects where best practice in digital transformation has been implemented to achieve positive results. These could be used to sell digital transformation to clients.
- Sell digital transformation as part of the whole life cost (WLC) of the building; this is the total cost of construction (CAPEX) as well as operating and maintaining the building. That way, the benefits of digital transformation can be seen in the long-term, which is where the greatest benefit is, not in the short-term (CAPEX).
- Having dedicated personnel to address digital transformation and measure performance of projects/business operations; experts say that an organisation that does not have a dedicated function is likely to lose credibility.

Resources and Capability

For any strategy to be implemented in an organisation, resources have to be allocated to support that strategy. Furthermore, the organisation needs to have the necessary competencies in its makeup to give it the capability to execute the strategy. The overriding resource required is knowledge. Expert technical knowledge that would not only enable understanding of what digital transformation entails but also the ability to use digital technologies to design, build, and operate buildings that could meet the requirements of sustainable development. There is also the need for an understanding of what market requirements, expectations, and opportunities are and where the best results can be achieved to increase business profitability.

Change Management

With digital transformation, built environment organisations needed to define what they want to achieve with digital transformation in which areas. They need to define where it is, where it wants to go and how it will get there. Having done this, the organisation can then define what changes needed to be made internally to help it achieve the defined goal. One can take a mixed view as to whether there would be a need for a new management structure in an organisation. One view could be that there would not necessarily be a change in the organisational management structure to address digital transformation as digital transformation would be part of an overall strategy. On the other hand, another view could be that to tackle digital transformation, there would be a need for the right structure to manage the change in strategy. However, in both views there is a recognition that there would be a need for some investment in digital transformation in terms of management of the digital transformation strategy formulation and implementation process.

To manage the change in the business processes, built environment organisation employees need to be informed and educated on digital transformation. As such, organisations should advance and design digital transformation awareness processes that can be used to give information to clients and employees alike. Over and above this, the organisations should be able to execute and roll out digital transformation processes or training courses to be given to employees. Having done this, the business processes and systems need to be developed to support the new approach to digital transformation. Current systems in literature cover general business environmental issues that the built environment organisation is involved in. These will need to be expanded to cover in greater detail the digital transformation attributes as a competitive tool.

Digital Transformation as a Business Differentiator in Sustainable Built Environment

Digital transformation can be a powerful strategy for built environment organisations to achieve business differentiation. By leveraging digital technologies, built environment organisations can streamline their operations, enhance their competitive advantage, and deliver greater value to their clients. Some ways digital transformations can be used as a strategy for achieving business differentiation and as a competitive advantage tool in the sustainable built environment include the following:

- **Innovation and Differentiated Offerings**

  Digital transformation can enable built environment organisations to innovate and offer differentiated services or products. For example, firms can leverage technologies such as BIM to create virtual models of construction projects, use AR and VR for immersive experiences, or implement smart building technologies for energy management and sustainability. By offering innovative and differentiated solutions, built environment organisations can stand out in the market and attract clients who are looking for cutting-edge solutions.

- **Data-Driven Decision-Making**

  Digital transformation generates a wealth of data that can be used for data-driven decision-making. Built environment organisations can leverage data analytics and business intelligence tools to analyse data and gain insights into various aspects of their operations, including project performance, resource allocation, risk management, and client preferences. Data-driven decision-making can lead to better strategic planning, improved operational efficiency, and informed decision-making, which can differentiate built environment organisations in terms of data-driven insights and evidence-based decision-making.

- **Enhanced Customer Experience**

  Digital transformation can enable contracting firms to deliver an enhanced customer experience. For instance, digital tools can provide real-time project updates, allow clients to access project information and documentation online, and facilitate online communication and collaboration. A superior customer experience can lead to increased client satisfaction, positive word-of-mouth, and repeat business, setting contracting firms apart from competitors in terms of customer-centricity.

- **Improved Project Management**

  Digital tools and platforms can greatly enhance project management processes for contracting firms. With project management software, firms can efficiently manage tasks, timelines, resources, and budgets, resulting in improved project delivery and client satisfaction. Digital project management solutions can also provide real-time visibility into project progress, enabling firms to make data-driven decisions and quickly adapt to changing project requirements, which can set them apart from competitors (Kapogiannis et al., 2021).

- **Enhanced Collaboration and Communication**

  Digital transformation can facilitate seamless collaboration and communication among team members, subcontractors, and clients. Cloud-based collaboration tools, such as project management platforms, document management systems, and communication applications, can enable real-time communication, document sharing, and collaboration across different stakeholders. Efficient communication and collaboration can lead to faster decision-making, increased transparency, and improved customer engagement, which can differentiate built environment organisations in terms of responsiveness and client satisfaction.

- **Automation and Efficiency Gains**

  Digital technologies, such as robotics, AI, and data analytics, can automate manual tasks, optimise processes, and improve overall operational efficiency. For example, drones can be used for site inspections and data collection, AI-powered analytics can automate data analysis for better insights, and data-driven decision-making can lead to more accurate cost estimates and resource allocation. Automation and efficiency gains can result in cost savings, faster project delivery, and improved quality, which can set contracting firms apart from competitors in terms of speed and accuracy.

Limitations of Digital Transformation in Business Opportunities for Built Environment Organisations

While digital transformation can offer numerous benefits to built environment organisations, there are also certain limitations and challenges that may impact business opportunities. Some potential limitations of digital transformation in the built environment are shown in Table 2.1.

Table 2.1 Limitations in Digital Transformation in the Built Environment

	Factor	Description
1.	Cost of Implementation	Implementing digital transformation initiatives can require significant upfront investment in technology infrastructure, software, training, and change management. Small and medium-sized built environment organisations may face financial constraints in terms of upfront costs, which can limit their ability to adopt and implement digital technologies.
2.	Skills and Knowledge Gap	Digital transformation requires skilled professionals who are knowledgeable in emerging technologies and can effectively implement and manage digital initiatives. Built environment organisations may face challenges in finding and retaining qualified personnel with the necessary skills and expertise, particularly in niche areas such as BIM, AI, and data analytics.
3.	Resistance to Change	Resistance to change from employees and stakeholders can pose a challenge in the adoption of digital transformation. There may be resistance to learning new technologies, changing established processes, and embracing new ways of working, which can hinder the successful implementation of digital initiatives.
4.	Data Privacy and Security Concerns	Digital transformation often involves the collection, storage, and analysis of large amounts of data. Built environment organisations need to ensure that they have robust data privacy and security measures in place to protect sensitive information from unauthorised access, data breaches, and cyber threats. Compliance with data protection regulations, such as the General Data Protection Regulation (GDPR), can also pose challenges.
5.	Fragmented Industry Landscape	The built environment can generally be fragmented, with numerous stakeholders involved in a project, including contractors, subcontractors, architects, engineers, and clients. This can result in challenges in terms of interoperability and data exchange between different digital tools and systems, hindering seamless collaboration and integration of digital technologies.
6.	Industry Regulations and Standards	The built environment is generally subject to various regulations and standards related to health and safety, building codes, and environmental requirements. Built environment organisations need to ensure that their digital transformation initiatives comply with these regulations and standards, which can add complexity and challenges to the adoption of digital technologies.
7.	Market Competition	The built environment can be highly competitive, with numerous firms vying for projects. While digital transformation can offer business differentiation, other firms may also be adopting similar technologies, reducing the competitive advantage of early adopters.

Conclusion

Digital transformation can bring significant benefits to built environment organisations. However, there are also limitations and challenges that may impact business opportunities for built environment organisations. These challenges include the cost of implementation, skills and knowledge gap, resistance to change, data privacy and security concerns, fragmented industry landscape, industry regulations and standards, and market competition. Therefore, built environment organisations need to carefully navigate these limitations and challenges to ensure successful implementation of digital transformation initiatives and realise their full potential.

Overall, digital transformation can be a powerful strategy for built environment organisations to achieve business differentiation. By leveraging digital technologies, built environment organisations can improve their project management processes, enhance collaboration and communication, automate tasks, optimise processes, enhance customer experience, offer differentiated solutions, and make data-driven decisions. Embracing digital transformation can set built environment organisations apart from competitors, drive operational excellence, and deliver greater value to their clients.

References

Balaguer, C., Abderrahim, M. (2008), *Robotics and Automation in Construction*, Intechopen. ISBN 978-953-7619-13-8.

Behzadi, A. (2016), Using Augmented and Virtual Reality Technology in the Construction Industry, *American Journal of Engineering Research (AJER)*, 5(12), 350–353, e-ISSN: 2320-0847 p-ISSN: 2320-0936.

Buckley, P. J., Casson, M. (2010), *The Multinational Enterprise Revisited: The Essential Buckley and Casson* (15 January 2010), Palgrave Macmillan, ISBN 978-0-230-51599-4.

Burns, B. (2017), *Managing Change*, 7th Edition, Pearson.

Cummings, T., Huse, E. (2018), *Organization Development and Change*, 11th Edition, South West College ISE.

Frost, D. (2021), *Strategic Management*, 1st Edition, Kendall Hunt Publishing, ISBN-10: 1792456301.

Grant, R.M. (2006), *Contemporary Strategy Analysis*, 5th Edition, Blackwell Publishing.

Grant, R.M. (2021), *Contemporary Strategy Analysis*, 11th Edition, Wiley.

Kapogiannis, G., Jiequn Guo, G., Hancock, C.M. (2021), *Digitial Supply Chain and Port Information Modelling (PIM), ICT Solutions and Digitalisation in Ports and Shipping.* https://doi.org/10.1049/PBTR030E_ch10

Mullin, J.L., Rees, G. (2023), *Management and Organisational Behaviour*, 13th Edition, Pearson.

Peters, T., Waterman, R., Jr. (2004), *In Search of Excellence: Lessons from America's Best-Run Companies*, HarperCollins Publishers Inc.

Pettigrew, A.M., Ferlie, E., McKee, L. (1992), *Shaping Strategic Change*, Sage.

Pitts, A. (2004), *Planning and Design Strategies for Sustainability and Profit*, Architectural Press/Elsevier.

Porter, M. (1998), *The Competitive Advantage of Nations*, The Free Press.

Porter, M. (2004), *Competitive Advantage: Creating and Sustaining Superior Performance*, The Free Press.

Romanelli, E., Tushman, M.L. (1994), Organisational Transformation as Punctuated Equilibrium: An Empirical Test, *Academy of Management Journal*, 37(5), 1141–1166.

Sacks, R., Eastman, C., Ghang, L., Teichholz, P. (2018), *BIM Handbook: A Guide to Building Information Modeling for Owners, Designers, Engineers, Contractors, and Facility Managers*, 3rd Edition, Wiley.

Thompson, A.A. Jr., Williams, M.P.L.E., Gamble, J.E., Strickland, A. J. III (2022), *Crafting and Executing Strategy – The Quest for Competitive Advantage – Concepts and Cases*, 23rd Edition, McGraw-Hill Irwin.

Thorne, P. (1991), *Mastering Change Management*, McGraw-Hill.

United Nations (2023a), *COP26: Together for Our Planet*, www.un.org/en/climatechange/cop26 (accessed 04/03/2023).

United Nations (2023b), *COP27: Delivering for People and the Planet*, www.un.org/en/climatechange/cop27 (accessed 04/03/2023).

Whittington, R., Angwin, D., Regner, P., Johnson, G., Scholes, K. (2023), *Exploring Corporate Strategy – Text & Cases*, 13th Edition, Pearson.

Wilson, D. (1992), *A Strategy of Change – Concepts and Controversies in the Management of Change*, Routledge.

World Business Council for Sustainable Development (WBCSD) (2021), *Digitalization of the Built Environment: Towards a More Sustainable Construction Sector*, Geneva, Beijing, Delhi, London, New York, Singapore.

3 Building Information Modelling Adoption in the New Zealand Construction Industry

Insights from BIM Experts

Dat Tien Doan, Yan Ma, and Ali Ghaffarianhoseini

Introduction

The construction industry is crucial to economic development, contributing 8.7% of New Zealand's GPD and employing over 292,000 people (MBIE, 2022). However, the industry's rapid growth also increases the burden on natural resources and energy, adding to the harmful effects of carbon emissions (Du et al., 2018). Several new techniques and models have been tested and commercialised to address this issue, including building information modelling (BIM), which provides the construction industry with design sustainability and enhanced productivity.

BIM is useful for the geometric modelling of building performance and construction project management (Bryde et al., 2013). Evidence suggests that BIM benefits include improving building design, increasing construction industry productivity (Wong & Fan, 2013), enabling better communication and collaboration, facilitating the organisation of project schedules and budgets, and allowing rapid analysis of different scenarios and feedback (Bryde, Broquetas & Volm, 2013).

BIM can bring down environmental concerns in the industry, reducing extra costs associated with design changes in subsequent phases of the construction process (Begić & Galić, 2021). However, low innovation, unsafe working practices, slow delivery, and high cost are common criticisms of the construction industry (Oraee et al., 2019). Despite its benefits, certain differences in the understanding of BIM application and limitations in awareness and training create a gap in BIM adoption. Effective plans are needed to analyse cost reduction and control (Lu et al., 2018), and BIM software is not feasible in all construction projects (Santos et al., 2019).

The challenges of BIM adoption are related to staff training, lack of BIM national standards, and software interoperability, with the BIM model not being entirely integrated on a large scale by enterprises in the construction industry (Criminale & Langar, 2017). To address these challenges, this chapter aims to examine the BIM adoption situation in the New Zealand construction industry, providing insights into its benefits, challenges, and solutions based on data collected from BIM projects in New Zealand.

DOI: 10.1201/9781003380559-3

This research is significant because an in-depth understanding of the elements linked with the application and adoption of BIM in New Zealand can be attained and discussed, extending engineering knowledge associated with BIM and its application in the New Zealand construction industry. It will help attain improved and defined design and improve the practice associated with the construction process. The research will also help in understanding what BIM is and how it should be integrated into the construction process in New Zealand. Evaluating aspects of research work involving a review of BIM literature will ensure that the practice of integrating BIM can be thoroughly understood, along with its possible associated challenges. The research work would be beneficial and provide context or a framework for BIM's application in future endeavours.

Literature review

The construction industry continually needs to develop new buildings and designs to contribute to the economy, which requires effective technology and knowledge (Olanrewaju et al., 2021). However, the negative impact of construction processes on societies and people has led to a need to integrate green design into buildings, resulting in the emergence and application of BIM. Rahman and Sainati (2021) concluded that BIM is one of the critical strategies to improve productivity in the construction industry.

Despite modern technology and world-class tools and experts, the construction industry is limited by low productivity, frequently facing changes in the delivery schedule, leading to project delays and increased costs (Olanrewaju et al., 2021). This results in a lack of achievement of construction project objectives and meeting government goals, with cost overruns being a common issue faced by governments and clients worldwide (Amri & Marey-Pérez, 2020). Nair and Suresh (2021) highlighted the need for well-planned and documented construction projects, as a well-planned building model could have prevented project extensions and delays due to unforeseen circumstances.

To overcome these issues, technological advancements such as BIM are being used. BIM involves the digital representation of the physical and functional characteristics of a facility designed or proposed to be designed. Rowlinson (2017) indicated that BIM is a medium of improvement in the construction industry that can make great leaps forward. BIM can bring about a fundamental change in the construction industry that can improve the designs of the buildings, the way they are constructed, and the operating standards they adopt (Ryan et al., 2021).

Although BIM has extensive benefits, there is a lack of integration and acceptance of the model, with key issues surrounding high implementation costs, weak leadership, resistance to change, lack of collaboration, lack of awareness, absence of training, unclear ownership data, and incomplete standards of BIM systems. The four critical issues in BIM implementation include execution of BIM, information and management processing, cost-based issues, and issues associated with legal technicalities (Dixit et al., 2019). Insufficient direction from the government, organisational issues, and insufficient external motivation are critical challenges to BIM adoption (Zhou et al., 2019), with unclear ownership data and incomplete standards of BIM systems also requiring attention (Sun et al., 2020).

While previous studies have provided valuable insights into the benefits and challenges of BIM adoption in the construction industry, it is essential to note that the context of each country or region may vary in terms of their unique challenges, regulatory environment, and industry norms. Therefore, it is necessary to conduct research focusing on the New Zealand construction industry to better understand the barriers to BIM adoption in this context and identify strategies for overcoming them. By conducting research on BIM adoption in the New Zealand construction industry, this research aims to fill a critical gap in the existing literature and provide practical recommendations for policymakers, industry stakeholders, and construction professionals to improve BIM adoption and leverage its potential benefits for sustainable construction.

Research methodology

This research aims to understand the effectiveness of adopting BIM in New Zealand's construction industry. A qualitative approach was used to gain familiarity with the topic and generate insights for future research. Specifically, this research adopted the semi-structured interview method, providing deep, rich observational data (Doan et al., 2021).

The sampling process involved a combination of purposive sampling and snowball sampling. Purposive sampling was used to ensure that the participants met the desirable criteria (Campbell et al., 2020). Due to the shortage of individuals and BIM specialists in New Zealand's construction industry, snowball sampling was also used to identify critical stakeholders of the construction industry. The interviewees were required to have at least five years of experience in the construction industry and be involved in a BIM project in New Zealand.

Table 3.1 presents the demographics of the five interviewees. They hold critical positions at large companies with at least five years of experience. Two out of five are managers; the rest are construction workers, designers, and site supervisors. They were chosen for the research because of their rich experience in the construction industry, representing different roles and responsibilities.

Table 3.1 Details of the interviewees

Interviewee	*Size of Company*	*BIM Experience*	*Position*
Interviewee 1	Large	5 years	Construction worker
Interviewee 2	Large	7 years	Manager
Interviewee 3	Large	10 years	Designer
Interviewee 4	Large	8 years	Manager
Interviewee 5	Large	10 years	Site supervisor

The data collected in this research was analysed using Nvivo software, which is a common and efficient tool for qualitative data analysis. This software has been shown to provide benefits such as increased efficiency, multiplicity, and transparency in the analysis process (Feng & Behar-Horenstein, 2019). The thematic analysis was conducted in six steps. The first step involved familiarising the researchers with the collected data and generating codes. The third step was to search for and review themes, then define and name the themes and produce the final report (Vaismoradi et al., 2016). To ensure the validity and reliability of the thematic analysis, appropriate measures were taken, including transcribing, checking codes, peer review, and triangulation (Castleberry & Nolen, 2018; Nowell et al., 2017).

Results and discussion

BIM perception

Most interviewees in this research were familiar with the concept of BIM and provided similar definitions when asked to define it. Interviewee 1 stated, "BIM is a technology used to digitally present the functional characteristics of a building. It also provides the layout of the building, making it efficient during the development of new construction". The interviewees agreed that BIM should be mandatory in the New Zealand construction industry. However, they acknowledged that time is necessary for construction companies to prepare and train their staff on BIM use. Interviewee 2 stated, "BIM should be made mandatory as it reduces delay and resonates with New Zealand's sustainable development goals on sustainable cities and climate action". The interviewees' shared interest is to prioritise BIM implementation and make it mandatory for all construction stakeholders.

All interviewees expressed positive views towards BIM, indicating bright prospects for BIM application in the New Zealand construction industry. This finding is consistent with the study conducted by Latiffi et al. (2017) despite the differences in context between Malaysia and New Zealand. In other words, this research confirms potential benefits of BIM adoption that can transform the industry (Abanda et al., 2015).

BIM adoption in New Zealand

The interviewees revealed that BIM has mainly been used in the design phase and has been adopted primarily by designers in the New Zealand construction industry. Contractors and suppliers have not fully integrated the BIM model. The interviewees also provided their perspectives on the different dimensions of BIM. Interviewee 4 expressed that "4D BIM facilitates risk mitigation as it improved team coordination, but it was not effective due to having no available classification systems for the 4D BIM". Interviewee 2 noted that "6D is more efficient, 5D is reliable, and 4D is affordable". Interviewee 3 shared that "unlike 5D and 6D, the construction practitioners are aware of the implementation and use of 4D". Interviewee 4 highlighted that "4D BIM is easily accessible, easy to use, and can be easily integrated into the system. For both 5D and 6D, they are efficient but very costly to implement".

The research findings indicate that BIM implementation in New Zealand is inadequate, with only a few stakeholders using it and low levels of adoption. Elmualim and Gilder (2014) suggested that a significant barrier to BIM adoption is the lack of capital, which means that implementation costs outweigh the benefits. This could explain why the BIM adoption level in New Zealand is not high. Moreover, the current status quo of inadequate BIM adoption in New Zealand could also be due to high initial costs, a lack of training, and cultural resistance (Harrison & Thurnell, 2015).

Advantages of BIM adoption in New Zealand

The research on adopting BIM in the New Zealand construction industry revealed several benefits. According to the interviewees, BIM adoption reduces the number of reworks, improves productivity and efficiency, saves time, and enhances collaboration and communication among project teams.

The findings indicate that using BIM reduces the number of reworks and facilitates the timely delivery of project objectives and goals. According to interviewee 1, "I have always been discouraged with the time wasted in redoing the same project over and over". "Using BIM reduces those changes as it can be easily projected, and proper plans implemented" (interviewee 2). Also, "when using BIM, it is easier to increase the accuracy of the information shared, which makes it easier to manage and reduce errors associated with it" (interviewee 5). This sentiment is supported by Gamil and Rahman (2019), who emphasised the significance of efficiency and reducing reworks among construction workers.

Improving productivity and efficiency is another benefit associated with BIM adoption. According to the interviewees, using BIM improves the overall productivity of construction projects, facilitating the timely transfer of construction information among stakeholders. "I think with the use of BIM, the management will increase its overall productivity" (interviewee 3). The efficiency of construction is also improved through leveraging different design optimisation models. Interviewee 4 believed that "the use of BIM was undertaken in improving the efficiency in the operations of the organisation". Using graphical and non-graphical information management and advanced engineering analysis enhances the project's success in the long run. Moreover, interviewee 5 emphasised that "the staff is motivated where there is efficiency of the operations and enhancing their commitment to cost reduction".

These findings are consistent with Xu's (2017) study, which showed that BIM adoption can improve productivity and efficiency in construction projects. By implementing BIM, the speed and efficiency of the construction process can be improved, leading to cost and time savings. Xu (2017) also noted that integrating BIM into New Zealand's construction sector can improve engineers' real-time operation. Xu et al. (2021) reiterated the positive impact of BIM on productivity and efficiency, highlighting its ability to reduce energy consumption and greenhouse gas emissions. According to their study, using BIM resulted in a 32% reduction in greenhouse gas emissions, which can be cost effective for companies that need to meet emissions-related legal requirements. Therefore, BIM adoption can improve productivity and efficiency in construction projects and help companies address environmental regulations.

Timesaving is another advantage of BIM adoption, with interviewees linking the efficiency of BIM use with reduced errors, which is critical when improving the timely delivery of operations. Delays are a common issue in the construction industry, often linked to scheduling. Interviewee 1 noted, "I have always been fascinated with the effectiveness and flexibility of using BIM in scheduling of the operations". "Every project manager is interested in being productive and realising the timely delivery of its operations, which can be achieved with BIM adoption" (interviewee 3). "With BIM, it is easier to increase the accuracy of the information shared, which makes it easier to manage and reduce errors associated with it" (interviewee 4). The importance of reducing construction reworks can be addressed by implementing technology that aligns with BIM. Qin et al. (2020) highlighted some of the technologies that should be implemented, including the technology acceptance model (TAM) and technology organisation environment (TOM), which are essential in improving the efficiency of operations.

The adoption of BIM in construction projects has been shown to improve collaboration and communication among project teams, leading to greater efficiency and productivity. According to interviewee 5, "I understand that BIM necessitates communication in the departments, which facilitates engagement of the team in every stage of construction". Improved collaboration can also enhance the team's commitment to developing a framework for achieving strategic goals in the New Zealand construction industry. As interviewee 5 explained, "I find it effective for the construction team to be involved in the implementation of the project, which can be done where there is online engagement". By sharing information on asset assessment and operational efficiencies, construction teams can improve business continuity and facilitate project success.

Understanding the engagement process of key stakeholders is necessary to realise the project's strategic objectives. This finding is consistent with the New Zealand Government Procurement's (2019) recommendation for workflow improvement and transparency in the construction sector. Effective communication and collaboration among stakeholders are critical in improving productivity and efficiency. BIM provides a platform for addressing issues related to functional and operational areas in the construction process. As such, documenting achievements in the industry and understanding the policy framework for the New Zealand construction industry can help capitalise on emerging trends and strategies in achieving strategic goals.

Challenges to BIM adoption in New Zealand

Based on the interviews, various challenges to BIM adoption were highlighted, including the high cost of implementation, weak leadership, resistance to change, lack of collaboration, and lack of awareness and training.

One of the primary challenges identified by the interviewees was the high cost of implementing BIM. The interviewees noted that many construction projects operate within limited budgets, and the cost of acquiring and implementing BIM can be a significant obstacle for new construction companies. For example, interviewee 2 stated that "the major challenge for BIM is the cost of acquisition and implementation", while interviewee 1 noted that "annual licence fees and subscriptions for BIM can be unsustainable for new construction companies". The interviewees suggested that proper planning could help to reduce the cost of implementing BIM. However, they also emphasised that the high cost of implementation remains a significant barrier to adoption. Interviewee 4 stated that "construction companies need to allocate more funding to the acquisition of BIM software". Reducing operational costs has always been a priority for most construction companies, and efficiency in resource usage is crucial for their success. However, cost overruns are common in construction projects, and implementing new technology like BIM can further increase construction costs.

This finding is consistent with Amri and Marey-Pérez's (2020) research, which emphasises the importance of addressing cost overruns in construction projects. In summary, the high cost of implementing BIM is a major challenge for construction companies, particularly those with limited budgets. Proper planning can help reduce the implementation cost, but construction companies may need to allocate more funding to acquire BIM software. The implications of cost overruns in construction projects further highlight the need to address the challenges associated with BIM adoption.

Weak leadership is identified as a challenge to BIM adoption in the construction industry. Interviewee 2 stated, "I know that weak leadership is associated with lack of proper implementation of policies and guiding principles for the organisation. This can be detrimental to the success of the project management". Interviewee 3 added, "in the implementation of BIM, consideration should be on effective allocation of resources and funding, and I think weak leadership can derail this process". The interviewees suggested that improving leadership can lead to strategic success in the construction industry. Interviewee 1 commented, "weak leadership can prevent the effective implementation of BIM". The concern with weak leadership is the challenge of providing training and direction for employees. As Rowlinson (2017) noted, integrating and accepting BIM is possible when there are strong leadership strategies involving planning, budgeting, controlling, and managing operations. Construction workers should be allowed to develop ideas that increase their commitment and realise the efficiency of the whole process. The leadership strategy employed in the New Zealand construction industry should be the basis for achieving strategic goals and minimising the cost implications of the project. When stakeholders are aligned with the requirements of the New Zealand construction industry, they can enhance their commitment and reduce issues that result in a lack of trust among employees.

Most interviewees highlighted that there is resistance to change among employees, which can hinder the adoption of BIM. Interviewee 3 stated, "I believe that the implementation of the BIM can be affected by employees not willing to accept change as they are used to traditional methods of construction in New Zealand". Some employees may be resistant to BIM adoption and prefer traditional construction methods, which can impede the achievement of project goals, as reported by interviewee 4: "Change is inevitable, but some employees might not accept change, either due to being unskilled or interested in their traditional construction methods". This resistance to change can have long-term implications for developing the construction industry in New Zealand, affecting its credibility and efficiency.

Construction companies may be reluctant to embrace technology and stick to outdated strategies. However, addressing this challenge can enhance the efficiency of operations in the construction industry and help realise strategic goals in the long run. This research's findings align with Sun et al. (2015), who found that resistance to change can delay the adoption of BIM. Juszczyk et al. (2015) also suggested that new approaches can be implemented to encourage innovation and change in the industry. Despite the extensive benefits of BIM, its acceptance and integration into the industry can be challenging. Effective leadership and team commitment can help address this challenge. Leaders should encourage their teams to be committed to meeting emerging demands in the New Zealand construction industry.

Lack of collaboration among employees and stakeholders is a significant challenge to the implementation of BIM adoption. The interviewees pointed out that effective collaboration is necessary for project success and achieving strategic goals. However, some managers can be inflexible in their approach, hindering collaboration and jeopardising BIM adoption. According to interviewee 3, "this cannot be achieved where there are departmental differences, and there is lack of understanding of the roles that they have to undertake within the organisation". The lack of collaboration can also affect employees' commitment towards implementing BIM. "Different departments may not commit towards the strategic goals of the organisation, which can be challenging for BIM adoption", as mentioned by interviewee 1.

The commitment of all stakeholders is essential to achieving timely delivery of products and services and realising intended goals. The study by Durdyev et al. (2021) supports this finding, highlighting that a lack of collaboration among stakeholders delays the implementation of BIM. Thus, effective leadership is necessary to encourage collaboration and minimise lack of commitment to the operations of the New Zealand construction industry. Leaders can provide direction on the significant measures to achieve strategic objectives and encourage collaboration among stakeholders.

Interviewees highlighted the lack of awareness of BIM, which can result in high costs for the construction company. Interviewee 1 stated that "some employees might lack the knowledge and skills to implement the new technology". The construction industry can benefit from implementing different technologies to achieve strategic goals, but the interviewees noted that some employees struggle to distinguish between the various BIM technologies and elements that can be used in construction (interviewee 5). Interviewees also emphasised the lack of training for employees. "On-job training of the employees tends to be prioritised in the implementation of BIM" (interviewee 4).

The interviewees recognised that the effectiveness of any technology implementation requires proper training on its use and functioning. However, some interviewees claimed that on-job training may not always provide enough information to realise strategic goals, which are critical in adopting new technologies. Interviewee 2 added, "I think lack of proper training and resources influences the desire of the employees in using BIM". Commitment from employees is necessary to achieve success in BIM. The lack of awareness among stakeholders can also limit their appreciation of the contribution of BIM to the New Zealand construction industry. Leaders can provide materials and case studies on how to implement BIM to improve stakeholder understanding. The findings support Dixit et al.'s (2019) views on the lack of awareness in executing BIM, including cost-based challenges, information and management processing, and issues with legal technicalities. Lack of training exacerbates these challenges, making it difficult to implement the model with ease.

Solutions for BIM adoption in New Zealand

According to the interviewees, the design of BIM can be critical in improving the implementation of all employees involved in BIM projects. Interviewee 1 stated, "I believe that the design can be improved by including the requirements of the New Zealand construction". Similarly, interviewee 2 suggested that "construction companies should be aware of the design to be implemented". By understanding the requirements of New Zealand's construction industry, mechanisms can be implemented that improve the use of BIM. Interviewee 1 further expressed, "I think aligning the design with the requirements of the construction legislation benefits the companies implementing BIM". Understanding the changes reported in New Zealand's construction industry, BIM can effectively improve building designs that align with the requirements and commitment of relevant stakeholders in enacting the strategy.

Shirowzhan et al. (2020) indicated that providing an allowance for the construction industry to have fixed software infrastructure for each stakeholder can help commit them to meeting the emerging trends in the industry. According to Shirowzhan et al. (2020) more powerful and efficient tools can be used to link scheduling and computer-aided design (CAD), which is essential in improving the functionality for object splitting and object granularity. By implementing these solutions, building designs can be done efficiently and reliably, improving the team's overall success.

Interviewees claimed that understanding the legal contracts existing in New Zealand's construction industry helps them to be in sync with the requirements. Interviewee 1 stated, "I believe it can be a challenge to change the contracts, but understanding the New Zealand construction industry's regulations and the possibility of aligning the current contract with the existing regulations helps in realising success". The intention is to reduce the cost of implementation, which can be achieved by aligning the legal contracts with the requirements of the construction industry. The interviewees highlighted the significance of aligning the existing contracts with New Zealand's construction industry requirements.

The findings also indicate that understanding the legality of the contracts associated with New Zealand's construction industry can effectively reduce the barriers to implementing BIM. The findings supported the views of Ryan et al. (2021), who highlighted the importance of understanding the issues with minor modifications undertaken in the contract forms. Ryan et al. (2021) stated that it is essential to design the BIM protocol by aligning it with New Zealand's construction industry requirements. As such, the team should be involved in an in-depth analysis of New Zealand's construction industry to formulate strategies that can implement BIM successfully.

This finding is crucial as it provides information to the construction companies for understanding the mechanism of reducing the inefficiencies associated with implementing BIM. This offers an insight into the success requirements for the construction companies involved in New Zealand's construction industry.

Interviewees pointed out the significance of training employees to understand new technologies and their effectiveness in the construction industry. Interviewees reported on the relevance of training, which is integral to the design process as employees need to understand the new measures of implementing strategies relevant to New Zealand's construction industry. "Some of the technology changes can be beyond the employees, requiring training for them to be successful in the implementation of the strategy" (interviewee 2). Training employees helps them understand and implement BIM in the construction industry. McGraw Hill Construction (2014) similarly found that training construction workers on the tools used in BIM can minimise the time used in scheduling and implementing the strategy in the long run.

Implementing BIM in the construction industry is crucial for improving efficiency and reducing costs. The design of BIM plays a critical role in the successful implementation of BIM projects. According to the interviewees, aligning the BIM design with the requirements of the New Zealand construction industry can significantly improve the use of BIM. Understanding the legal contracts and regulations associated with New Zealand's construction industry can also reduce the barriers to implementing BIM. Furthermore, training employees on using new technologies and BIM is essential for successful implementation.

The findings of this research suggest that construction companies should consider aligning their BIM design and legal contracts with New Zealand's construction industry requirements to improve the effective use of BIM. They should also invest in employee training to reduce the time to implement the strategy and achieve long-term success. Overall, the research underscores the significance of understanding the requirements of the construction industry and the need for strategic planning to implement BIM effectively in the industry.

Conclusion

The study aimed to explore the advantages and challenges of BIM adoption in the New Zealand construction industry, as well as to identify potential solutions to these challenges. The findings highlighted that BIM can significantly improve the accuracy and efficiency of construction projects in New Zealand, resulting in fewer reworks, increased productivity, timesaving, and better collaboration among stakeholders. However, there are several challenges that construction practitioners face in implementing BIM, including the high cost of implementation, resistance to change, lack of collaboration, awareness and training, weak leadership, and legal and contractual issues.

To address these challenges, the study recommends prioritising training for construction practitioners and aligning building design with the requirements of the New Zealand construction industry. Additionally, it is essential to understand the legal contracts prevailing in New Zealand and determine minor adjustments that will not have legal ramifications for the construction company.

The study demonstrates the effectiveness of implementing BIM and the benefits that construction practitioners can gain from adopting it in New Zealand's construction industry. The study provides insights into construction practitioners' challenges in implementing BIM and suggests potential solutions to overcome them. It is hoped that the findings of this study will contribute to the adoption and successful implementation of BIM in the New Zealand construction industry.

To establish the appropriate mechanisms for implementing BIM in New Zealand's construction industry, a joint research effort with the government is recommended. This effort would shed light on the contractual legal issues and adoption of BIM, focusing on the construction management process. Several key issues must be addressed, including procurement management measures, contract management, and information management initiatives. Furthermore, it is crucial to understand the government's and construction clients' perceptions of the proposed solutions for New Zealand's construction industry, particularly about building design improvements and their alignment with regulations.

Future research in this area should also assess current legal contracts and consider minor modifications that could be implemented to facilitate BIM adoption. Additionally, it is critical to evaluate the significance of training for construction practitioners, as their level of training could impact the credibility and reliability of BIM adoption in the construction industry. Understanding the effects of these changes in the technology world is vital for future researchers seeking to advance this field.

References

Abanda, F. H., Vidalakis, C., Oti, A. H., & Tah, J. H. (2015). A critical analysis of Building Information Modelling systems used in construction projects. *Advances in Engineering Software*, *90*, 183–201. https://doi.org/10.1016/j.advengsoft.2015.08.009

Amri, T., & Marey-Pérez, M. (2020). Towards a sustainable construction industry: Delays and cost overrun causes in construction projects of Oman. *Journal of Project Management, 5*, 87–102. https://doi.org/10.5267/j.jpm.2020.1.001

Begić, H., & Galić, M. (2021). A systematic review of construction 4.0 in the context of the BIM 4.0 premise. *Buildings, 11*(8), 337. http://doi.org/10.3390/buildings11080337

Bryde, D., Broquetas, M., & Volm, J. (2013). The project benefits of Building Information Modelling (BIM). *International Journal of Project Management, 31*(7), 971–980. https://doi.org/10.1016/j.ijproman.2012.12.001

Campbell, S., Greenwood, M., Prior, S., Shearer, T., Walkem, K., & Young, S. et al. (2020). Purposive sampling: Complex or simple? Research case examples. *Journal of Research in Nursing, 25*(8), 652–661. https://doi.org/10.1177/1744987120927206

Castleberry, A., & Nolen, A. (2018). Thematic analysis of qualitative research data: Is it as easy as it sounds? *Currents in Pharmacy Teaching and Learning, 10*(6), 807–815. https://doi.org/10.1016/j.cptl.2018.03.019

Criminale, A., & Langar, S. (2017). Challenges with BIM implementation: A review of literature. In 53rd ASC Annual International Conference Proceedings (pp. 329–335).

Dixit, M., Venkatraj, V., Ostadalimakhmalbaf, M., Pariafsai, F., & Lavy, S. (2019). Integration of facility management and Building Information Modeling (BIM). *Facilities, 37*(7/8), 455–483. https://doi.org/10.1108/f-03-2018-0043

Doan, D., Hoseini, A., Naismith, N., Hoseini, A., Zhang, T., & Tookey, J. (2021). Examining critical perspectives on Building Information Modelling (BIM) adoption in New Zealand. *Smart and Sustainable Built Environment, 10*(4), 594–615. https://doi.org/10.1108/sasbe-04-2020-0043

Du, Q., Lu, X., Li, Y., Wu, M., Bai, L., & Yu, M. (2018). Carbon emissions in China's construction industry: Calculations, factors and regions. *International Journal of Environmental Research and Public Health, 15*(6), 1220. https://doi.org/10.3390/ijerph15061220

Durdyev, S., Mbachu, J., Thurnell, D., Zhao, L., & Hosseini, M. (2021). BIM adoption in the Cambodian construction industry: Key drivers and barriers. *ISPRS International Journal of Geo-Information, 10*(4), 215. https://doi.org/10.3390/ijgi10040215

Elmualim, A., & Gilder, J. (2014). BIM: Innovation in design management, influence and challenges of implementation. *Architectural Engineering and Design Management, 10*(3–4), 183–199.

Feng, X., & Behar-Horenstein, L. (2019). Maximising NVivo utilities to analyse open-ended responses. *The Qualitative Report, 24*(3), 563–571. https://doi.org/10.46743/2160-3715/2019.3692

Gamil, Y., & Rahman, I. (2019). Awareness and challenges of Building Information Modelling (BIM) implementation in the Yemen construction industry. *Journal of Engineering, Design and Technology, 17*(5), 1077–1084. https://doi.org/10.1108/jedt-03-2019-0063

Harrison, C., & Thurnell, D. (2015). BIM implementation in a New Zealand consulting quantity surveying practice. *International Journal of Construction Supply Chain Management, 5*(1), 1–15. https://doi.org/10.14424/ijcscm501015-01-15

Juszczyk, M., Výskala, M., & Zima, K. (2015). Prospects for the use of BIM in Poland and the Czech Republic – Preliminary research results. *Procedia Engineering, 123*, 250–259. https://doi.org/10.1016/j.proeng.2015.10.086

Latiffi, A. A., Brahim, J., & Fathi, M. S. (2017). Building Information Modelling (BIM) after ten years: Malaysian construction players' perception of BIM. *IOP Conference Series: Earth and Environmental Science, 81*(1), 012147.

Lu, W., Lai, C. C., & Tse, T. (2018). *BIM and Big Data for Construction Cost Management.* Routledge.

MBIE (Ministry of Business, Innovation & Employment) (2022). *Building and Construction Sector Trends Biannual Snapshot: May 2022.* www.mbie.govt.nz/building-and-energy/building/building-system-insights-programme/sector-trends-reporting/biannual-snapshots/may-2022/

McGraw Hill Construction (2014). *The Business Value of BIM in Australia and New Zealand: How Building Information Modeling is Transforming the Design and Construction Industry* [Ebook] (1st ed., pp. 18–23).

Nair, G., & Suresh, M. (2021). Challenges faced by construction organisations during Covid-19 era. *IOP Conference Series: Earth and Environmental Science*, *796*(1), 012004. https://doi.org/10.1088/1755-1315/796/1/012004

New Zealand Government Procurement (2019). *Building Information Modelling (BIM): Construction Procurement Guidelines* [Ebook] (2nd ed., pp. 1–10). New Zealand Government.

Nowell, L., Norris, J., White, D., & Moules, N. (2017). Thematic analysis. *International Journal of Qualitative Methods*, *16*(1), 160940691773384. https://doi.org/10.1177/1609406917733847

Olanrewaju, O., Kineber, A., Chileshe, N., & Edwards, D. (2021). Modelling the impact of Building Information Modelling (BIM) implementation drivers and awareness on project lifecycle. *Sustainability*, *13*(16), 8887. https://doi.org/10.3390/su13168887

Oraee, M., Hosseini, M. R., Edwards, D. J., Li, H., Papadonikolaki, E., & Cao, D. (2019). Collaboration barriers in BIM-based construction networks: A conceptual model. *International Journal of Project Management*, *37*(6), 839–854. https://doi.org/10.1016/j.ijproman.2019.05.004

Qin, X., Shi, Y., Lyu, K., & Mo, Y. (2020). Using a TAM-TOA model to explore factors of Building Information Modelling (BIM) adoption in the construction industry. *Journal of Civil Engineering and Management*, *26*(3), 259–277. https://doi.org/10.3846/jcem.2020.12176

Rahman, A., & Sainati, T. (2021). Adapting standard forms of contract to facilitate Building Information Modelling. *Proceedings of the Institution of Civil Engineers – Management, Procurement and Law*, *175*(3), 110–118. https://doi.org/10.1680/jmapl.21.00017

Rowlinson, S. (2017). Building Information Modelling, integrated project delivery and all that. *Construction Innovation*, *17*(1), 45–49. https://doi.org/10.1108/ci-05-2016-0025

Ryan, A., Miller, G., & Wilkinson, S. (2021). Successfully Implementing Building Information Modelling in New Zealand: Maintaining the Relevance of Contract Forms and Procurement Models [Ebook] (pp. 1–11). University of Auckland. www.library.auckland.ac.nz/external/finalproceeding/Files/Papers/46530final00150.pdf.

Santos, R., Costa, A. A., Silvestre, J. D., & Pyl, L. (2019). Informetric analysis and review of literature on the role of BIM in sustainable construction. *Automation in Construction*, *103*, 221–234. https://doi.org/10.1016/j.autcon.2019.02.022

Shirowzhan, S., Sepasgozar, S., Edwards, D., Li, H., & Wang, C. (2020). BIM compatibility and its differentiation with interoperability challenges as an innovation factor. *Automation in Construction*, *112*, 103086. https://doi.org/10.1016/j.autcon.2020.103086

Sun, C., Jiang, S., Skibniewski, M., Man, Q., & Shen, L. (2015). A literature review of the factors limiting the application of BIM in the construction industry. *Technological and Economic Development of Economy*, *23*(5), 764–779. https://doi.org/10.3846/20294913.2015.1087071

Sun, C., Xu, H., & Jiang, S. (2020). Understanding the risk factors of BIM technology implementation in the construction industry: An interpretive structural modelling (ISM) approach. *Engineering, Construction and Architectural Management*, *27*(10), 3289–3308. https://doi.org/10.1108/ecam-09-2019-0508

Vaismoradi, M., Jones, J., Turunen, H., & Snelgrove, S. (2016). Theme development in qualitative content analysis and thematic analysis. *Journal of Nursing Education and Practice*, *6*(5), 100–110. https://doi.org/10.5430/jnep.v6n5p100

Wong, K., & Fan, Q. (2013). Building Information Modelling (BIM) for sustainable building design. *Facilities*, *31*(3/4), 138–157. https://doi.org/10.1108/02632771311299412

Xu, J. (2017). Research on application of BIM 5D technology in central grand project. *Procedia Engineering*, *174*, 600–610. https://doi.org/10.1016/j.proeng.2017.01.194

Xu, X., Mumford, T., & Zou, P. (2021). Life-cycle building information modelling (BIM) engaged framework for improving building energy performance. *Energy and Buildings*, *231*, 110496. https://doi.org/10.1016/j.enbuild.2020.110496

Zhou, Y., Yang, Y., & Yang, J. (2019). Barriers to BIM implementation strategies in China. *Engineering, Construction and Architectural Management*, *26*(3), 554–574. https://doi.org/10.1108/ecam-04-2018-0158

4 Assessing the Awareness and Adoption of Robotics/Automation in Foreign and Indigenous Firms in Nigeria

Oluwadamilare Olamide Ilesanmi, Ayodeji Emmanuel Oke, John Aliu, and John Oluwaseun Atolagbe

Introduction

The awareness and adoption of robotics and automation in the construction industry vary significantly worldwide (Aghimien et al., 2024). Developed nations, driven by technological advancements and investment, incorporate these innovations to enhance productivity, quality, and safety across construction processes (Jaiswal et al., 2024). In contrast, developing countries face challenges such as limited infrastructure, high costs, and a lack of technical expertise, which often result in slower adoption rates (Ilesanmi et al., 2024). Globally, the construction industry continues to shift toward robotics and automation, recognizing their potential to address labor shortages, increase efficiency, and improve precision (Umoh et al., 2024).

In Nigeria, the construction sector partially reflects this global trend but encounters notable constraints (Oke et al., 2024). Awareness of robotics and automation remains low, especially among local construction firms (Chan et al., 2024). While foreign firms operating in Nigeria may bring advanced technologies from their home countries, local firms face barriers that limit both awareness and adoption (Babalola and Harinarain, 2024). As a result, indigenous firms often lag in adopting robotics and automation, unlike their foreign counterparts, who sometimes utilize these technologies to gain a competitive advantage (Wang et al., 2024).

Increased awareness and adoption of these technologies in Nigeria's construction industry could yield substantial benefits, including enhanced safety, cost-effectiveness, and project efficiency (Ebekozien et al., 2023). Strategies such as government incentives, industry training programs, and collaborations between academia and industry are essential to encourage more widespread technological integration (Moyanga et al., 2024). Understanding the current state of adoption and exploring effective strategies could help bridge the gap between local and foreign firms in Nigeria.

This study aims to assess and compare the levels of awareness and adoption of robotics and automation between foreign and indigenous construction firms in Nigeria. Three main research questions guide this study:

DOI: 10.1201/9781003380559-4

1 What is the level of awareness of robotics and automation in construction firms?
2 What is the level of adoption of robotics and automation in construction firms?
3 How do foreign and indigenous firms differ in their levels of awareness and adoption of robotics and automation?

Previous studies have extensively examined the use of automation and robotics in construction industries across countries like China (Yan et al., 2023), the United States (Al Bashar et al., 2024), Malaysia (Almatari et al., 2024), South Africa (Ikuabe et al., 2023), and several European nations (Schöberl et al., 2023; Damiani et al., 2023; Piras et al., 2024). Most research has focused on specific technologies, such as three-dimensional (3D) printing (Volpe et al., 2024), the Internet of Things (Dosumu and Uwayo, 2023), digital twin technology (Arowoiya et al., 2024), and cyber-physical systems (Adeosun and Oke, 2024). However, few researchers have evaluated awareness and adoption levels specific to robotics and automation, and none have directly compared adoption between foreign and indigenous firms. By examining the degree of awareness and adoption within both types of firms, this study fills a notable gap in the literature. This research contributes valuable insights into the distinct challenges and opportunities each type of firm faces in adopting these technologies. Highlighting disparities and proposing solutions to enhance adoption, the study underscores the importance of increased awareness and adoption. The originality of this study lies in its comparative approach, providing a nuanced understanding of technological adoption across diverse organizational contexts within Nigeria.

Literature Review

Issues in Awareness and Adoption of Robotics and Automation in the Nigerian Construction Industry

The Nigerian construction industry has experienced limited adoption of robotics and automation technologies, a trend that reflects significant challenges in achieving higher productivity and efficiency (Oke et al., 2024). Although digitization is increasingly recognized as essential to modernizing the industry, barriers such as high costs, inadequate infrastructure, and limited technical expertise hinder the widespread adoption of advanced technologies (Ilesanmi et al., 2024). Many firms lack sufficient awareness of the benefits robotics and automation offer, including improved precision, reduced labor dependency, and enhanced safety (Babalola and Harinarain, 2024). Consequently, the industry remains largely reliant on traditional methods, leading to slower project completion times, increased safety risks, and higher costs (Arowoiya et al., 2024). Currently, the adoption of technology within Nigeria's construction sector is relatively low, with digital tools like Building Information Modeling and some automation tools gaining ground only in recent years (Banmgbose et al., 2024). However, awareness and acceptance of robotics and automation remain minimal, particularly among local firms, which are further constrained by budget limitations and limited access to technical training (Oke et al., 2023). The consequences of this low adoption rate are evident in project inefficiencies, frequent delays, and quality inconsistencies (Kineber et al., 2022). Embracing robotics and automation could fundamentally transform the Nigerian construction industry, offering substantial benefits in terms of productivity, quality, and cost-effectiveness (Adepoju and Aigbavboa, 2020). With increased awareness and strategic investment, the successful adoption of these technologies could drive a much-needed shift, addressing industry setbacks and enhancing the overall competitiveness of Nigeria's construction sector (Ebekozien et al., 2023).

An Overview of Foreign and Indigenous Construction Firms

Globally, the construction sector shows a clear divide between foreign and indigenous firms, with varying levels of technological adoption and project execution (Yusuf et al., 2022). In developed countries, foreign construction firms often integrate advanced robotics and automation technologies to enhance productivity and efficiency, benefiting from well-established infrastructure and skilled labor pools (Abishov and Mamutova, 2024). However, in developing countries, these technologies are often limited to foreign firms with access to global capital and resources (Selcuk et al., 2024). Indigenous construction firms in these regions face limited funding, outdated practices, and a shortage of skilled labor, which hinder the adoption of advanced technologies like robotics and automation . In Nigeria, foreign construction firms have been active since colonial times, establishing companies like Cappa and D'Alberto in 1932 (Yusuf et al., 2022). These firms have maintained a strong presence in large-scale construction projects, often employing advanced technologies (Oke et al., 2023). Their impact on Nigeria's construction landscape is significant, as they handle the majority of high-value projects (Kineber et al., 2022). However, while they bring technical expertise, their dominance limits the growth and competitiveness of indigenous firms (Kumar et al., 2024). Indigenous construction firms in Nigeria are typically smaller and less equipped to adopt robotics and automation due to financial and technical constraints (Denny-Smith et al., 2024). They control only a fraction of large projects, completing around 5% of civil engineering work and 25% of general construction (Kumar et al., 2024). Local firms struggle with performance issues, such as poor project management and inadequate technical capabilities, often resulting in project delays and cost overruns (Rajput et al., 2024). This reliance on foreign firms hampers indigenous growth and reduces local expertise (Fu et al., 2024). The successful adoption of robotics and automation across foreign and indigenous firms could drive efficiency, reduce costs, and improve project timelines, ultimately transforming Nigeria's construction industry (Oke et al., 2023).

Robotics and Automation Technologies in Construction

The global construction industry has increasingly embraced robotics and automation to address challenges related to labor shortages, efficiency, safety, and cost-effectiveness (Umoh et al., 2024). In developed countries, these technologies have gained substantial traction, contributing significantly to productivity and quality improvements (Chan et al., 2024). However, in developing countries like Nigeria, adoption remains limited due to technological and economic barriers (Oke et al., 2023). Nonetheless, rising awareness of these technologies presents a pathway to harness their potential benefits (Kineber et al., 2022).

Demolition robots, 3D printing robots, robotic drones, bricklaying robots, welding robots, exoskeletons, forklift robots, and roadwork robots are some of the new robotic technologies making a mark in the construction industry (Chang and Hasanzadeh, 2024). Each of these technologies offers distinct potential to improve construction processes in terms of quality, productivity, and safety (Aghimien et al., 2024).

The following sections discuss various robotic technologies currently shaping the construction industry, with a focus on their use, adoption levels, and potential impact in Nigeria.

Demolition Robots

Demolition robots are mechanized tools used to dismantle building components, such as walls and concrete structures while allowing operators to maintain a safe distance (Dindorf and Wos, 2024). In countries like China and the United States, demolition robots are valued for their efficiency in handling hazardous tasks (Zhang et al., 2024). However, in Nigeria, the high cost of procurement and maintenance restricts widespread adoption (Iyiola et al., 2024). Additionally, the lack of trained operators poses a barrier to efficient utilization (Oke et al., 2023). Despite these challenges, demolition robots present an opportunity for safer demolition processes in high-risk environments, particularly for complex deconstruction projects that require precision and safety (Dindorf and Wos, 2024).

3D Printing (Additive Manufacturing)

3D printing, particularly in the form of contour crafting, has transformed construction by enabling rapid, cost-effective production of building components (Parisi et al., 2024). In the Netherlands, China, and the United Arab Emirates, large-scale 3D printers have been used to construct houses and bridges with reduced material waste and construction time (Placzek and Schwerdtner, 2024). While awareness of 3D printing is growing in Nigeria, adoption remains in its infancy, largely due to high investment costs and a lack of locally accessible printing materials (Akande et al., 2023). However, for Nigeria, this technology holds the potential to address affordable housing challenges, as 3D printing can significantly reduce labor and material costs, making it a viable solution for low-cost housing projects (Opawole et al., 2022).

Drones

Drones are widely used in surveying, site monitoring, and transporting lightweight materials (Oyebode et al., 2023). The ability of drones to capture aerial data in real time makes them invaluable for precision mapping, reducing manual surveying time and labor costs (Wahab, 2020). Drones are commonly used in the United States and Europe for project monitoring and safety inspections (Yan et al., 2023). In Nigeria, while awareness of drone technology is high, regulatory constraints and operational costs have limited their use (Patrick et al., 2020). Expanding drone usage in Nigeria's construction industry could streamline surveying processes, improve site safety, and facilitate efficient project monitoring, particularly in urban areas with complex topographies (Onososen et al., 2023).

Bricklaying Robots

Bricklaying robots, such as Semi-Automated Mason (SAM), are designed to enhance construction efficiency by layering bricks at a faster rate and with higher precision than manual labor (Dindorf and Wos, 2024). Widely adopted in Australia and parts of Europe, these robots are used in large-scale infrastructure projects that demand consistent and speedy bricklaying (Usmanov et al., 2021). However, in Nigeria, the high upfront costs and maintenance requirements of bricklaying robots have hindered their adoption (Adepoju and Aigbavboa, 2020). Training the local workforce to operate and maintain these machines remains a challenge (Oke et al., 2023). Despite these obstacles, bricklaying robots could accelerate housing projects and improve construction quality in Nigeria, especially in large-scale residential developments (Ikuabe et al., 2023).

Welding Robots

Welding robots are particularly useful in tasks requiring precision, such as joining steel components in skyscrapers or large infrastructure projects (Guan, 2023). In the United States, welding robots are extensively used in industrial construction, ensuring high-quality welds with reduced human error (Lee et al., 2024). In Nigeria, however, the adoption of welding robots is minimal, primarily due to limited availability and high operational costs (Oke et al., 2023). Enhancing access to welding robots could benefit Nigeria's infrastructure sector by ensuring stronger, more reliable structural connections, which is crucial for high-rise construction and other complex projects (Ikuabe et al., 2023).

Exoskeletons

Exoskeletons are wearable robotic suits that amplify human strength, enabling workers to handle heavy materials with less physical strain (Kineber et al., 2020). In Japan and the United States, construction workers use exoskeletons to perform physically demanding tasks, thereby reducing the risk of injuries (Golabshi et al., 2023). In Nigeria, the application of exoskeletons is still novel, with limited awareness and significant cost barriers (Okpala et al., 2022). Nevertheless, integrating exoskeletons could address labor shortages and enhance worker safety, particularly in tasks requiring heavy lifting, which are common in Nigeria's construction sites (Nnaji et al., 2023).

Forklift Robots

These autonomous robots streamline material handling, moving heavy materials across construction sites without human intervention. In regions such as North America, forklift robots have gained popularity for their ability to reduce reliance on manual labor and optimize logistics (Licardo et al., 2024). In Nigeria, however, the lack of advanced infrastructure and high importation costs have hindered their adoption (Ikuabe et al., 2023). Forklift robots could support Nigeria's construction sector by automating material transportation, enhancing productivity, and reducing risks associated with manual handling of heavy items (Aghimien et al., 2024).

Roadwork Robots

Roadwork robots are designed for tasks such as line painting, surface finishing, and minor road repairs (Zhang et al., 2024). Commonly used in Europe and North America, these robots can quickly and precisely complete tasks with minimal disruption to traffic (Melenbrink et al., 2020). In Nigeria, where road maintenance is critical due to rapid urbanization, roadwork robots could contribute to efficient and safer road construction. However, high costs and technical limitations currently impede their widespread use (Aghimien et al., 2024). Nevertheless, increased awareness and targeted investments could make roadwork robots a viable option for improving road quality and longevity in Nigeria (Adepoju and Aigbavboa, 2020).

Humanoid Robots

Though still in the early stages, humanoid robots are being developed to perform complex tasks in environments that are hazardous to human workers (Okagbue et al., 2024). In South Korea and the United States, humanoid robots are used experimentally in high-risk environments, such as disaster zones (Pradhananga et al., 2021). In Nigeria, humanoid robots remain largely unfamiliar due to their cost and complexity (Ikuabe et al., 2023). However, as awareness of robotics grows, there may be future opportunities for these robots in Nigeria's construction industry, especially in high-risk operations such as confined-space inspections or work in hazardous conditions (Oke et al., 2023).

The Level of Awareness and Adoption of Robotics and Automation in Construction between Foreign and Indigenous Firms

The adoption of robotics and automation in construction spans all phases of the project lifecycle, from initial design through on-site construction to post-completion maintenance, and eventual demolition (Pirisi et al., 2024). This technology covers an array of applications, including computer-aided design for architectural automation, cost estimation, project planning tools, and on-site smart-controlled machinery that enhances precision and efficiency (Arowoiya et al., 2024). The extent of robotics and automation use varies widely across different construction processes and between developed and developing countries. In developed countries, such as Japan, Germany, and the United States, advanced automation and robotics are widely integrated into construction (Pradhananga et al., 2021; Golabshi et al., 2023). For example, Japan has embraced robotics for tasks like welding and bricklaying, which reduces labor demands and boosts project efficiency (Cisneros-Gonzalez et al., 2024).

The use of Industrialized Building Systems has also led to greater output, as prefabricated components can be efficiently assembled on-site (Dindorf and Wos, 2024). This shift from conventional construction methods to prefabrication has made processes faster, safer, and more cost-effective (Oke et al., 2023). In civil engineering, robotic applications in bridge and tunnel construction are now common, with advanced machinery enhancing both productivity and safety (Kineber et al., 2022). However, the level of awareness and adoption in developing countries remains significantly lower. Factors such as high equipment costs, limited technical expertise, and resistance to new technologies hinder adoption (Oke et al., 2023). In Nigeria, the awareness of robotics and automation is rising, yet adoption is still restricted mainly to high-value projects managed by foreign firms (Ikuabe et al., 2023). Indigenous firms often lack the resources to invest in these technologies, limiting their competitiveness in the market (Rajput et al., 2024). Nevertheless, the growing recognition of automation's benefits, including cost savings and enhanced safety, is gradually driving interest among local firms (Yusuf et al., 2022). Despite these challenges, the global construction industry increasingly sees robotics and automation as essential for productivity and safety gains (Denny-Smith et al., 2024). The adoption gap between foreign and indigenous firms underscores the need for targeted initiatives to enhance awareness and reduce the barriers that developing nations face in fully embracing these transformative technologies (Kumar et al., 2024).

Research Methodology

Research Approach and Compilation of Robotic and Automation Technologies, Awareness, and Adoption

This study assesses and compares the levels of awareness and adoption of robotics and automation between foreign and indigenous construction firms in Nigeria. By identifying the unique challenges and opportunities each type of firm encounters, this analysis provides insights into enhancing the adoption of these technologies within the Nigerian construction industry. The cities of Lagos and Abuja were chosen as focal points due to their prominent roles in construction. Lagos is a major urban hub with extensive infrastructure projects, while Abuja, as the capital, hosts significant government and private sector developments (Balogun et al., 2020). Recent data highlight Lagos and Abuja as two of Nigeria's fastest-growing regions (Obia, 2016). To generate an up-to-date understanding of robotic and automation technologies and their adoption, a comprehensive literature review was conducted using databases such as IEEE Xplore, Scopus, Web of Science, Google

Scholar, JSTOR, Emerald Insight, Elsevier, and Taylor & Francis. The selection of the nine robotics and automation technologies utilized in this study was based on their current relevance and visibility within the Nigerian construction industry. While other technologies exist, these nine represent the most recognized and widely discussed in local contexts, providing a focused and manageable scope for the study's objectives.

A questionnaire was selected as the primary tool for data collection, offering a structured approach to capture participants' insights, and enabling detailed analysis of awareness and adoption trends (Moyanga et al., 2023). Target respondents included professionals from contracting and consulting firms as well as government organizations.

Respondent Selection and Sample Sizing Strategy

This study employed a convenience sampling, which selects respondents based on accessibility, availability, and proximity, aligning with Golzar et al.'s (2022) definition of this technique. The sample comprises construction firms in Lagos and Abuja, with the population drawn from 81 registered firms in Abuja's FCT and 193 in Lagos. Sources include professional bodies such as NIQS, NIA, NSE, NIOB, and PMI (2021). The sample included 60 firms in Lagos and 57 in Abuja, totaling 117. The Yamane (1967) formula was used to determine the appropriate sample size for the study.

Questionnaire Design and Data Collection Process

Before distributing the questionnaires, a pilot study was conducted to test the clarity and effectiveness of the survey instrument (Quiles et al., 2019). This pilot involved 15 experienced project and construction managers with expertise in digital systems and smart technologies, whose feedback was instrumental in refining the questionnaire. The final questionnaire consisted of three sections: the first section gathered respondent background information, the second section evaluated the level of awareness of robotics and automation, and the third one assessed adoption levels among construction professionals, distinguishing between foreign and indigenous firms. A five-point Likert scale was used to measure awareness and adoption levels, with awareness ratings from 5 (very high awareness) to 1 (very low awareness), and adoption ratings from 5 (fully adopted) to 1 (not adopted at all). Google Forms was used for survey distribution due to its ease of use and efficiency in data collection. Of the 117 questionnaires distributed, 85 were returned, resulting in a 73% response rate.

Evaluation of Data and Statistical Techniques

Data were analyzed using the Statistical Package for the Social Sciences software, widely used for statistical analysis. Descriptive statistics, including frequency and percentage distributions, were applied to analyze respondents' background information. Additional statistical techniques included calculating mean values and standard deviations to rank and compare data, offering insights into central tendencies and trends. The mean item score tool was used to assess the awareness and adoption level of robotics and automation between foreign and indigenous firms. To test differences in adoption levels, the Mann-Whitney U test and the Wilcoxon signed-rank test were used, with a significance level of 5% ($p < 0.05$). A p-value below 0.05 indicated statistically significant differences between the two groups (Pallant & Bailey, 2005).

Results and Findings

This study assesses the awareness and adoption levels of robotics and automation among Nigerian construction firms. It explores the awareness and adoption of various robotic technologies, comparing indigenous and foreign firms' engagement with these innovations. The findings highlight differences in awareness and adoption between the two groups, offering insights into how each type of firm integrates robotics and automation in construction practices.

Respondents' Background Information

A total of 105 questionnaires were distributed to construction firms in Abuja and Lagos, with 85 successfully retrieved. Table 4.1 presents the types of firms that participated in the survey. Contractors account for 50.6% of the respondents, while 34.1% are consultants or professional service providers, 12.9% are subcontractors, and 2.4% are developers. The second factor in Table 4.1 illustrates the qualifications of the respondents. Among them, 16.5% hold an HND/PGD, 42.2% have a BTech/BSc, 36.5% possess an MTech/MSc, 3.5% hold a PhD, and 1.2% fall under "Others." The third factor in Table 4.1 outlines the areas of specialization of the respondents. Of the respondents, 37.6% had specialization in building, 3.5% in non-residential projects, 35.3% in civil engineering, 4.7% in infrastructure, and 18.8% were categorized as "Others." The fourth factor examines the years of professional experience among the respondents: 35.3% had less than 5 years of experience, 36.5% had 6–10 years of experience, 15.3% had 11–15 years of experience, 7.1% had 16–20 years of experience, and 5.9% had more than 20 years of professional experience. The fifth factor identifies the respondent firms' categories: 87.1% were indigenous firms, while 12.9% were foreign firms.

Level of Awareness of Robotics and Automation in Construction Firms

Table 4.2 presents the mean scores and standard deviations for the level of awareness of nine robotics and automation technologies in construction firms, using a threshold of 3.4 as established by Velasco. Based on this threshold, technologies with a mean score above 3.4 indicate a higher level of awareness, while those below 3.4 show a lower level of awareness. Technologies like drones (4.41), 3D printing and contour crafting robots (4.39), and demolition robots (4.25) have

Table 4.1 Respondents' Background Information

Factor	*Variables*	*Frequency Result*	*Percent (%)*
Firm's Type	*Contractor*	*43*	*50.6*
	Consultant or professional services	29	34.1
	Sub-contractor	11	12.9
	Developer	2	2.4
	Total	**85**	**100**
Academic Qualifications of Respondent	HND/PGD	14	16.5
	BTECH/BSC	36	42.4
	MTECH/MSC	31	36.5
	PhD	3	3.5
	Others	1	1.2
	Total	**85**	**100**
Areas of Specialization	Building	32	37.6
	Non-residential	3	3.5
	Civil engineering	30	35.3
	Infrastructure	4	4.7
	Others specify	16	18.8
	Total	**85**	**100**
Years of Professional Experience	1–5 years	30	35.3
	6–10 years	31	36.5
	11–15 years	13	15.3
	16–20 years	6	7.1
	Above 20 years	5	5.9
	Total	**85**	**100**
Category of Firm	Indigenous firm	74	87.1
	Foreign firm	11	12.9
	Total	**85**	**100**

Table 4.2 Level of Awareness

	Frequency	*Percent*	*Cumulative Percent*
Very high	8	9.4	9.4
High	25	29.4	38.8
Average	42	49.4	88.2
Low	7	8.2	96.5
Very low	3	3.5	100.0
Total	85	100.0	

mean scores above 3.4, reflecting a strong awareness within the construction firms. These technologies are more widely recognized and known by firms, suggesting that they are gaining traction in the industry. Welding robots (4.22), forklift robots (4.21), and bricklaying robots (4.19) also fall above the 3.4 threshold, indicating moderate to high awareness levels, though slightly less so compared to the top three.

On the other hand, technologies like exoskeletons (3.19), roadwork robots (3.16), and humanoids (3.08) have mean scores below the 3.4 threshold, suggesting that these technologies are less known and adopted within the Nigerian construction industry. The higher standard deviations for these technologies (ranging from 0.27 to 1.39) also imply greater variability in awareness, with some firms being more aware than others. This indicates that while some firms might be familiar with these technologies, the general awareness is still quite low. In summary, technologies like drones, 3D printing, and demolition robots are highly recognized in the industry, while exoskeletons, roadwork robots, and humanoids are the least known, according to the 3.4 threshold.

Level of Adoption of Robotics and Automation in Construction Industry

Table 4.3 presents the mean scores and standard deviations for the level of adoption of nine robotics and automation technologies in construction firms, using a threshold of 3.4 as established by Velasco. According to this threshold, technologies with a mean score above 3.4 indicate higher adoption levels, while those below 3.4 suggest lower adoption. Drones (4.11) and 3D printing and contour crafting robots (4.09) have the highest mean scores, both well above the 3.4 threshold, indicating that these technologies have a higher level of adoption in the Nigerian construction industry. Demolition robots (3.85) also show a relatively high level of adoption, though slightly lower compared to drones and 3D printing. Technologies like welding robots (3.67) are also above the 3.4 threshold, suggesting a moderate adoption level in the industry.

In contrast, technologies like roadwork robots (3.21), exoskeletons (3.19), and humanoids (3.09) have mean scores below 3.4, reflecting lower adoption levels.

Table 4.3 Robotics Technology

	Mean	*Std. Deviation*	*Frequency*
Drones	.61	.490	1
3D printing and contour crafting	.39	.490	2
Roadwork robot	.25	.434	3
Welding robots	.22	.419	4
Demolition robots	.21	.411	5
Exoskeleton	.19	.393	6
Forklift robot	.19	.393	7
Bricklaying robots	.16	.373	8
Humanoids	.08	.277	9

These technologies are less frequently used in construction firms, indicating that they are not yet widely implemented. The higher standard deviations for these technologies, particularly humanoids (1.23) and roadwork robots (0.51), suggest significant variability in their adoption, with some firms adopting them more than others. The technologies with the lowest adoption levels are forklift robots (2.86) and bricklaying robots (2.08), both of which fall far below the 3.4 threshold. These technologies have minimal adoption within the Nigerian construction industry, with forklift robots being slightly more adopted than bricklaying robots, but both showing very limited usage. In summary, drones and 3D printing have the highest adoption levels, while forklift robots and bricklaying robots have the least adoption in the industry, according to the 3.4 threshold.

Comparative Evaluation of Robotics and Automation Awareness with the Adoption Levels in Local and Foreign Companies

Table 4.4 presents the Mann-Whitney U and Wilcoxon W test statistics comparing the awareness and adoption levels of robotics and automation technologies between indigenous and foreign firms. For demolition robots, 3D printing and contour crafting, bricklaying robots, and welding robots, both foreign firms and indigenous firms show significant differences in awareness and adoption, with foreign firms generally having higher levels. The p-values for these technologies are less than 0.05, indicating that foreign firms are more aware and have adopted these technologies at higher levels than indigenous firms. The Z-values range from –3.672 to –5.948, confirming this significant difference.

For drones, there is no significant difference in the level of awareness and adoption between indigenous and foreign firms, as shown by the p-value of 0.553, meaning that both types of firms have similar levels of awareness and adoption of drones. For exoskeletons, forklift robots, roadwork robots, and humanoids, foreign firms have higher awareness and adoption compared to indigenous firms, with the p-values all being less than 0.05, indicating significant differences. In summary, foreign firms generally show higher levels of awareness and adoption across most robotics and automation technologies, while indigenous firms lag in adopting and

Table 4.4 Level of Awareness of Automation and Robotics

Category of Firm		*N*	*Mean Rank*	*Sum of Ranks*	*Median*
Level of Awareness of Automation and Robotics	Indigenous firm	74	47.60	3522.50	3.00
	Foreign firm	11	12.05	132.50	1.00
	Total	85			
	Test Statistics, Grouping Variable: Category of firm				
Mann-Whitney U	66.500				
Wilcoxon W	132.500				
Z	−4.828				
Asymp.Sig. (2-tailed)	.000				

being aware of these technologies, with drones being the only exception where both types of firms show similar levels.

Mann-Whitney U Test and the Wilcoxon Signed-Rank Test for the Comparative Analysis

Table 4.5 presents the mean ranks and sum of median ranks for the different types of robotic and automation technologies across indigenous and foreign firms. These values reflect how each firm category ranks the awareness and adoption of these technologies. For demolition robots, 3D printing and contour crafting, welding robots, and bricklaying robots, foreign firms have higher mean ranks than indigenous firms, indicating that foreign firms are more familiar with and likely to adopt these technologies.

The mean ranks for foreign firms range from 61.05 to 70.27, while the mean ranks for indigenous firms range from 38.74 to 42.47. For drones, exoskeletons,

Table 4.5 Ranks

Which Category Is Your Firm		*N*	*Mean Rank*	*Sum of Ranks*	*Median*
Demolition robots	Indigenous firm	74	40.32	2983.5	0
	Foreign firm	11	61.05	671.5	1
	Total	85			
3D printing and contour crafting	Indigenous firm	74	39.51	2923.5	0
	Foreign firm	11	66.5	731.5	1
	Total	85			
Drones	Indigenous firm	74	42.47	3143	1
	Foreign firm	11	46.55	512	1
	Total	85			
Bricklaying robots	Indigenous firm	74	38.95	2882	0
	Foreign firm	11	70.27	773	1
	Total	85			
Welding robots	Indigenous firm	74	39.74	2941	0
	Foreign firm	11	64.91	714	1
	Total	85			
Exoskeleton	Indigenous firm	74	39.59	2930	0
	Foreign firm	11	65.91	725	1
	Total	85			
Forklift robot	Indigenous firm	74	39.67	2935.5	0
	Foreign firm	11	65.41	719.5	1
	Total	85			
Roadwork robot	Indigenous firm	74	38.74	2867	0
	Foreign firm	11	71.64	788	1
	Total	85			
Humanoids	Indigenous firm	74	41.15	3045	0
	Foreign firm	11	55.45	610	0
	Total	85			

and forklift robots, foreign firms again hold higher mean ranks, indicating better awareness and adoption. The mean ranks for foreign firms in these categories are around 55.45 to 71.64, while indigenous firms have lower ranks in these technologies, signaling limited awareness or adoption. In summary, foreign firms consistently rank higher across all robotic and automation technologies, suggesting greater awareness and adoption. Indigenous firms generally rank lower, showing a gap in awareness and adoption compared to their foreign counterparts. This indicates a significant disparity in the use and understanding of these technologies between indigenous and foreign firms in the Nigerian construction industry.

Test Statistics

Table 4.6 presents the results of a statistical analysis comparing the awareness and adoption of various robotic technologies between foreign and indigenous firms in Nigeria. The analysis employs the Mann-Whitney U and Wilcoxon W tests, which are non-parametric tools suitable for comparing two independent groups. The technologies assessed include demolition robots, 3D printing and contour crafting robots, drones, bricklaying robots, welding robots, exoskeletons, forklift robots, roadwork robots, and humanoid robots. The results indicate statistically significant differences in awareness and adoption levels for all technologies, as all p-values (asymptotic significance, two-tailed) are less than 0.05. The negative Z-scores suggest that indigenous firms consistently rank lower in awareness and adoption compared to their foreign counterparts. Technologies such as humanoid robots, roadwork robots, and forklift robots exhibit the most significant disparities, as reflected by higher Mann-Whitney U and Wilcoxon W values. In contrast, drones and bricklaying robots show relatively smaller differences, indicating closer levels of awareness and adoption between the two groups. These findings highlight a substantial technological gap between foreign and indigenous firms in Nigeria. While foreign firms demonstrate higher awareness and adoption rates, indigenous firms require targeted interventions to enhance their capacity, particularly for advanced robotics such as humanoid robots and roadwork robots. Addressing this gap is crucial for fostering equitable technological advancement in the Nigerian construction industry.

Discussion and Implication of Findings

This section presents the findings based on data collected from the questionnaires completed by respondents. The results are compared with previous studies related to the research topic, and the implications of these findings are discussed in detail.

Level of Awareness of Robotics and Automation in the Construction Firm

The findings reveal that technologies such as drones, 3D printing, and demolition robots are highly recognized within the Nigerian construction industry, while exoskeletons, roadwork robots, and humanoids remain the least known. In

Table 4.6 Test Statistics (A Grouping Variable: Which Category Is Your Firm)

	Demolition Robots	*3D Printing and Contour Crafting*	*Drones*	*Bricklaying Robots*	*Welding Robots*	*Exoskeleton*	*Forklift Robot*	*Roadwork Robot*	*Humanoids*
Mann-Whitney U	208.500	148.500	368.000	107.000	166.000	155.000	160.500	92.000	270.000
Wilcoxon W	2983.500	2923.500	3143.000	2882.000	2941.000	2930.000	2935.500	2867.000	3045.000
Z	−3.672	−3.930	−.593	−5.948	−4.459	−4.873	−4.658	−5.613	−4.043
Asymp.Sig. (2-tailed)	.000	.000	.553	.000	.000	.000	.000	.000	.000

comparison, a study in South Africa showed that similar to Nigeria, drones and 3D printing are well known, while awareness of exoskeletons and humanoids is limited (Jaiswal et al., 2024). In Germany, however, a study found that exoskeletons are more recognized compared to other countries, which is attributed to higher investments and research in robotic technologies for construction (Al-Khiami et al., 2024). In China, widespread awareness of drones and 3D printing contrasts with the limited recognition of roadwork robots, indicating a trend similar to Nigeria (Yan et al., 2023). These findings highlight the need for increased awareness efforts in Nigeria and other developing countries. Policymakers and construction stakeholders should focus on promoting lesser-known technologies like exoskeletons and roadwork robots. Enhanced awareness will foster innovation, drive adoption, and improve competitiveness in the construction sector.

Level of Adoption of Robotics and Automation in the Construction Firm

The study findings show that drones and 3D printing exhibit the highest adoption levels in Nigeria's construction industry, while forklift robots and bricklaying robots have the lowest adoption according to the 3.4 threshold. In a study conducted in the United Kingdom, drones and 3D printing also emerged as widely adopted technologies in construction, driven by efficiency and cost-effectiveness (Newman et al., 2021). In contrast, in Malaysia, the adoption of 3D printing was higher than drones, with bricklaying robots seeing limited implementation (Almatari et al., 2024). These findings highlight that while adoption rates in Nigeria are similar to those in other countries, the low adoption of forklift and bricklaying robots suggests that these technologies face barriers, such as cost and lack of awareness. Policymakers and construction stakeholders should consider strategies to increase adoption by addressing these challenges and promoting the benefits of automation.

Comparative Evaluation of Robotics and Automation Awareness and Adoption Levels in the Construction Industry

The findings reveal that foreign firms in Nigeria exhibit higher levels of awareness and adoption of robotics and automation technologies compared to indigenous firms. Foreign firms lead in adopting and being aware of these technologies, with drones being the only technology where both foreign and indigenous firms show similar levels of awareness and adoption. Indigenous firms consistently rank lower across all technologies, indicating a noticeable gap in both awareness and adoption.

A study conducted in Italy by Aggogeri et al. (2024) highlighted that foreign firms tend to adopt advanced robotics more quickly, whereas local companies face challenges due to a lack of infrastructure and capital. This contrasts with Akpuokwe et al.'s (2024) research that describes multinational corporations and foreign firms as being responsible for the transfer of technology and knowledge to local firms. In China, foreign firms also outperform indigenous ones in both awareness and adoption of robotics and automation, although drones show equal recognition across both groups (Wong and Yap, 2012). These findings indicate that foreign firms have more access to resources, infrastructure, and information, which contributes to their higher levels of awareness and adoption. In contrast, indigenous firms in Nigeria face challenges in keeping up with technological advancements. Policymakers and construction stakeholders should focus on bridging this gap by providing targeted support and incentives to indigenous firms to boost their adoption of robotics and automation.

Conclusions and Recommendations

The development of sustainable communities through digital transformation is essential, as the adoption of robotics and automation can drive efficiency, reduce environmental impacts, and address labor shortages. This study seeks to understand the awareness and adoption levels of various robotic technologies within Nigeria's construction sector and how these technologies contribute to digital transformation. Using a quantitative research design, data were collected from 85 respondents through structured questionnaires. Analysis tools included frequency and percentage calculations, mean item scores, standard deviation, Mann-Whitney U test, and Wilcoxon signed-rank test. Key findings reveal that technologies like drones, 3D printing, and demolition robots are highly recognized, while exoskeletons, roadwork robots, and humanoids are the least known. In terms of adoption, drones and 3D printing show the highest levels, while forklift and bricklaying robots are the least adopted. Foreign firms generally exhibit higher awareness and adoption across most technologies, while indigenous firms lag, with drones as the only technology showing similar levels of adoption across both groups. This indicates a significant gap in awareness and adoption levels between foreign and indigenous firms.

This study recommends that the Nigerian construction industry should include increasing awareness and training for workers and professionals on the benefits of robotics and automation. Structured training programs can foster an in-depth understanding of these technologies. Furthermore, the Nigerian Electricity Regulatory Commission should ensure a stable power supply to facilitate the adoption of advanced technologies. This study's limitations include its focus on the two cities of Nigeria (Lagos and Abuja), where construction is centralized. Future studies could examine other regions and countries, especially the developing ones that are hesitant to adopt robotics and automation, in order to gain broader insights into awareness and adoption levels across different settings.

References

Abishov, M. and Mamutova, A. (2024). Features of taxation of construction companies in foreign countries and their use in our country. *Modern Science and Research*, *3*(2), pp. 296–304.

Adeosun, F.E. and Oke, A.E. (2024). Examining the awareness and usage of cyber physical systems for construction projects in Nigeria. *Journal of Engineering, Design and Technology*, *22*(1), pp. 281–294.

Adepoju, O.O. and Aigbavboa, C.O. (2020). Implementation of construction 4.0 in Nigeria: Evaluating the opportunities and threats on the workforce. *Academic Journal of Interdisciplinary Studies*, *9*(5), pp. 254–264.

Aggogeri, F., Pellegrini, N. and Taesi, C. (2024). Towards industrial robots' maturity: An Italian case study. *Robotics*, *13*(3), p. 42.

Aghimien, D., Ikuabe, M., Aghimien, L.M., Aigbavboa, C., Ngcobo, N. and Yankah, J. (2024), PLS-SEM assessment of the impediments of robotics and automation deployment for effective construction health and safety. *Journal of Facilities Management, 22*(3), pp. 458–478. doi:10.1108/JFM-04-2022-0037

Akande, I., Alagbe, O.T., Adebayo, O.A., Olaosun, S. and Oyedele, I.O. (2023). Prospects and challenges of emerging technologies in architectural design practice within the Nigerian context. *Journal of Namibian Studies: History Politics Culture*, *35*(1), pp. 4594–4610.

Akpuokwe, C.U., Bakare, S.S., Eneh, N.E. and Adeniyi, A.O. (2024). Corporate law in the era of globalization: A review of ethical implications and global impacts. *Finance & Accounting Research Journal*, *6*(3), pp. 304–319.

Al Bashar, M., Taher, M.A., Islam, M.K. and Ahmed, H. (2024). The impact of advanced robotics and automation on supply chain efficiency in industrial manufacturing: A comparative analysis between the US and Bangladesh. *Global Mainstream Journal of Business, Economics, Development & Project Management, 3*(03), pp. 28–41.

Al-Khiami, M.I., Lindhard, S.M. and Wandahl, S. (2024). Charting the exoskeleton industry: A comprehensive insight into dynamics and trends. In *IOP Conference Series: Earth and Environmental Science* (Vol. 1389, No. 1, p. 012014). IOP Publishing.

Almatari, H.A.Q., Chan, M. and Masrom, M.A.N. (2024). Factors inhibiting the adoption of industrial revolution 4.0 in Malaysian construction industry. *Smart and Sustainable Built Environment*, *13*(4), pp. 1041–1065.

Arowoiya, V.A., Oke, A.E., Ojo, L.D. and Adelusi, A.O. (2024). Driving factors for the adoption of digital twin technology implementation for construction project performance in Nigeria. *Journal of Construction Engineering and Management*, *150*(1), p. 05023014.

Babalola, A. and Harinarain, N. (2024). Policy barriers to sustainable construction practice in the Nigerian construction industry: An exploratory factor analysis. *Journal of Engineering, Design and Technology, 22*(1), pp. 214–234.

Balogun, O.L., Olumide, M.D., Gbaiye, O.O., Ayo-Bello, T.A., Akinwole, O.T., Ayantoye, K. (2020). Consumers' willingness to pay for packaged chicken eggs in Lagos State, Nigeria. *AMIS Agriculture and Food, 5*(2), pp. 204–217.

Bamgbose, O.A., Ogunbayo, B.F. and Aigbavboa, C.O. (2024). Barriers to building information modelling adoption in small and medium enterprises: Nigerian construction industry perspectives. *Buildings, 14*(2), p. 538.

Chan, D.W., Olawumi, T.O., Saka, A.B. and Ekundayo, D. (2024). Comparative analysis of the barriers to smart sustainable practices adoption in the construction industry between Hong Kong and Nigeria. *International Journal of Construction Management*, *24*(14), pp. 1499–1509.

Chang, W.C. and Hasanzadeh, S. (2024). Toward a framework for trust building between humans and robots in the construction industry: A systematic review of current research and future directions. *Journal of Computing in Civil Engineering*, *38*(3), p. 03124001.

Cisneros-Gonzalez, J.J., Rasool, A. and Ahmad, R. (2024). Digital technologies and robotics in mass-timber manufacturing: a systematic literature review on construction 4.0/5.0. *Construction Robotics, 8*(2), pp. 1–26.

Damiani, M., Pompei, F. and Kleinknecht, A. (2023). Robots, skills and temporary jobs: Evidence from six European countries. *Industry and Innovation, 30*(8), pp. 1060–1109.

Denny-Smith, G., Williams, M., Loosemore, M., Sunindijo, R.Y. and Piggott, L. (2024). What social value do Indigenous contractors create?. *Construction Management and Economics*, *42*(1), pp. 16–34.

Dindorf, R. and Wos, P. (2024). Challenges of robotic technology in sustainable construction practice. *Sustainability*, *16*(13), p. 5500.

Dosumu, O.S. and Uwayo, S.M. (2023). Modelling the adoption of Internet of Things (IoT) for sustainable construction in a developing economy. *Built Environment Project and Asset Management, 13*(3), pp. 394–411.

Ebekozien, A., Aigbavboa, C. and Samsurijan, M.S. (2023). An appraisal of blockchain technology relevance in the 21st century Nigerian construction industry: Perspective from the built environment professionals. *Journal of Global Operations and Strategic Sourcing, 16*(1), pp. 142–160.

Fu, X., Zanello, G., Contreras, C. and Ding, X. (2024). Innovation under constraints: The role of open innovation in Ghana. *Industry and Innovation, 31*(4), pp. 444–474.

Golabchi, A., Jasimi Zindashti, N., Miller, L., Rouhani, H. and Tavakoli, M. (2023). Performance and effectiveness of a passive back-support exoskeleton in manual material handling tasks in the construction industry. *Construction Robotics*, *7*(1), pp. 77–88.

Golzar, J., Noor, S. and Tajik, O. (2022). Convenience sampling. *International Journal of Education & Language Studies*, *1*(2), pp. 72–77.

Guan, T. (2023). Research on the application of robot welding technology in modern architecture. *International Journal of System Assurance Engineering and Management*, *14*(2), pp. 681–690.

Ikuabe, M., Aigbavboa, C. and Kissi, E. (2023). Potential applications and benefits of humanoids in the construction industry: A South African perspective. *International Journal of Building Pathology and Adaptation*, *41*(6), pp. 254–268.

Ilesanmi, O.O., Oke, A.E., Moyanga, D.T. and Aliu, J. (2024). Unveiling the imperatives: Drivers for incorporating smart building technologies in the construction sector in Nigeria. In: Chakir, A., Bansal, R., and Azzouazi, M. (eds.), *Industry 5.0 and Emerging Technologies*. Studies in Systems, Decision and Control, vol. 565. Springer, Cham. doi:10.1007/978-3-031-70996-8_10

Iyiola, C.O., Shakantu, W. and Daniel, E.I. (2024). Digital technologies for promoting construction and demolition waste management: A systematic review. *Buildings*, *14*(10), p. 3234.

Jaiswal, S.V., Hunt, D.V. and Davies, R.J. (2024). Construction 4.0: A systematic review of its application in developing countries. *Applied Sciences, 14*(14), p. 6197.

Kineber, A.F., Oke, A.E., Hamed, M.M., Rached, E.F., Elmansoury, A. and Alyanbaawi, A. (2022). A partial least squares structural equation modeling of robotics implementation for sustainable building projects: A case in Nigeria. *Sustainability, 15*(1), p. 604.

Kumar, V., Singh, R. and Pandey, A. (2024). Multiple stakeholders' critical success factors scale for success on large construction projects. *Asian Journal of Civil Engineering, 25*(2), pp. 1691–1705.

Lee, D., Nie, G.Y. and Han, K. (2024). Automatic and real-time joint tracking and three-dimensional scanning for a construction welding robot. *Journal of Construction Engineering and Management*, *150*(3), p. 04023165.

Licardo, J.T., Domjan, M. and Orehovački, T. (2024). Intelligent robotics—A systematic review of emerging technologies and trends. *Electronics*, *13*(3), p. 542.

Melenbrink, N., Werfel, J. and Menges, A. (2020). On-site autonomous construction robots: Towards unsupervised building. *Automation in Construction, 119*(1), p. 103312.

Moyanga, D.T., Ojo, L.D., Ilesanmi, O.O. and Elyamany, A. (2024). Characterisation of the effects of coronavirus pandemic on construction projects delivery. *International Journal of Business Governance and Ethics, 18*(2), pp. 169–187.

Newman, C., Edwards, D., Martek, I., Lai, J., Thwala, W.D. and Rillie, I. (2021). Industry 4.0 deployment in the construction industry: A bibliometric literature review and UK-based case study. *Smart and Sustainable Built Environment, 10*(4), pp. 557–580.

Nnaji, C., Okpala, I., Gambatese, J. and Jin, Z. (2023). Controlling safety and health challenges intrinsic in exoskeleton use in construction. *Safety Science, 157*(1), p. 105943.

Obia, A.E. (2016). Emerging Nigerian megacities and sustainable development: Case study of Lagos and Abuja. *Journal of Sustainable Development, 9*(2), pp. 27–42.

Okagbue, E.F., Muhideen, S., Anulika, A.G., Nchekwubemchukwu, I.S., Chinemerem, O.G., Tsakuwa, M.B., Achaa, L.O., Adarkwah, M.A., Funmi, K.B., Nneoma, N.C. and Mwase, C. (2024). An in-depth analysis of humanoid robotics in higher education system. *Education and Information Technologies, 29*(1), pp. 185–217.

Oke, A.E., Aliu, J. and Onajite, S.A. (2024). Barriers to the adoption of digital technologies for sustainable construction in a developing economy. *Architectural Engineering and Design Management, 20*(3), pp. 431–447.

Oke, A.E., Aliu, J., Oni, S.B. and Ilesanmi, O.O. (2023). The complexities of embracing mechatronics in the Nigerian construction industry. *Construction Innovation*. doi:10.1108/CI-07-2023-0154

Okpala, I., Nnaji, C., Ogunseiju, O. & Akanmu, A. (2022). Assessing the role of wearable robotics in the construction industry: Potential safety benefits, opportunities, and implementation barriers. *Automation and Robotics in the Architecture, Engineering, and Construction Industry, 7*(3) pp. 165–180.

Onososen, A.O., Musonda, I., Onatayo, D., Tjebane, M.M., Saka, A.B. and Fagbenro, R.K. (2023). Impediments to construction site digitalisation using unmanned aerial vehicles (UAVs). *Drones, 7*(1), p. 45.

Opawole, A., Olojede, B.O. and Kajimo-shakantu, K. (2022). Assessment of the adoption of 3D printing technology for construction delivery: A case study of Lagos State, Nigeria. *Journal of Sustainable Construction Materials and Technologies, 7*(3), pp. 184–197.

Pallant, J.F. and Bailey, C.M. (2005). Assessment of the structure of the hospital anxiety and depression scale in musculoskeletal patients. *Health and Quality of Life Outcomes*, 3, pp. 1–9. doi: 10.1186/1477-7525-3-82

Parisi, F., Sangiorgio, V., Parisi, N., Mangini, A.M., Fanti, M.P. and Adam, J.M. (2024). A new concept for large additive manufacturing in construction: Tower crane-based 3D printing controlled by deep reinforcement learning. *Construction Innovation, 24*(1), pp. 8–32.

Patrick, O.O., Nnadi, E.O. and Ajaelu, H.C. (2020). Effective use of quadcopter drones for safety and security monitoring in a building construction sites: Case study Enugu Metropolis Nigeria. *Journal of Engineering and Technology Research, 12*(1), pp. 37–46.

Piras, G., Muzi, F. and Tiburcio, V.A. (2024). Digital management methodology for building production optimization through digital twin and artificial intelligence integration. *Buildings, 14*(7), p. 2110.

Placzek, G. and Schwerdtner, P. (2024). A global snapshot of 3D-printed buildings: Uncovering robotic-oriented fabrication strategies. *Buildings, 14*(11), p. 3410.

Pradhananga, P., ElZomor, M. and Santi Kasabdji, G. (2021). Identifying the challenges to adopting robotics in the US construction industry. *Journal of Construction Engineering and Management, 147*(5), p. 05021003.

Quiles, N.N., McCullough, A.K. and Piao, L. (2019). Validity and reliability of the exercise vital sign questionnaire in an ethnically diverse group: A pilot study. *Journal of Primary Care and Community Health,* 10, 2150132719844062. doi: 10.1177/2150132719844062

Rajput, B., Kore, S. and Agarwal, A. (2024). Competencies of civil engineering graduates: Indian construction professionals' perspective. *International Journal of Construction Management, 24*(11), pp. 1240–1248.

Schöberl, M., Huber, A., Kreppold, S., Dirnaichner, J., Kessler, S. and Fottner, J. (2023). Cobot uptake in construction: Embedding collaborative robots in digital construction processes. *Construction Robotics, 7*(1), pp. 89–103.

Selcuk, O., Turkoglu, H., Polat, G. and Hajdu, M. (2024). An integrative literature review on the causes of delays in construction projects: Evidence from developing countries. *International Journal of Construction Management, 24*(6), pp. 610–622.

Umoh, A.A., Nwasike, C.N., Tula, O.A., Adekoya, O.O. and Gidiagba, J.O. (2024). A review of smart green building technologies: Investigating the integration and impact of AI and IOT in sustainable building designs. *Computer Science & IT Research Journal, 5*(1), pp. 141–165.

Usmanov, V., Illetško, J. and Šulc, R. (2021). Digital plan of brickwork layout for robotic bricklaying technology. *Sustainability, 13*(7), p. 3905.

Volpe, S., Sangiorgio, V., Fiorito, F. and Varum, H. (2024). Overview of 3D construction printing and future perspectives: A review of technology, companies and research progression. *Architectural Science Review, 67*(1), pp. 1–22.

Wahab, A.B. (2020). Applications of drone technology in the management of disaster and risk associated with COVID-19 pandemic in the built environment: The Nigerian experience. *International Journal of Real Estate Studies, 14*(S1), pp. 73–81.

Wang, T., Zhang, Y. and Liu, C. (2024). Robot adoption and employment adjustment: Firm-level evidence from China. *China Economic Review, 84*(1), p. 102137.

Wong, C.Y. and Yap, X.S. (2012). Mapping technological innovations through patent analysis: A case study of foreign multinationals and indigenous firms in China. *Scientometrics, 91*(3), pp. 773–787.

Yan, J.K., Zheng, Z., Zhou, Y.C., Lin, J.R., Deng, Y.C. and Lu, X.Z. (2023). Recent research progress in intelligent construction: A comparison between China and developed countries. *Buildings, 13*(5), p. 1329.

Yusuf, S.O., Aribisala, A.F., Muhammed, A.O., Mobayo, J.O. and Belgore, U. (2022). Competitive advantage among indigenous construction firms in competing with expatriate contractors in Nigeria. *International Journal of Real Estate Studies, 16*(2), pp. 11–22.

Zhang, N., Konyalıoğlu, A.K., Duan, H., Feng, H. and Li, H. (2024). The impact of innovative technologies in construction activities on concrete debris recycling in China: A system dynamics-based analysis. *Environment, Development and Sustainability, 26*(6), pp. 14039–14064.

5 Sustainable and Lean Healthcare Facility Management

Gökhan Demirdöğen, Yusuf Arayici, and Zeynep Işık

Introduction

Facility management (FM) is becoming more important due to its contribution to economies. The studies showed that the turnover of FM companies and related services reached nearly £115 billion in the UK (Atkin and Bildsten, 2017). Additionally, it has been revealed that companies related to FM contribute 3% to the national workforce (Atkin and Bildsten, 2017). Beside the economic contribution of FM companies, facility-related activities have negative consequences for the environment and resource consumption. Another study showed that facilities are responsible for 30% of energy consumption (Costa et al., 2013). Depending on other activities and energy consumption, studies showed that facilities have a considerable impact on CO_2 emission directly. Also, according to other statistics, FM-related costs correspond to 60% of building lifecycle costs (Koch et al., 2014; Edirisinghe et al., 2017).

Even though FM is very important for both the economy and the environment, there are many productivity-related issues that need to be overcome, such as price increases, reactive interventions, loss of time due to problems in accessing data, data loss, and so on. In other words, access to data in-situ and validation of data are the main issues that cause loss of time for facility managers due to traditional FM approaches. Other examples include data labelling issues, data storage, data volume, and storage of unnecessary data and its management (Terreno, Asadi, and Anumba, 2019). To overcome these issues, some FM systems have been produced, such as computer-aided facility management, maintenance management systems, building automation systems, and so on (Lu et al., 2018).

The abovementioned issues and process bottlenecks in FM necessitate a change in FM activities towards more lean activities (Gao, Sui Pheng, and Tay, 2020). Therefore, in the literature, lean management (LM) principles were implemented on FM to review processes and eliminate waste. The main aim of LM principles is to obtain more effective and efficient activities by considering customer demands (Gao, Sui Pheng, and Tay, 2020). However, LM practices in FM are not at the intended level. Within this context, Demirdöğen, Işık and Arayici (2020) stated that non-involvement of facility managers in the design process; lack of data analysis and knowledge generation; productivity issues during activity monitoring and

DOI: 10.1201/9781003380559-5

reporting; and lack of valorisation technologies are the most important observed issues in FM.

Within this context, building information modelling (BIM) gives an opportunity to store, visualise, and manage data produced from the design stage to the end of the building lifecycle. In addition to building lifecycle management, BIM enables FM, navigation, marketing, personnel training, space management, and maintenance management (Becerik-Gerber et al., 2011; Oskouie et al., 2012; Ilter and Ergen, 2015; Liu and Issa, 2015; Terreno et al., 2015; Nicał and Wodyński, 2016; Oti et al., 2016; Davtalab, 2017; Edirisinghe et al., 2017; Wong, Ge and He, 2018). However, the BIM deficits include the query of FM data, FM data monitoring, and FM data storage. Therefore, this issue necessitates the integration of BIM with data storage and data analysis (Demirdöğen, Işik and Arayici, 2020).

In recent years, with the web-based applications and sensors to facilities, big data analytics (BDA) has gained importance (Demirdöğen, Işik and Arayici, 2020). BDA provides proactive operation with the analysis of users/customers/end users' data (Kambatla et al., 2014). Enormous data volume and a variety of data have been observed during FM. Within this context, BDA enables faster data analysis and integration into the decision-making process by eliminating productivity issues (Aibinu, Koch, and Ng, 2019; Ahmed et al., 2020; Gao, Sui Pheng, and Tay, 2020). Therefore, the aim of the chapter is to show the development processes of an FM framework by employing design science research methodology, which embraces BIM, BDA, building energy performance simulation (BEPS), and computational fluid dynamics (CFD) as technological digital concepts. In addition to these tools, FM information types and FM KPIs were considered together.

Identification of Facility Management Performance Indicators

FM necessitates effective and efficient resource management due to the complexity and variety of FM activities, since facility managers need various FM tools to conduct FM activities. However, existing FM systems either are designed to query limited and predefined queries or present a data analysis capability for FM data (Hopland and Kvamsdal, 2019). Additionally, as stated earlier, knowledge stays hidden in big FM data masses due to a lack of data analytics (Ahmed et al., 2018). Therefore, in the literature, the researchers stated that data analytics can be used effectively in the analysis and estimation of FM performance indicators. Further, it was stated that the existing performance management indicators lag behind current technological developments, and there is a need for the discovery of new indicators through the use of data analytics (Burak Gunay, Shen, and Newsham, 2019; Dutta, Burak Gunay, and Bucking, 2020).

Demirdöğen, Işık, and Arayici (2022) stated that ageing in the population, emerging pandemics, and society's changing behaviour in energy consumption have made facilities management in hospitals more critical. Moreover, healthcare facilities have financial problems due to the high cost of FM expenditures. Within the study, it was stated that the existing studies are performance- and maintenance-oriented

(Shohet and Lavy, 2017); considerations of limited FM factors such as financial, functionality, physical, and user satisfaction (Róka-Madarász, Mályusz, and Tuczai, 2016); and identification of FM indicators specific to healthcare FM (Lai and Yuen, 2019). However, the existing FM practices are not helpful to generate and use knowledge by analysing data stack produced during the FM process (Demirdöğen, Işık, and Arayici, 2022). Within this context, Demirdöğen, Işık, and Arayici (2022) identified healthcare FM performance indicators that can be discovered through business intelligence and analytics. Additionally, the authors determined the priorities of FM performance indicators specific to healthcare facilities in Türkiye. The flow chart employed in the study is depicted in Figure 5.1.

According to the results of the study, 98 FM performance indicators were identified and categorised under 6 categories. The study also categorised healthcare FM performance indicators to identify measurement frequencies of indicators, such as daily, weekly, monthly, quarterly, semi-annually, and annually. As a research

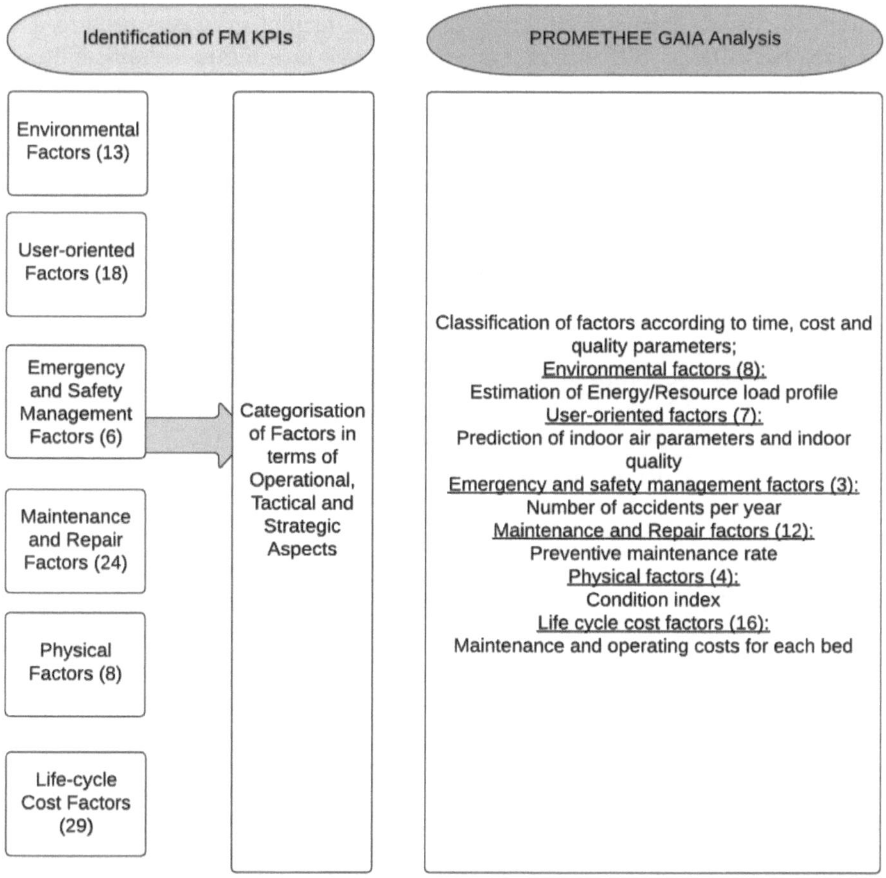

Figure 5.1 Identification of FM KPIs.

methodology, PROMETHEE II was used to enable prioritisation of FM performance indicators. In the study, identified FM performance indicators were evaluated against time, cost, and quality, which are contradictory criteria. Figure 5.1 gives the highest priority indicators in this context (Demirdöğen, Işık, and Arayici, 2022).

Development of FM Information Taxonomy Framework

The equipment and operation rooms found in healthcare facilities must provide uninterrupted service. Therefore, identification of FM information requirements is important both to facilitate decision-making processes and to overcome unavailability of data. Moreover, the failure to define information requirements for FM leads to incomplete collection of information (Demirdöğen, Işık, and Arayici, 2021). Nowadays, much FM software enables us to collect FM-related data and data queries. These software tools can be computer-aided FM systems, computerised maintenance management systems, building automation systems, and so on. However, the main aim of these systems is to manage information that focuses on specific management areas. Therefore, facility managers need to use and integrate different FM systems (Lu et al., 2018). In addition, identification of required FM information types helps to eliminate redundant data collection. So, the data stack can be reduced with the identification of FM information types (Kang and Hong, 2015).

FM information types were not identified depending on a plan; facility managers and staff need both to control all documents and models to access the required information and to validate obtained information through site visits (Sacks et al., 2018). Within this context, the studies showed that BIM comes into prominence for managing all FM-related information. Therefore, researchers proposed BIM-integrated building automation systems, maintenance management systems, and big data systems (Demirdöğen, Işık, and Arayici, 2021). Furthermore, interoperability between BIM and these systems helps to enable required information to calculate FM performance indicators. To summarise briefly, BIM can be used in the determination of the depreciation value of assets, access to building data, material data, the management of competent staff data, complaint management, management of work orders, cost management, emergency management, access to energy consumption data, access to equipment and system lists, maintenance schedule, failure detection and resolution, and the management of FM information types (Demirdöğen, Işık, and Arayici, 2022).

Demirdöğen, Işık, and Arayici (2022) stated that there was no compilation of BIM-based FM information types, and the FM information types were not identified comprehensively in the literature. Therefore, the authors of the chapter identified BIM-based FM information types and FM information types. The authors then validated FM information types identified via literature review. In the validation, the authors used case studies in which the information transfer process was depicted with business process modelling notation. The research methodology employed in the study is depicted in Figure 5.2.

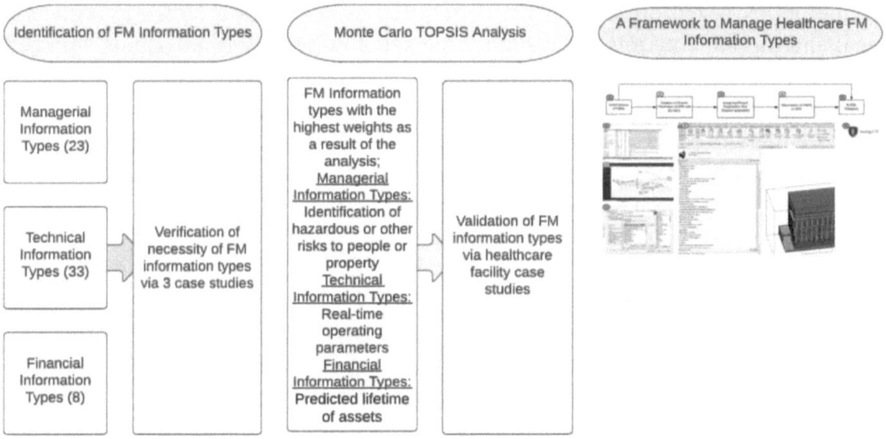

Figure 5.2 Identification of FM Information Types.

Lean Healthcare FM Framework

In FM, there are many inefficiencies related to activities and processes, as in every field. Within this context, the dynamic and static data needed for FM should be prepared and stored, while the existing data stack needs to be converted to a form in which the FM process is supported. In other words, FM and LM should be merged to obtain more efficient and effective FM activities and processes. While LM is helpful for obtaining more effective and efficient activities and processes, it eliminates activities and processes that do not create value for businesses. Moreover, the LM philosophy puts the customer at the centre of all processes. The concept of LM contributes to companies in many areas, such as collaboration, schedule performance management, occupational health and safety, the environmental impact of buildings, sustainable development, quality, and customer satisfaction (Demirdöğen, Işik, and Arayici, 2020).

As mentioned earlier, BIM is used in many areas related to FM. Some of these areas are energy consumption monitoring, space management, retrofitting, maintenance and asset management, emergency management, interoperability with other FM software, warranty information, location information of building components, data storage, and so on (Brooks and Lucas, 2014; Terreno et al., 2015; Oti et al., 2016; Davtalab, 2017; Edirisinghe et al., 2017; Reza et al., 2018).

However, Demirdöğen, Işik, and Arayici (2020) stated that both BIM and FM systems need to be integrated with data analytics to facilitate decision-making. Furthermore, this integration helps to store FM data. Within this context, the authors asserted that FM systems including the integration of BDA should be used during the FM. According to the literature review conducted by the authors, there are many studies about the use of BIM in FM, the integration of BIM and building energy performance simulation, and BDA-integrated BIM systems. However, the authors explored the idea that the proposed systems don't consider organisational and implementation aspects and also stated that many FM systems are used together, since all FM software aims to solve different FM issues. Therefore, it was stated that the relationship between data sources and knowledge generation are hidden due to the use of many software tools. Within this context, the authors integrated an LM philosophy into FM to eliminate data management, organisational concerns, and managerial waste issues.

In this context, a framework integrating BIM, BEPS, and BDA is presented in the study (research) considered in the chapter. This framework is depicted in Figure 5.3.

- **Process 1:** Process 1, depicted in this framework, involves a process related to the identification of FM information types since facility managers do not need all information related to the facility. In addition, redundant data collection and management cause inefficiencies in the FM process. Within this context, identified information types, need to be evaluated by facility managers for each facility. The authors proposed the use of multi-criteria decision-making methods to find out priority orders of FM information types. At the last step, the authors emphasised that the facility manager or client needs to approve the identified information types. One of the main objectives of this process is to design data specifications and query algorithms for the data to be managed within the FM systems to be developed or used (Demirdöğen, Işik, and Arayici, 2020).
- **Process 2:** Building component–related information, such as reflection properties of materials, thickness, or heat convection coefficient, are needed to perform BIM-based energy analysis. Therefore, this information and FM information

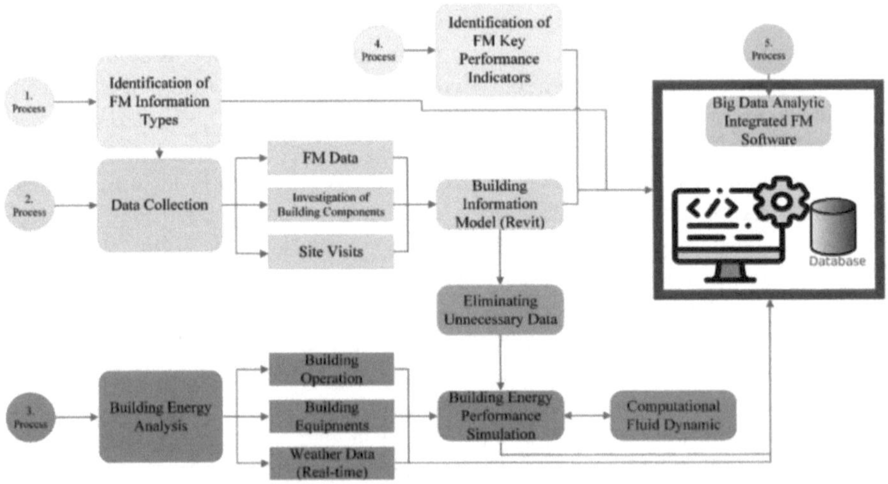

Figure 5.3 Lean Management–Integrated FM Framework.
Source: Demirdöğen, Işik, and Arayici (2020).

types should be identified in building components in the BIM. The created model helps both to perform energy analysis and to create a base for NoSQL databases to manage data sources (Demirdöğen, Işik, and Arayici, 2020).

- **Process 3:** This includes the generation of energy consumption data with the use of BEPS. Weather data is an important parameter to obtain reliable facility energy consumption data. Therefore, access to weather data where a facility is located is enabled and needs to be entered into simulations. Furthermore, equipment and systems installed in the facility, facility operation hours, building geometry, building materials, and the heat convection coefficient should be discovered and used in the energy analysis.

  The authors proposed the use of CFD to obtain reliable results for energy consumption. CFD helps to divide spaces into smaller parts so that temperature or other parameters can be calculated more realistically than using other methodologies. For example, design builder simulation results accept that every point in the room has the same temperature value. However, CFD can calculate indoor air parameters for a given mesh size. However, calibration of the building model needs to be performed before the analysis results are used. Within this context, the authors proposed the use of root mean square error (RMSE), mean absolute error (MAE), regression error (R Squared:R2), and coefficient of variance (CV). In the study, the authors proposed to combine the analysis results with machine learning to estimate building energy consumption and to develop scenarios by changing operational and weather parameters (Demirdöğen, Işik, and Arayici, 2020). Finally, the authors aimed to store analysis results in the NoSQL database, enabling the querying previous scenarios and adapting them to new conditions.
- **Process 4:** This includes the identification of healthcare facility performance indicators. Within this context, the performance indicators designed to be monitored can be different for each facility. Therefore, they can be identified by using multi-criteria decision-making methods. This method will reveal the priority orders and facilitate the selection process of FM performance indicators (Demirdöğen, Işik, and Arayici, 2020).
- **Process 5:** FM information types and FM performance indicators found as a result of Process 1 and Process 4 are converted to query codes, and data generated in Process 2 and Process 3 is stored in the set-up database. The authors integrated the database with a MapReduce algorithm (with which data stack is distributed to the computer slaves to process and filter data). With the use of the MapReduce algorithm, data analysis and filtering can be performed faster. Moreover, BDA enables the use of data mining, text mining, and visual analytics (Demirdöğen, Işik, and Arayici, 2020).

The proposed framework was validated in terms of both doability and interaction with LM principles. The expert review was followed for the validation. The expert views are summarised below.

- **Standardisation:** The framework will contribute to standardisation efforts as it enables the elicitation and management of the types of information required.
- **Alignment between short- and long-term goals:** Thanks to the database within the proposed framework, previous decisions and their results will be recorded, and this will form a basis for future decisions.
- **Process bottlenecks:** The Plan–Do–Act–Check cycle was embedded in the framework. The experts stated that the cycle will be helpful to eliminate process bottlenecks.
- **Continuous improvement:** Storing historical data and information will pave the way for continuous improvement, and forecasts that can be made through the system will benefit the simplification of processes.
- **Optimisation of FM activities and processes:** Data management is needed to plan optimal processes and activities within FM. The experts stated that the proposed FM framework includes a database. With the use of databases, previous actions and their results will be ready to use in the decision-making process.
- **Idle time for healthcare equipment and systems:** Equipment and systems should give uninterrupted service in healthcare facilities since the conditions of equipment and systems directly affect patients' health. In addition, the fact that the equipment is in working condition when there are no employees or patients in the hospital causes further waste. Many activities, such as ageing of equipment, can be realised with the proposed framework.
- **Excessive inventory:** A LM philosophy requires non-storage of materials, spare parts, and equipment since this is seen as a waste. Within this context, the experts stated that estimations used in maintenance management will be helpful to determine excessive inventories in the facilities.
- **Labour and time loss:** According to expert evaluations, finding necessary FM information is a time-consuming activity for facility managers. It was also stated that thanks to the framework, planning can be carried out more efficiently. Another expert stated that not analysing the existing data is an additional inefficiency that will be eliminated thanks to the framework.
- **Rehandling and idle time for staff:** Rehandling, which is one of the LM principles, will be eliminated when standardisation is achieved. In addition to this, duplication of activities and unnecessary or unorganised employee irregularities can be tracked through the proposed framework as staff assigned to work orders and maintenance activities can be monitored with the analysis of maintenance or work order history.
- **Improvement in organisational structure:** Healthcare FM professionals as participants stated that the presented framework provides an opportunity to monitor the FM process. In addition, the managerial activities that will be integrated into the FM system will enable a leaner philosophy to be followed in processes such as business management, reporting, monitoring activities, and so on.
- **On-site review:** According to expert opinions, building information models presented in the proposed framework and integration of building information models with AR-VR applications will be helpful to comprehend FM-related

issues and to facilitate the decision-making process during the stakeholder meetings.

- **Identification of improvement opportunities:** One of the experts stated that during the design process, the parameters are assumed to calculate the energy consumption of the building. With the use of the proposed framework, it is believed that, with the use of actual weather data and operational data, excessive and actual energy consumption can be detected.
- **Use of visual management tools:** It has been stated that visualising the FM system on A3-sized windows and presenting the data analysis results with graphics will simplify the FM processes.
- **Defining the cause–effect relationship:** The participant stated that energy consumption may vary depending on the air temperature and the number of people in the room, so it will contribute to the determination of cause and effect.
- **Reduction of diversity and cycle:** The experts believe that performing preventive and predictive maintenance through the proposed systems will help to reduce maintenance cycle times.
- **Establishing and measuring FM performance indicators:** One of the participants stated that each process presented in the framework is a performance indicator, while another stated that Process 4 in the study would be useful in this regard.
- **Data use obtained from FM systems:** It was noted that the database presented in the framework provides an opportunity to achieve leaner activities.
- **Awareness raising:** The lack of awareness-raising training activities in the study was identified as a missing aspect of the framework.

Development of a BIM–BDA-based FM System for Healthcare Facilities

BIM can be used as a database for the storage of data that does not change depending on time. BIM is also used to depict dynamic data such as sensors on the model. However, BIM does not allow for the storing of historical data. In other words, if the modeller changes the data content of objects, the previous data content will disappear. Therefore, BIM needs to be supported with a database. Within this context, Demirdöğen, Işık, and Arayici (2023) proposed a FM system that illustrates queries of FM information types and FM performance indicators by integrating BIM, BDA, and NoSQL databases.

The authors stated that studies focusing on the integration of BIM into the FM phase are limited in the literature. Additionally, Dixit et al. (2019) stated that storage of misinformation, non-updating of building models, file exchange formats, interoperability issues between FM systems and BIM, and large file size are the biggest obstacles for adapting BIM into the FM process. To overcome interoperability issues, industry foundation class (IFC), information delivery manual (IDM), model view definition (MVD), and construction operations building information exchange (COBie) formats have been developed in the context of industrial standardisation efforts. However, these formats were designed for data exchange between software tools. In other words, data exchange for BIM objects is not possible. Furthermore, it was stated that studies on BIM and BDA, literature reviews, BIM-integrated BDA system proposals, dynamic data analysis with BDA, and performance management were carried out (Demirdöğen, Işık, and Arayici, 2023). The authors revealed that IFC and COBie data exchange formats are used to transfer data from BIM to relational databases. As mentioned earlier, this structure helps to transfer whole building model data rather than object-based data.

It was stated that existing studies focus on a limited number of FM information types and FM performance indicators. Establishing a BDA system without revealing the information types and FM performance indicators requires coding knowledge, coding of MapReduce files within the system and their analysis. Data transfer is also required due to the lack of a common database for storing and analysing data in the BDA part (Chen, Chang, and Lin, 2016; Aziz, Riaz, and Arslan, 2017; Mawed and Aal-Hajj, 2017).

In this context, the following objectives were targeted in the study.

- Preparation of FM data in BIM
- Establishing a semi-automatic data transfer mechanism to transfer asset data to the database
- Saving data in the FM system
- The ability to query and analyse FM information with the proposed FM system

To achieve the abovementioned objectives, the flow presented in Figure 5.4 was followed in the study (Demirdöğen, Işık and Arayici, 2023).

Step 1: Identification of FM performance indicators. In this process, the authors used the most important FM performance indicators that were determined in the first process.

Step 2: Identification of FM information types. The FM information types were identified in Process 2. In the prototype testing, the most important FM information types were queried.

Step 3: Development of shared parameters with Dynamo. All geometric and non-geometric information types were identified during Process 2. However, these information types are not defined in BIM by default. Therefore, Dynamo visual script was used to identify them in all building components, such as furniture, equipment, system, and so on.

Step 4: Data entry in BIM. As a result of Process 3, information types were identified in the objects. However, corresponding information related to FM information types should be identified by responsible parties such as modellers, facility managers, and so on. Therefore, the responsible parties should gather all information related to FM information types and validate the information by making site visits.

Step 5: Creation of visual script for the retrieval of FM data from BIM. The IFC data schema, which is widely used in professional life, is very general and difficult to interpret and consists of many property sets. For this reason, this data schema is difficult to customise according to needs (Hu et al., 2016; Farghaly et al., 2017; Lu et al., 2018). Within this context, Afsari, Eastman, and Castro-Lacouture (2017) developed the ifcJSON data exchange structure. The main purpose here is to take advantage of the more efficient data parsing and resource utilisation of the JSON data exchange scheme. However, the developed ifcJSON data schema and transfer approach is intended to ensure that all data is transferred. It is also compatible with the data logging infrastructure of the MongoDB (NoSQL) database that can be shared with BDA. For this reason, a Dynamo visual code was written to be used to retrieve FM information, i.e., geometric and non-geometric data from each object.

Step 6: Setting up MongoDB Database. Information retrieved from BIM should be stored in the database to be used and analysed. Therefore, historical data can be observed via the MongoDB database, which presents a document-based database. With the use of MongoDB, the user can create different collections for every FM area. Under collections, documents can be created for every asset or BIM object. Therefore, the data can be managed dynamically.

Step 7: The connection between Hadoop and MongoDB. Apache Hadoop is one of the BDA engines with MapReduce infrastructure. In this context, MongoDB has developed a connector for Apache Hadoop and their database to work together. MapReduce and MongoDB have emerged because of massive volumes of internet and data records. The MongoDB database can support many big data systems and real-time data analysis both offline and online. Scaling is also possible (Dede et al., 2013; MongoDB, 2021).

Step 8: Development of user interface to illustrate the results of FM information query and FM performance indicators. In the process of developing the interfaces in the study, PyQt5 library was used in the Python coding program. Separate modules for CRUD operation can be directly used through MongoDB when the data size is low and a system that can work together with the BDA engine when the data size is very high.

In the study, the proposed prototype was also evaluated by four experienced facility managers. Within the evaluation, 14 questions were asked under three different headings and the efficiency, practicality and usability of the system were evaluated. According to the evaluation results, the efficiency and practicality of the system was high, while the usability was low due to the need for training (due to ICM). However, the usability value was still above average.

Conclusion

The services and operations provided in healthcare buildings are different from other facilities. This is because the 24-hour service provided in healthcare facilities, the possibility of negative impact of the decisions taken regarding FM on patients sensitive to environmental conditions, and the complexity of FM operations make the FM of healthcare facilities special. The currently used FM systems do not meet the needs of FM in terms of data analytics. In addition, the failure to define and integrate the types of information required for FM into the systems; the inability to monitor FM success indicators on the systems, and their non-existence in some of them; the very fast data growth rate in FM in health facilities; and the need for rapid decision-making constitute the starting point of the study.

As a result of the study, 50 FM success indicators were defined under the environmental, user-oriented, emergency and safety management, maintenance and renovation, and physical and lifecycle-related factor groups. In addition to these success indicators, the types of knowledge required for the FM were identified under the managerial, financial, and technical factor groups. In total, 64 knowledge types were defined. Subsequently, a framework to integrate building energy performance simulation, BIM, and BDA, while considering both FM information types and FM KPIs, was developed and evaluated from the perspective of the LM philosophy. In light of the findings, a prototype was developed to analyse the most important FM information types and KPIs. According to the findings, 100% of the respondents confirmed that the proposed system would be efficient in solving the FM information queries and KPI analysis, while 97% stated that it is practical in carrying out FM activities. In terms of usage, 60% of the participants stated that the proposed system would be practical to use during the FM process.

It is thought that the proposed system will be useful for expanding the use of BIM in the FM phase, which is one of the lifecycle processes of the building, in addition to the design phase process.

References

Afsari, K., Eastman, C.M. and Castro-Lacouture, D. (2017) 'JavaScript Object Notation (JSON) data serialization for IFC schema in web-based BIM data exchange', *Automation in Construction*, 77, pp. 24–51. Available at: https://doi.org/10.1016/j.autcon.2017.01.011

Ahmed, V. et al. (2018) 'Challenges and drivers for data mining in the AEC sector', *Engineering, Construction and Architectural Management*, 25(11), pp. 1436–1453. Available at: https://doi.org/10.1108/ECAM-01-2018-0035

Ahmed, V. et al. (2020) 'The future of Big Data in facilities management: opportunities and challenges', *Facilities*, 35(13), pp. 725–745. Available at: https://doi.org/10.1108/F-06-2016-0064

Aibinu, A.A., Koch, F. and Ng, S.T. (2019) 'Data analytics and big data in construction project and asset management', Built Environment Project and Asset Management, 9(4), pp. 474–475. Available at: https://doi.org/10.1108/BEPAM-09-2019-139

Atkin, B. and Bildsten, L. (2017) 'Editorial: A future for facility management', *Construction Innovation*, 17(2), pp. 116–124. Available at: https://doi.org/10.1108/CI-11-2016-0059

Aziz, Z., Riaz, Z. and Arslan, M. (2017) 'Leveraging BIM and Big Data to deliver well maintained highways', *Facilities*, 35(13–14), pp. 818–832. Available at: https://doi.org/10.1108/F-02-2016-0021

Becerik-Gerber, B. et al. (2011) 'Application areas and data requirements for BIM-enabled facilities management', *Journal of Construction Engineering and Management*, 138(3), pp. 431–442. Available at: https://doi.org/10.1061/(asce)co.1943-7862.0000433

Brooks, T.J. and Lucas, J.D. (2014) 'A study to support BIM turnover to facility managers for use after construction', in *American Society of Civil Engineers (ASCE)*, pp. 243–250. Available at: https://doi.org/10.1061/9780784413616.031

Burak Gunay, H., Shen, W. and Newsham, G. (2019) 'Data analytics to improve building performance: A critical review', *Automation in Construction* [Preprint], 97, pp. 96–109. Available at: https://doi.org/10.1016/j.autcon.2018.10.020

Chen, H.M., Chang, K.C. and Lin, T.H. (2016) 'A cloud-based system framework for performing online viewing, storage, and analysis on big data of massive BIMs', *Automation in Construction*, 71, pp. 34–48. Available at: https://doi.org/10.1016/j.autcon.2016.03.002

Costa, A. et al. (2013) 'Building operation and energy performance: Monitoring, analysis and optimisation toolkit', *Applied Energy*, 101, pp. 310–316. Available at: https://doi.org/10.1016/j.apenergy.2011.10.037

Davtalab, O. (2017) 'Benefits of real-time data driven BIM for FM departments in operations control and maintenance', *Computing in Civil Engineering*, 3, pp. 326–334. Available at: https://doi.org/doi:10.1061/9780784407943

Dede, E. et al. (2013) 'Performance evaluation of a MongoDB and Hadoop platform for scientific data analysis', in *HPDC'13: The 22nd International Symposium on High-Performance Parallel and Distributed Computing*. New York, NY, USA: Elsevier. Available at: https://doi.org/10.1016/J.IJAR.2021.07.004

Demirdöğen, G., Işik, Z. and Arayici, Y. (2020) 'Lean management framework for healthcare facilities integrating BIM, BEPS and big data analytics', *Sustainability (Switzerland)*, 12(17). Available at: https://doi.org/10.3390/su12177061

Demirdöğen, G., Isik, Z. and Arayici, Y. (2021) 'Facility management information taxonomy framework for queries in healthcare buildings', *Journal of Building Engineering*, 44, p. 102510. Available at: https://doi.org/10.1016/j.jobe.2021.102510

Demirdöğen, G., Işık, Z. and Arayici, Y. (2022) 'Determination of business intelligence and analytics-based healthcare facility management key performance indicators', *Applied Sciences (Switzerland)*, 12(2). Available at: https://doi.org/10.3390/APP12020651

Demirdöğen, G., Işık, Z. and Arayici, Y. (2023) 'BIM-based big data analytic system for healthcare facility management', *Journal of Building Engineering*, 64, p. 105713. Available at: https://doi.org/10.1016/J.JOBE.2022.105713

Dixit, M.K., Venkatraj, V. and Ostadalimakhmalbaf, M. (2019) 'Integration of facility management and building information modeling (BIM): A review of key issues and challenges', 37(7), pp. 455–483. Available at: https://doi.org/10.1108/F-03-2018-0043

Dutta, S., Burak Gunay, H. and Bucking, S. (2020) 'A method for extracting performance metrics using work-order data', *Science and Technology for the Built Environment*, 26(3), pp. 414–425. Available at: https://doi.org/10.1080/23744731.2019.1693208

Edirisinghe, R. et al. (2017) 'Building information modelling for facility management: Are we there yet?' *Engineering, Construction and Architectural Management*, 24(6), pp. 1119–1154. Available at: https://doi.org/10.1108/ECAM-06-2016-0139

Farghaly, K. et al. (2017) 'BIM Big Data System Architecture for Asset Management: A conceptual framework', July, pp. 289–296. Available at: https://doi.org/10.24928/jc3-2017/0163

Gao, S., Sui Pheng, L. and Tay, W. (2020) 'Lean facilities management: Preliminary findings from Singapore's international schools', *Facilities*, 38(7), pp. 539–558. Available at: https://doi.org/10.1108/F-07-2019-0076

Hopland, A.O. and Kvamsdal, S.F. (2019) 'Critical success factors for maintenance of local public buildings: The role of contingencies', *Journal of Facilities Management*, 17(5), pp. 412–427. Available at: https://doi.org/10.1108/JFM-10-2018-0065

Hu, Z.Z. et al. (2016) 'Construction and facility management of large MEP projects using a multi-scale building information model', *Advances in Engineering Software*, 100, pp. 215–230. Available at: https://doi.org/10.1016/j.advengsoft.2016.07.006

Ilter, D. and Ergen, E. (2015) 'BIM for building refurbishment and maintenance: Current status and research directions', *Structural Survey*, 33(3), pp. 228–256. Available at: https://doi.org/10.1108/SS-02-2015-0008

Kambatla, K. et al. (2014) 'Trends in big data analytics', *Journal of Parallel and Distributed Computing*, 74(7), pp. 2561–2573. Available at: https://doi.org/10.1016/j.jpdc.2014.01.003

Kang, T.W. and Hong, C.H. (2015) 'A study on software architecture for effective BIM/GIS-based facility management data integration', *Automation in Construction*, 54, pp. 25–38. Available at: https://doi.org/10.1016/j.autcon.2015.03.019

Koch, C. et al. (2014) 'Natural markers for augmented reality-based indoor navigation and facility maintenance', Automation in Construction [Preprint], 48, pp. 18–30. Available at: https://doi.org/10.1016/j.autcon.2014.08.009

Lai, J. and Yuen, P.L. (2019) 'Performance evaluation for hospital facility management: Literature review and a research methodology', *Journal of Facility Management Education and Research*, 3(1), pp. 38–43. Available at: https://doi.org/10.22361/jfmer/96267

Liu, R. and Issa, R.R.A. (2015) 'Survey: Common knowledge in BIM for facility maintenance', *Journal of Performance of Constructed Facilities*, 30(3), p. 04015033. Available at: https://doi.org/10.1061/(asce)cf.1943-5509.0000778

Lu, Q. et al. (2018) 'Activity theory-based analysis of BIM implementation in building O&M and first response', *Automation in Construction*, 85(October 2017), pp. 317–332. Available at: https://doi.org/10.1016/j.autcon.2017.10.017

Mawed, M. and Aal-Hajj, A. (2017) 'Using big data to improve the performance management: A case study from the UAE FM industry', *Facilities*, 35(13–14), pp. 746–765. Available at: https://doi.org/10.1108/F-01-2016-0006

MongoDB (2021) *What is Big Data? | MongoDB*. Available at: www.mongodb.com/big-data-explained (Accessed: 12 June 2021).

Nicał, A.K. and Wodyński, W. (2016) 'Enhancing Facility Management through BIM 6D', Procedia Engineering, 164, pp. 299–306. Available at: https://doi.org/10.1016/j.proeng.2016.11.623

Oskouie, P., Gerber, D.J., Alves, T. & Becerik-Gerber, B. (2012). 'Extending the Interaction of Building Information modelling and lean construction', *20th Annual Conference of the International Group for Lean Construction*, San Diego, CA, USA.

Oti, A.H. et al. (2016) 'A framework for the utilization of Building Management System data in building information models for building design and operation', *Automation in Construction*, 72, pp. 195–210. Available at: https://doi.org/10.1016/j.autcon.2016.08.043

Reza, H.M. et al. (2018) 'Integrating BIM into facility management: Typology matrix of information handover requirements', *International Journal of Building Pathology and Adaptation*, 36(1), pp. 2–14. Available at: https://doi.org/doi:10.1108/IJBPA-08-2017-0034

Róka-Madarász, L., Mályusz, L. and Tuczai, P. (2016) 'Benchmarking facilities operation and maintenance management using CAFM database: Data analysis and new results', *Journal of Building Engineering*, 6, pp. 184–195. Available at: https://doi.org/10.1016/j.jobe.2016.03.007

Sacks, R., Eastman, C., Lee, G., & Teicholz, P. (2018) *BIM Handbook: A Guide to Building Information Modeling for Owners, Designers, Engineers, Contractors, and Facility Managers*. John Wiley & Sons.

Shohet, I.M. and Lavy, S. (2017) 'Facility maintenance and management: A health care case study', *International Journal of Strategic Property Management*, 21(2), pp. 170–182. Available at: https://doi.org/10.3846/1648715X.2016.1258374

Terreno, S., Asadi, S. and Anumba, C. (2019) 'An exploration of synergies between lean concepts and BIM in FM: A review and directions for future research', *Buildings*, 9(6), p. 147. Available at: https://doi.org/10.3390/BUILDINGS9060147

Terreno, S. et al. (2015) 'The benefits of BIM integration with facilities management: A preliminary case study', *Computing in Civil Engineering* [Preprint]. Available at: https://doi.org/10.1061/9780784479247.084

Wong, J.K.W., Ge, J. and He, S.X. (2018) 'Digitisation in facilities management: A literature review and future research directions', *Automation in Construction*, 92(March), pp. 312–326. Available at: https://doi.org/10.1016/j.autcon.2018.04.006

6 Circular Economy for Sustainable Built Environment Through Building Information Modelling

Bahriye Ilhan Jones

Introduction

As the hub of social, cultural and economic activities, cities contribute significantly to natural resource consumption and discharge considerable volumes of waste (Huang et al., 2020).

According to UNEP (2016), urban precincts are responsible for about 75% of the world's energy consumption and over 70% of global greenhouse gas emissions. Among the primary elements in the urban built environment, buildings are responsible for 40% of global energy consumption and a similar proportion of greenhouse gas emissions (Dabous et al., 2022).

Moreover, identified as the largest consumer of raw materials and other resources, the engineering and construction industry is using more than 3 billion tonnes of raw materials (WEF, 2016). The built environment sector in London alone consumes 400 million tonnes of material each year, a quarter of which is wasted, accounting for 54% of waste in the capital (UK Green Building Council, 2022a). Putting such pressure on the natural environment, traditional building practices and construction activities need to be transformed toward more sustainable systems and procedures. The way cities are developed can mitigate the negative impacts by designing and constructing buildings for low energy and resource consumption, since within cities the built environment, where the energy transition and circular transition come together, has the potential to play a key role in reducing environmental impacts (Redevco, 2018).

Owing to the colossal nature of the construction industry, a gradual approach is required to achieve sustainability that encompasses triple bottom-line aspects of environmental conservation, social equality and economic security (Mhatre et al., 2021). Any successful adaptation and adoption of modern innovative approaches will create a positive impact on the industry. The circular economy is viewed as a condition for sustainability, a beneficial relation, or a trade-off (Geissdoerfer et al., 2017). Even though it is perceived as a new paradigm in the built environment, the concept of a circular economy has its foundations in the principles of regenerative design (Lyle, 1996), the cradle-to-cradle school of thought (Braungart, McDonough and Bollinger, 2007), industrial ecology (Stahel, 2016) and biomimicry (Blok and

DOI: 10.1201/9781003380559-6

Gremmen, 2016). It is a sustainable development model that offers an alternative to the traditional 'take-make-waste' model (Despeisse et al., 2017).

The fundamental idea of the circular economy revolves around replacing the end-of-life (EoL) concept with redesign, reduce, reuse, refurbish, repair, material recovery and recycling in production and consumption processes at micro, meso and macro levels (Kirchherr, Reike and Hekkert, 2017). The circular economy is a restorative and regenerative system in which resource use, waste and emissions are minimised by narrowing (efficient resource use), slowing (temporally extended use) and closing (cycling) material loops (Reike, Vermeulen and Witjes, 2018). It can eliminate waste by adopting appropriate resource-efficient methods from a sustainability viewpoint (Ghisellini, Cialani and Ulgiati, 2016).

Despite rising interest in circular economy in recent years (Circular Economy Promotion Law of the People's Republic of China, 2008; Advisory Board for Cirkulær Økonomi, 2017; European Commission, 2017), its adoption in the built environment is still in its early stages. Even though there have been some attempts at the implementation of circular economy strategies in building design and construction, the process has so far been incoherent and without a commonly acknowledged or established direction across the industry (Eberhardt, Birkved and Birgisdottir, 2022). Some examples of circular economy principles applied to the built environment include mining the physical infrastructure through the reuse of building components (Iacovidou and Purnell, 2016) and increasing the value of building materials (BAMB, 2020). However, the lack of cross-sector communication and coordination tools needs to be addressed to enable the broad implementation of a feasible circular approach strategy in current construction practice (De Wolf, Hoxha and Fivet, 2020).

Digital technologies are an essential enabler of the circular economy in various industries; however, to date, very few studies have investigated which digital technologies could enable the circular economy in the built environment (Cetin, De Wolf and Bocken, 2021). Building information modelling (BIM), as a promising technology, can be an instrument suitable to perform assessments (environmental and/or economic) that take into consideration the life cycle of a building and may be required multiple times over the life cycle (Di Biccari et al., 2019). As yet, the main contribution of BIM has often been in the design, bidding, construction and facility management stages, and only a few works – which focus on the recoverability of materials in various design alternatives through BIM platforms and add-ins (Akanbi et al., 2018; Akanbi et al., 2019; Honic, Kovacic and Rechberger, 2019) – have considered EoL phase issues (Akinade et al., 2017). Along with circularity, the vision of digital transformation for the future of the built environment not only empowers people and businesses but also supports the transition to a climate-neutral, circular and resilient economy (Europe's Digital Decade, 2021).

Clearly, there is a need for a holistic approach to utilise circular economy principles throughout the life cycle phases of buildings. The main motivation of this chapter is to provide a comprehensive understanding of the circularity concept within the built environment through comparison of the current dominant linear model and investigation of digital aid by exploring the capabilities of BIM technology to enhance the adoption of circular economy principles. For a thorough and successful adoption, the study proposes a framework to implement a circularity strategy focusing on each phase of a building's life cycle.

The proposed framework aims to elaborate on the great scheme of circular economy principles within the built environment. Based on current practices within the industry, the framework determines drivers and barriers and offers circular economy solutions as a remedy together with indicators. Then, the role of BIM use is addressed to achieve the proposed indicators where possible to help realise the impact of digitalisation on circularity acceptance.

Background

Circular Economy Principles in the Built Environment

The evolution of the global economy has been dominated by a linear model of production and consumption in which goods are manufactured from raw materials, sold, used and then discarded or incinerated as waste (Ellen MacArthur Foundation, 2015). On the contrary, the circular economy is an economic paradigm where resources are kept in use as long as possible, with maximum value extracted from them; the paradigm has its conceptual root in industrial ecology, emphasising the benefits of recycling waste materials and by-products (Jacobsen, 2006).

The Ellen MacArthur Foundation (2021) defines the circular economy as

> a systems solution framework that tackles global challenges like climate change, biodiversity loss, waste, and pollution, and it is based on three principles, driven by design: eliminate waste and pollution, circulate products and materials (at their highest value), and regenerate nature.

The circular model, as in the linear economy, entails the extraction of raw materials and then the usage of these resources to manufacture new products consumed; however, different from the traditional business model, the waste is transformed into another resource (i.e., they are kept in use, retaining their value and contributing to a substantial reduction in the extraction of virgin materials) (Mangialardo and Micelli, 2018). Providing a pathway to more sustainable and resilient buildings and cities, circularity, similar to sustainable development, is considered as a development that meets the needs of the present without compromising the ability of future generations, based on social, economic and environmental pillars, with a focus on technological advances to solve the economic and environmental problem of finite resources by reducing consumption, as well as on the user behaviour that needs to change to close the cycle (Munaro, Tavares and Bragança, 2020). Where sustainability is the goal, the circular economy is a road map to a more sustainable economy (Anastasiades et al., 2020).

In the construction industry, the adoption of circular economy practices helps minimise the adverse impacts of the built environment and accelerate more

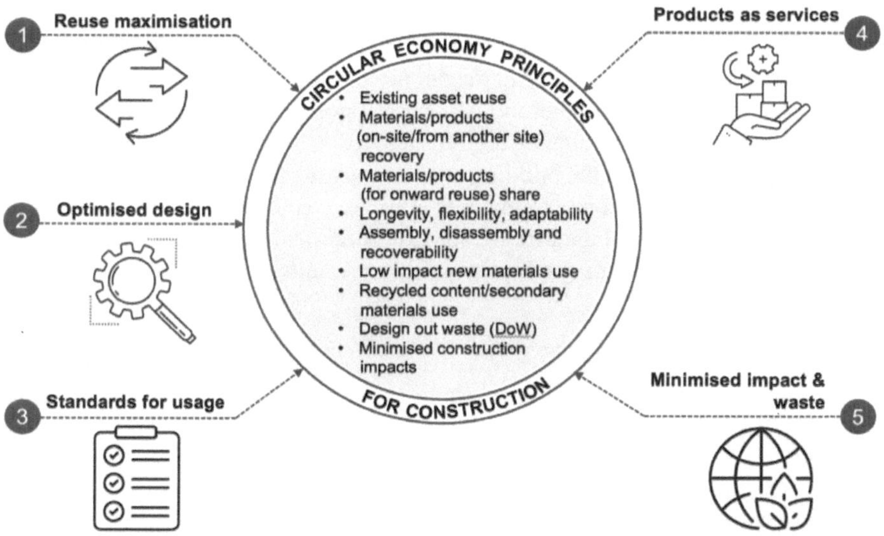

Figure 6.1 Circular economy principles for construction.

Source: Based on the circular economy guidance for construction clients by Surgenor et al. (2019) and the UK Green Building Council (2022b).

sustainable solutions. In their guide to construction clients, Surgenor et al. (2019) present the circular economy principles for construction. Figure 6.1 illustrates these, which include: reusing the sub- and superstructure; designing for optimisation, including longevity, flexibility, adaptability and assembly, disassembly and recoverability; using standardised elements, components, or modular design solutions for materials and products; transitioning from selling products to selling services through a payment structure where customers only pay for what they actually use; and encouraging use of new materials with little or no adverse effect, recycled or secondary materials, design out waste (DoW) and reduced construction site impacts (UK Green Building Council, 2022b). Scaling up the circular economy principles to buildings by keeping materials and components in use and guaranteeing their reuse when the building reaches its EoL will decrease the use of raw materials and reduce waste generation from the design phase (for instance, designing for a circular economy implies designing for adaptability, flexibility and disassembly to enable reversibility and salvage the value of the building's products) (Rahla, Mateus and Bragança, 2021).

The literature is also focusing on the implementation of circular economy practices in the built environment from these perspectives. In their systematic literature review, Benachio, Freitas and Tavares (2020) highlight that the reuse of materials, materials stocks, circular economy in the project design, life cycle assessment and material passports are the main areas discussed from most to least, respectively, and the most common approach is to address only one phase of the building life cycle: building EoL is the phase that most appeared in the articles, followed by project design, while operation is the least covered. Even though there is increasing research on circular economy in the construction industry, it still struggles for efficient and practical adoption of circular economy principles, which are successfully being implemented in many other industries (Pomponi and Moncaster, 2017). The change towards business circularity requires a focus on systemic thinking to understand the entire life cycle of the building and the construction value chain, involving better stakeholder integration (Zimmann et al., 2016). Due to their capacity to store data and practicability to the entire lifespan of buildings, BIM tools are attractive for achieving circular economy adoption within the built environment.

BIM for Circular Economy Adoption within the Built Environment

Utilised by many stakeholders within the architecture, engineering, construction and facility management domains for various uses, including design authoring, 3D coordination, design review, analysis, scheduling, cost estimation and asset management, BIM also has the potential to support circular economy implementation (Cetin, De Wolf and Bocken, 2021).

Along with these current BIM uses that influence achieving circular economy (e.g., structural design directly impacts the disassembly potential of a building), seven new BIM uses, including a digital model for sustainable EoL, material passport development, project database, data checking, circularity assessment, material recovery processes and material banks, are identified by Charef and Emmitt (2021) for a circular built environment. Table 6.1 summarises the recent BIM-based studies carried out for circular economy embracement within the built environment. Recent focus includes circularity indicators, reuse of building materials and components through industry foundation class (IFC) extensions, BIM-based design for deconstruction (DfD) and material passport.

Research Methodology and Process Plan

Focusing on incorporating the circular economy principles into a building life cycle through digitalisation, this chapter adopts an observe–orient–decide–act (OODA) loop (Boyd, 1995). The four-step research method consists of observation, orientation, decision and action stages (Figure 6.2).

The first step, which is important for data collection, is observing and understanding the current situation of the industry as well as future trends. The general overview of the built environment and the construction industry is based on various aspects, including the problems caused by the use of the linear economy approach and needs, demand and motivation for each phase of the building life cycle. The second step requires filtering and analysing the data collected in the first step with the circular economy orientation. By recognising the constraints and difficulties within the industry, interpreting the changing dynamics accurately – including both more digitalisation and adoption of modern approaches – is the key point of this step. The next step is where the optimum suggestions are made.

Table 6.1 Summary of BIM-based circular economy literature

Main focus	*Life cycle phase*	*BIM tool*	*Research method*	*Author and year*
BIM-based circularity indicators	Not applicable	°	Systematic literature review	Khadim et al. (2022)
Reusable materials and components visualisation through BIM and virtual reality	Prefabrication	•	Case study	O'Grady et al. (2021)
Reuse of building components	End-of-life	•	BIM standard proposal through IFC extensions	Sanchez et al. (2021)
Formation of new BIM uses for deconstruction	Use	•	Case study	van den Berg, Voordijk and Adriaanse (2021)
BIM-based material passport	End-of-life	•	Existing building (use case)	Honic et al. (2021)
Deconstructability of steel structures and DfD	Design	•	Case study	Basta, Serror and Marzouk (2020)
BIM-based end-of-life	End-of-life	°	Literature review	Akbarieh et al. (2020)
Material passport compilation for recycling potential	Design	•	Case study	Honic, Kovacic and Rechberger (2019)
BIM-based disassembly and deconstruction analytics	Design	•	Software development	Akanbi et al. (2019)
Potential construction waste of design alternatives	Design and Construction	•	Predictive model development	Akinade and Oyedele (2019)
Recoverability of materials	Design	•	Mathematical model	Akanbi et al. (2018)
Impact of DfD on design improvements	Design	°	Method development (Disassembly Network Analysis [DNA])	Denis, Vandervaeren and De Temmerman (2018)

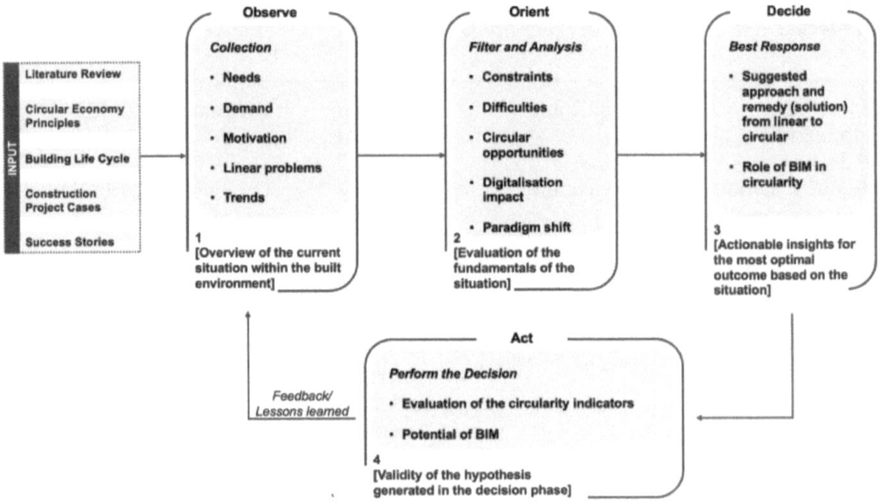

Figure 6.2 Research method.

Circular Building Life Cycle Approach

Successful adoption of circular economy principles can be achieved through a comprehensive assessment of each phase of the building life cycle. To facilitate the utilisation of circularity, this study proposes an overall approach for relevant stakeholders to apply circular principles to their processes. The proposed approach – based on the review of relevant literature, including Kubbinga et al. (2018); Heisel and Rau-Oberhuber (2020); Rahla, Mateus and Bragança (2021); Munaro and Tavares (2023) – aims to increase the awareness of the circular economy within the built environment as well as provide a guideline to employ a circular manner for the design, construction, use and EoL of a building life cycle.

Based on the drivers and barriers to implementing a circular economy, Figure 6.3 presents the fundamental principles to apply for each phase of the building life cycle along with the overall indicators. The increasing attention to net-zero, low carbon, innovative methods and systems, advanced technologies and new business models motivates the industry to embrace more circular lines within the practices. Despite these catalysts, the industry faces a lack of standardisation, knowledge and awareness of circularity, a fragmented and vulnerable supply chain, difficulty in keeping accurate material information for reuse/recover/recycle, cost of recovery of materials and their storage and lack of compatible tools/methods with advanced technologies. To increase and maintain the value of building components, all stakeholders within the building life cycle take an essential part in assessing the impact of their roles in circular economy adoption, and the need for collaboration throughout the value chain is crucial. Starting from the planning of the design

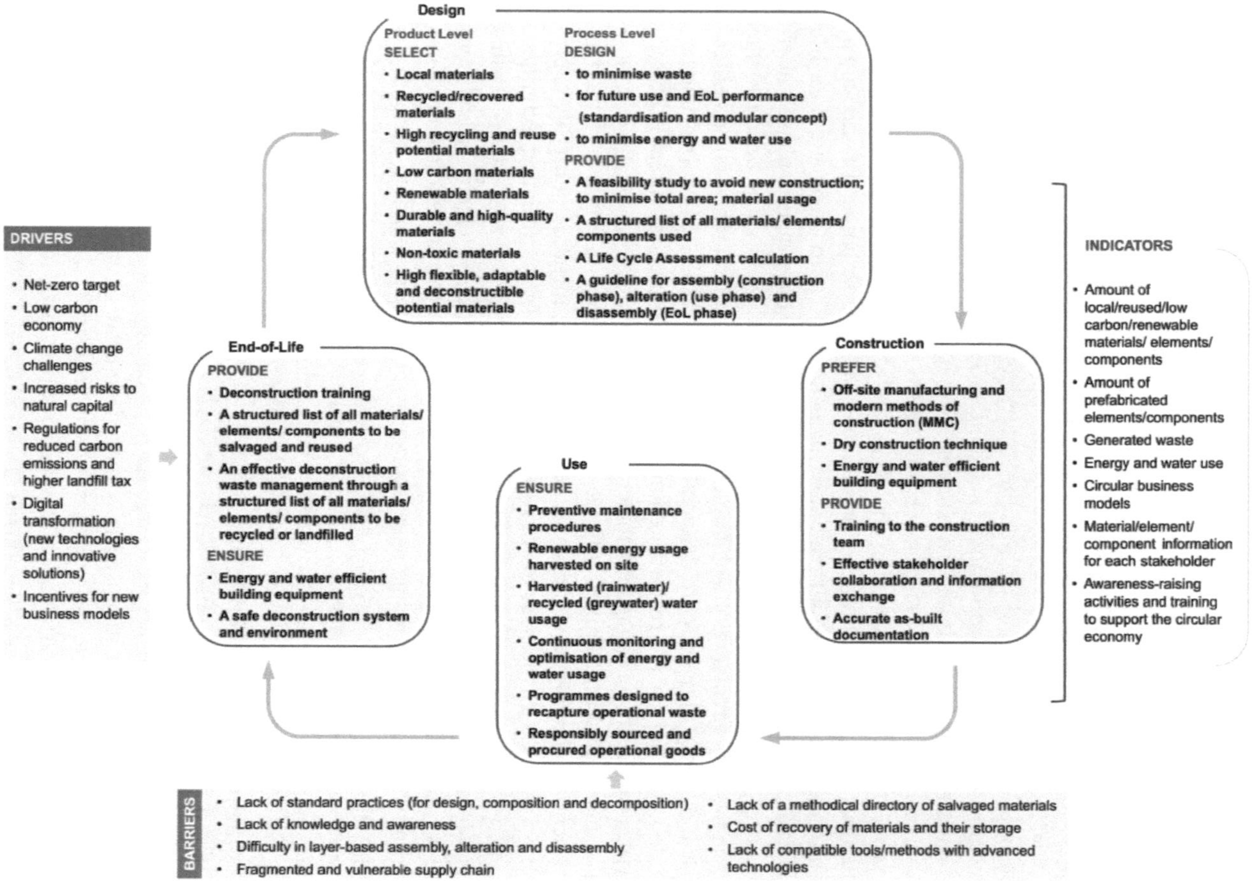

Figure 6.3 Circular building life cycle approach.

phase, a feasibility study for the optimum solution (avoid new construction or minimise total area and material usage) is the initial step. The main decisions at the product level should be based on the material features, focusing on not only the design aspect but also its performance during the use phase – which highlights the locality, recyclability, recoverability, renewability, flexibility, adaptability, durability and deconstructibility – to maximise their value.

The design should follow the principles of minimising waste, energy and water use and standardisation and modularity for future use and EoL performance. For a successful adaptation of the circular economy, the principles of longevity, quality and adaptability should be embedded in the final product from the early stages of design. More durable and high-quality materials and components are easier to maintain and renew with minimum contribution. Even though such an approach may initially increase the initial investment costs, it will be beneficial for not only the owners but also end users, considering the savings it will provide throughout the product life cycle. Designing materials and components to be reused also stimulates the economy by turning waste into a valuable resource. The documentation of a structured list of all materials, elements and components and their life cycle assessments as well as a guideline for assembly (construction phase), alteration (use phase) and disassembly (EoL phase) would be the key action of the design phase for circular approach implementation. This documentation and guideline will not only allow a more deliberate and methodical design process but also provide key input for the rest of the building life cycle phases so that necessary activities can easily be engaged with accurate data.

The main circular economy attitudes for the construction phase include both construction technology and process management. The utilisation of off-site manufacturing and modern methods of construction (MMC), dry construction techniques and energy- and water-efficient building equipment are listed as the considerable actions to adopt in this phase, which will help maximise productivity and minimise waste. Moreover, construction team training, effective stakeholder collaboration and information exchange and accurate as-built documentation increase the success of circularity. All of these aspects would require operative and functional construction process management, which includes planning, organising and controlling steps. Ensuring the construction team acquires the relevant information from the design phase for assembly and builds accordingly is one of the critical actions for this phase. The accuracy and quality of the construction work will affect not only the disassembly processes in the EoL phase but also the maintenance and possible modifications during the use phase. Training investments for the construction team, therefore, are inevitable for most cases; however, the return on this investment would be positive as this will result in more circular and less environmentally impact construction processes in the long term. Providing the as-built project documentation systematically with the correct data, to be transferred to the use and EoL phases, has a great impact on the success of those phases.

To provide the optimum building performance, the operation and maintenance phase should incorporate preventive maintenance procedures, harvested energy and water (rainwater, greywater) usage and their continuous monitoring and optimisation, operational waste recapture and responsibly sourced operational goods to minimise the environmental impact of a building in its use. Unlike traditional approaches (e.g., correct when it is broken), preventive maintenance requires regular activities, both time- and condition-based, to prevent critical issues. Energy and water usage as well as operational waste should be recorded, stored and observed to optimise for the best performance. Having a unique position in changing the environmental impact of a building, the use phase and facility management can minimise costs, waste and carbon and provide healthier living and working environments to improve well-being and quality of life.

As the key phase for eliminating waste and supporting reuse, the EoL phase should have a special focus towards circularity, unlike traditional methods used for the demolition of a building. As obstacles to circular economy adoption, deconstruction training to relevant parties and structured lists of all materials, elements and components to be salvaged, reused, recycled or landfilled should be the strategic areas to be focused on for the EoL phase. In addition, effective deconstruction waste management, energy- and water-efficient building equipment usage and providing a safe deconstruction system and environment are other principles to implement. The overall indicators listed to measure the success of a project include details about the materials, elements and components used; generated waste in construction, use and EoL phases of the building life cycle; energy and water use; circular business models; and awareness and training activities to support the circular economy.

The Role of BIM in Circularity

As one of the drivers for a circular economy, engaging advanced technologies in building production processes can accelerate the implementation of circularity within the built environment. BIM enables the storage of different types of data during the building life cycle and promotes circular economy principles. Key connections between BIM and circular economy within the building life cycle can be listed under adaptability and flexibility in the design phase, efficiency in the construction phase and control in the use and EoL phases. Using BIM as an effective collaboration tool among different disciplines not only increases efficiency and productivity but also can provide energy, environment and economy optimisation throughout the life of a building, while different stakeholders work in coordination from the beginning of the design phase.

Having the product and process data available in the BIM database helps stakeholders to fully control compliance with regulations while minimising construction errors and making the building production process faster and safer. By facilitating off-site production, BIM reduces mobility on the construction site and prevents incorrect and excessive orders, which results in less waste. Access to the data required for facility management and maintenance by monitoring and recording building performance during its use becomes easy through as-built models generated by BIM. At the EoL phase, through digital documentation, BIM can facilitate the process by providing an accurate list of materials and components and the exact location of those to be reused or recycled.

In this study, the role of BIM is tackled from the point of its potential to provide the indicators proposed in the circular building life cycle approach. Figure 6.4 presents the main steps for achieving the indicators, which include the input based on the design, construction, use and EoL phases; use of existing and extended property sets within the BIM software; acquiring the input data from the IFC file; and finally analysing the data to gauge the performance and reaching the indicators. Different colours are used to indicate the corresponding input for each phase and the relevant indicators.

The required input for BIM software is based on circular economy principles for each phase of the building life cycle, provided in the previous section. Accurate and detailed material, element and component data (e.g., including type, for design phase: local, reused, low carbon, renewable; for construction phase: prefabricated; for EoL: to be salvaged (for reuse, recycle etc.)) should be embedded in the BIM software. Detailed specifications regarding the construction technique, operation and maintenance, future use and deconstruction method for each component and/or activity provided at the design phase not only facilitate accurate data recording but also provide effective information exchange among the stakeholders. Furthermore, real-time updates with construction progress on site for effective communication and accurate as-built documentation and preventive maintenance data for both product and process levels allow each relevant stakeholder to obtain the essential information accurately and timely. Starting from the design phase, energy- and water-related aspects should also be provided so that robust monitoring and optimisation can be achieved.

To be able to measure the waste generated throughout the building life cycle, construction, operational and deconstruction waste data should be recorded separately. Data concerning different alternatives for early design appraisal, operational goods and real-time updates with use phase product and process and deconstruction progress on site would be helpful to gauge circular business model adoption. Finally, the data acquired from previous phases for knowledge sharing can be used for awareness-raising activities and training to support the circular economy within the built environment.

For the successful implementation of circular economy principles in the construction industry, the building should be addressed with a system approach and building parts should be easily replaced, adapted and renewed. As a technological and innovative solution, due to its benefits, BIM can be one of the biggest supporters of this process, which can be defined as 'from demolition to reuse'. In the BIM scenario, the main steps to be taken for circular economy principles for each phase of the building life cycle should be clarified with their corresponding indicators at the beginning of a project so that each stakeholder can perform their task in a timely and accurate manner.

The opportunity of embedding non-geometrical data into the model and acquiring them, when necessary, can not only provide periodical and project-based

Figure 6.4 BIM for circular economy indicators.

measurement and monitoring but also help more strategical decision-making for circular approaches within the industry.

Discussion and Conclusion

Adopting a circular economy approach in a high-growth, high-waste sector like the built environment presents a huge opportunity to capture more value, and organisations across the value chain – from city governments to developers, asset managers to construction product manufacturers – can redesign the sector using the principles of a circular economy to better manage resources and create (Ellen MacArthur Foundation, 2020). As a condition for sustainability, adapting the principles of the circular economy into built environment practices will help achieve net-zero targets and reduce carbon emissions. Even though changing the dynamics of the industry is not straightforward and requires considerable effort, there has been increasing attention to the transition from linear to circular, and driving factors such as the utilisation of new technologies and innovative business models can accelerate this change.

Focusing on the role of BIM in circularity, this chapter proposes a holistic circular building life cycle approach that includes the main steps and principles for design, construction, use and EoL phases along with the indicators. For successful circularity adoption, key stakeholders should embrace integrated and inclusive methods and systems of design, construction, use and deconstruction. Since most design decisions affect the rest of the building life cycle phases, the design phase requires particular attention, consideration, planning and programming at both product and process levels. Based on accurate design data and information flow for the succeeding processes, the construction, operation and maintenance and EoL phases – applying the phase-specific principle and generating the necessary data for the other phase and relevant stakeholders – can complete the loop.

Alongside incorporating circular economy principles into practice, continuous monitoring and improvement throughout the phases play an essential role in a thorough performance. BIM, offering an integrated database to exchange information among different parties through a common model, can easily incorporate circular economy principles into the practices of different phases of the building life cycle. Moreover, the circularity data embedded into the BIM software can then be used for measuring the performance (achieving the indicators) as well as strategic decisions to foster circular business models. To indicate its potential, this study summarises key BIM inputs for each phase for achieving the main circularity indicators. Whereas data entry can be achieved through existing and/or extended property sets, IFC could be the standard to extract the data for analysis. The provided basis for BIM integration with circular economy principles in this chapter can lead to further study that focuses on the details of the property set development and steps for data extraction, along with verification via a case study.

References

Advisory Board for Cirkulær Økonomi (2017) *Circular Economy Advisory Board. Recommendations for the Danish Government*. Available at: https://en.mfvm.dk/fileadmin/user_upload/MFVM/Miljoe/Cirkulaer_oekonomi/Advisory-Board-for-Circular-Economy-Report-2017-Content_Single_pages_WEB.pdf (Accessed: 12 December 2022).

Akanbi, L. A., Oyedele, L. O., Akinade, O. O., Ajayi, A. O., Delgado, M. D., Bilal, M. and Bello, S. A. (2018) 'Salvaging building materials in a circular economy: A BIM-based whole-life performance estimator', *Resources, Conservation and Recycling*, 129, pp. 175–186. Available at: https://doi.org/10.1016/j.resconrec.2017.10.026

Akanbi, L. A., Oyedele, L. O., Omoteso, K., Bilal, M., Akinade, O. O., Ajayi, A. O., Delgado, J. M. D. and Owolabi, H. A. (2019) 'Disassembly and deconstruction analytics system (D-DAS) for construction in a circular economy', *Journal of Cleaner Production*, 223, pp. 386–396. Available at: https://doi.org/10.1016/j.jclepro.2019.03.172

Akbarieh, A., Jayasinghe, L. B., Waldmann, D. and Teferle, F. N. (2020) 'BIM-based end-of-lifecycle decision making and digital deconstruction: Literature review', *Sustainability*, 12(7), p. 2670. Available at: https://doi.org/10.3390/su12072670

Akinade, O. O. and Oyedele, L. O. (2019) 'Integrating construction supply chains within a circular economy: An ANFIS-based waste analytics system (A-WAS)', *Journal of Cleaner Production*, 229, pp. 863–873. Available at: https://doi.org/10.1016/j.jclepro.2019.04.232

Akinade, O. O., Oyedele, L. O., Omoteso, K., Ajayi, S. O., Bilal, M., Owolabi, H. A., Alaka, H. A., Ayris, L. and Looney, J. H. (2017) 'BIM-based deconstruction tool: Towards essential functionalities', *International Journal of Sustainable Built Environment*, 6(1), pp. 260–271. Available at: https://doi.org/10.1016/j.ijsbe.2017.01.002

Anastasiades, K., Blom, J., Buyle, M. and Audenaert, A. (2020) 'Translating the circular economy to bridge construction: Lessons learnt from a critical literature review', *Renewable and Sustainable Energy Reviews*, 117, p. 109522. Available at: https://doi.org/10.1016/j.rser.2019.109522

Basta, A., Serror, M. H. and Marzouk, M. (2020) 'A BIM-based framework for quantitative assessment of steel structure deconstructability', *Automation in Construction*, 111, p. 103064. Available at: https://doi.org/10.1016/j.autcon.2019.103064

Benachio, G. L. F., Freitas, M. D. C. D. and Tavares, S. F. (2020) 'Circular economy in the construction industry: A systematic literature review', *Journal of Cleaner Production*, 260, p. 121046. Available at: https://doi.org/10.1016/j.jclepro.2020.121046

Blok, V. and Gremmen, B. (2016) 'Ecological innovation: Biomimicry as a new way of thinking and acting ecologically', *Journal of Agricultural and Environmental Ethics*, 29(2), pp. 203–217. Available at: https://doi.org/10.1007/s10806-015-9596-1

Boyd, J. R. (1995) *The Essence of Winning and Losing*. Available at: https://web.archive.org/web/20110324054054/ http://www.danford.net/boyd/essence.htm (Accessed: 12 January 2023).

Braungart, M., McDonough, W. and Bollinger, A. (2007) 'Cradle-to-cradle design: Creating healthy emissions – A strategy for eco-effective product and system design', *Journal of Cleaner Production*, 15(13–14), pp. 1337–1348. Available at: https://doi.org/10.1016/j.jclepro.2006.08.003

Buildings as Material Banks (BAMB) (2020) *Enabling a Circular Building Industry*. Available at: www.bamb2020.eu/ (Accessed: 11 November 2022).

Cetin, S., De Wolf, C. and Bocken, N. (2021) 'Circular digital built environment: An emerging framework', *Sustainability*, 13(11), p. 6348. Available at: https://doi.org/10.3390/su13116348

Charef, R. and Emmitt, S. (2021) 'Uses of building information modelling for overcoming barriers to a circular economy', *Journal of Cleaner Production*, 285, p. 124854. Available at: https://doi.org/10.1016/j.jclepro.2020.124854

China Circular Economy Promotion Law (2008) *The Circular Economy Promotion Law of the People's Republic of China.* Available at: https://ppp.worldbank.org/public-private-partnership/sites/ppp.worldbank.org/files/documents/China_CircularEconomyLawEngl ish.pdf (Accessed: 19 November 2022).

Dabous, S. A., Shanableh, A., Al-Ruzouq, R., Hosny, F. and Khalil, M. A. (2022) 'A spatio-temporal framework for sustainable planning of buildings based on carbon emissions at the city scale', *Sustainable Cities and Society*, 82, p. 103890. Available at: https://doi.org/ 10.1016/j.scs.2022.103890

Denis, F., Vandervaeren, C. and De Temmerman, N. (2018) 'Using network analysis and BIM to quantify the impact of design for disassembly', *Buildings*, 8(8), p. 113. Available at: https://doi.org/10.3390/buildings8080113

Despeisse, M., Baumers, M., Brown, P., Charnley, F., Ford, S. J., Garmulewicz, A., Knowles, S., Minshall, T. H. W., Mortara, L., Reed-Tsochas, F. P. and Rowley, J. (2017) 'Unlocking value for a circular economy through 3D printing: A research agenda', *Technological Forecasting and Social Change*, 115, pp. 75–84. Available at: https://doi.org/10.1016/ j.techfore.2016.09.021

De Wolf, C., Hoxha, E. and Fivet, C. (2020) 'Comparison of environmental assessment methods when reusing building components: A case study', *Sustainable Cities and Society*, 61, p. 102322. Available at: https://doi.org/10.1016/j.scs.2020.102322

Di Biccari, C., Abualdenien, J., Borrmann, A. and Corallo, A. (2019) 'A BIM-based framework to visually evaluate circularity and life cycle cost of buildings', IOP Conference Series: Earth and Environmental Science, 290(1), p. 012043. Available at: https://iopscie nce.iop.org/article/10.1088/1755-1315/290/1/012043/pdf

Eberhardt, L. C. M., Birkved, M. and Birgisdottir, H. (2022) 'Building design and construction strategies for a circular economy', *Architectural Engineering and Design Management*, 18(2), pp. 93–113. Available at: https://doi.org/10.1080/17452007.2020.1781588

Ellen MacArthur Foundation (EMF) (2015) Growth Within: A Circular Economy Vision for a Competitive Europe. Available at: https://ellenmacarthurfoundation.org/growth-wit hin-a-circular-economy-vision-for-a-competitive-europe (Accessed: 16 January 2023).

Ellen MacArthur Foundation (EMF) (2020) *Reimagining Our Buildings and Spaces for a Circular Economy.* Available at: https://ellenmacarthurfoundation.org/topics/built-envi ronment/overview (Accessed: 10 June 2023).

Ellen MacArthur Foundation (EMF) (2021) Circular Economy Glossary. Available at: https://emf.thirdlight.com/link/vj6i9k5yax0n-1fkyvu/@/preview/1?o (Accessed: 31 March 2023).

European Commission (EU) (2017) Circular Economy Action Plan. Available at: https:// environment.ec.europa.eu/strategy/circular-economy-action-plan_en (Accessed: 16 October 2022).

European Commission (EU) (2021) *Europe's Digital Decade: Commission Sets the Course Towards a Digitally Empowered Europe by 2030.* Available at: https://ec.europa.eu/com mission/presscorner/detail/en/IP_21_983 (Accessed: 14 May 2022).

Geissdoerfer, M., Savaget, P., Bocken, N. M. and Hultink, E. J. (2017) 'The circular economy – A new sustainability paradigm?' *Journal of Cleaner Production*, 143, pp. 757–768. Available at: https://doi.org/10.1016/j.jclepro.2016.12.048

Ghisellini, P., Cialani, C. and Ulgiati, S. (2016) 'A review on circular economy: The expected transition to a balanced interplay of environmental and economic systems',

Journal of Cleaner Production, 114, pp. 11–32. Available at: https://doi.org/10.1016/j.jclepro.2015.09.007

Heisel, F. and Rau-Oberhuber, S. (2020) 'Calculation and evaluation of circularity indicators for the built environment using the case studies of UMAR and Madaster', *Journal of Cleaner Production*, 243, p. 118482. Available at: https://doi.org/10.1016/j.jclepro.2019.118482

Honic, M., Kovacic, I., Aschenbrenner, P. and Ragossnig, A. (2021) 'Material Passports for the end-of-life stage of buildings: Challenges and potentials', *Journal of Cleaner Production*, 319, p. 128702. Available at: https://doi.org/10.1016/j.jclepro.2021.128702

Honic, M., Kovacic, I. and Rechberger, H. (2019) 'Improving the recycling potential of buildings through Material Passports (MP): An Austrian case study', *Journal of Cleaner Production*, 217, pp. 787–797. Available at: https://doi.org/10.1016/j.jclepro.2019.01.212

Huang, B., Xing, K., Pullen, S., Liao, L. and Huang, K. (2020) 'Ecological–economic assessment of renewable energy deployment in sustainable built environment', *Renewable Energy*, 161, pp. 1328–1340. Available at: https://doi.org/10.1016/j.renene.2020.08.004

Iacovidou, E. and Purnell, P. (2016) 'Mining the physical infrastructure: Opportunities, barriers and interventions in promoting structural components reuse', *Science of the Total Environment*, 557, pp. 791–807. Available at: https://doi.org/10.1016/j.scitotenv.2016.03.098

Jacobsen, N. B. (2006) 'Industrial symbiosis in Kalundborg, Denmark: A quantitative assessment of economic and environmental aspects', *Journal of Industrial Ecology*, 10(1–2), pp. 239–255. Available at: https://doi.org/10.1162/108819806775545411

Khadim, N., Agliata, R., Marino, A., Thaheem, M. J. and Mollo, L. (2022) 'Critical review of nano and micro-level building circularity indicators and frameworks', *Journal of Cleaner Production*, 357, p. 131859. Available at: https://doi.org/10.1016/j.jclepro.2022.131859

Kirchherr, J., Reike, D. and Hekkert, M. (2017) 'Conceptualizing the circular economy: An analysis of 114 definitions', *Resources, Conservation and Recycling*, 127, pp. 221–232. Available at: https://doi.org/10.1016/j.resconrec.2017.09.005

Kubbinga, B., Bamberger, M., Van Noort, E., Van den Reek, D., Blok, M., Roemers, G., ... & Faes, K. (2018) 'A framework for circular buildings – Indicators for possible inclusion in BREEAM', Circle Economy, DGBC, Metabolic and SGS, Netherlands, 52. Available at: https://assets.website-files.com/5d26d80e8836af2d12ed1269/5dea6b3713854714c4a8b755_A-Framework-For-Circular-Buildings-BREEAM-report-20181007-1.pdf (Accessed: 14 May 2023).

Lyle, J. T. (1996) *Regenerative Design for Sustainable Development*. John Wiley and Sons.

Mangialardo, A. and Micelli, E. (2018) 'Rethinking the construction industry under the circular economy: Principles and case studies', In *Smart and Sustainable Planning for Cities and Regions: Results of SSPCR 2017*, 2, pp. 333–344. Available at: https://link.springer.com/chapter/10.1007/978-3-319-75774-2_23

Mhatre, P., Gedam, V., Unnikrishnan, S. and Verma, S. (2021) 'Circular economy in built environment – Literature review and theory development', *Journal of Building Engineering*, 35, p. 101995. Available at: https://doi.org/10.1016/j.jobe.2020.101995

Munaro, M. R. and Tavares, S. F. (2023) 'A review on barriers, drivers, and stakeholders towards the circular economy: The construction sector perspective', Cleaner and Responsible Consumption, p. 100107. Available at: https://doi.org/10.1016/j.clrc.2023.100107

Munaro, M. R., Tavares, S. F. and Bragança, L. (2020) 'Towards circular and more sustainable buildings: A systematic literature review on the circular economy in the built environment', *Journal of Cleaner Production*, 260, p. 121134. Available at: https://doi.org/10.1016/j.jclepro.2020.121134

O'Grady, T. M., Brajkovich, N., Minunno, R., Chong, H. Y. and Morrison, G. M. (2021) 'Circular economy and virtual reality in advanced BIM-based prefabricated construction', *Energies*, 14(13), p. 4065. Available at: https://doi.org/10.3390/en14134065

Pomponi, F. and Moncaster, A. (2017) 'Circular economy for the built environment: A research framework', *Journal of Cleaner Production*, 143, p. 710–718. Available at: https://doi.org/10.1016/j.jclepro.2016.12.055

Rahla, K. M., Mateus, R. and Bragança, L. (2021) 'Implementing circular economy strategies in buildings – From theory to practice', *Applied System Innovation*, 4(2), p. 26. Available at: https://doi.org/10.3390/asi4020026

Redevco Foundation (2018) *A Framework for Circular Buildings*. Available at: https://assets.website-files.com/5d26d80e8836af2d12ed1269/5dea6b3713854714c4a8b755_A-Framework-For-Circular-Buildings-BREEAM-report-20181007-1.pdf (Accessed: 21 October 2022).

Reike, D., Vermeulen, W. J. V. and Witjes, S. (2018) 'The circular economy: New or refurbished as CE 3.0? – Exploring controversies in the conceptualization of the circular economy through a focus on history and resource value retention options', Resources, *Conservation and Recycling*, 135, pp. 246–264. Available at: https://doi.org/10.1016/j.resconrec.2017.08.027

Sanchez, B., Rausch, C., Haas, C. and Hartmann, T. (2021) 'A framework for BIM-based disassembly models to support reuse of building components', *Resources, Conservation and Recycling*, 175, p. 105825. Available at: https://doi.org/10.1016/j.resconrec.2021.105825

Stahel, W. R. (2016) 'The circular economy', *Nature*, 531(7595), pp. 435–438. Available at: https://doi.org/10.1038/531435a

Surgenor, A., Winch, R., Moodey, L. and Mant, A. (2019) *Circular Economy Guidance for Construction Clients: How to Practically Apply Circular Economy at the Project Brief Stage*. Available at: https://ukgbc.org/wp-content/uploads/2019/04/Circular-Economy-Report-singles.pdf (Accessed: 18 March 2023).

UK Green Building Council (2022a) *The Circular Score*. Available at: www.ukgbc.org/inside_innovation/the-circular-score/#:~:text=The%20built%20environment%20sector%20in,behavioural%20shift%20in%20the%20industry (Accessed: 18 December 2022).

UK Green Building Council (2022b) Insights on How Circular Economy Principles Can Impact Carbon and Value. Available at: https://ukgbc.org/wp-content/uploads/2022/08/Whole-Life-Carbon-Circular-Economy-Report.pdf (Accessed: 23 January 2023).

United Nations Environment Programme (UNEP) (2016) The Emissions Gap Report 2016. Available at: https://wedocs.unep.org/bitstream/handle/20.500.11822/10016/emission_gap_report_%202016.pdf (Accessed: 16 November 2022).

Van den Berg, M., Voordijk, H. and Adriaanse, A. (2021) 'BIM uses for deconstruction: An activity-theoretical perspective on reorganising end-of-life practices', *Construction Management and Economics*, 39(4), pp. 323–339. Available at: https://doi.org/10.1080/01446193.2021.1876894

World Economic Forum (WEF) (2016) Shaping the Future of Construction, *a* Breakthrough in Mindset and Technology. Available at: www3.weforum.org/docs/WEF_Shaping_the_Future_of_Construction_report_020516.pdf (Accessed: 10 December 2022).

Zimmann, R., O'Brien, H., Hargrave, J. and Morrell, M. (2016) The Circular Economy in the Built Environment. Available at: www.arup.com/perspectives/publications/research/section/circular-economy-in-the-built-environment (Accessed: 10 April 2023).

7 Human Augmentation Technology Adoption for the Construction Industry

Samuel Adeniyi Adekunle, Clinton Aigbavboa, Obuks Ejohwomu, Opeoluwa Akinradewo, Andrew Ebekozien, and Douglas Aghimien

Introduction

The construction industry is vital to any economy. It is a great employer of labour; it employs up to 7.70% of the global workforce and is critical to economic recovery post-Covid (Khanal et al., 2020). The construction industry's impact is felt across sectors and human endeavours; it has environmental, social, and economic influence. However, it has faced many challenges over the years. Some of the challenges facing the industry include productivity, labour shortages, and slow adoption of technology, among others. In the present age, the issue of labour shortages and the adoption of technology to improve productivity while making up for labour shortages has become a critical issue for discussion in the construction industry.

A critical look at labour shortage, as it affects the construction industry, reveals that the industry is suffering from an image problem that makes it unattractive to potential workers. Haupt and Harinarain (2017) state that the skill shortage in the construction industry is related to poor image, and this is not changing as the image has not improved from the prospective workers' perspective. But, again, skill shortage is observed to be dependent on technological level, as high technology firms struggle more to recruit (Haskel & Martin, 2001). This introduces another dimension to the skill shortage based on the technology implementation level of an organisation. Another identified challenge to skill shortage is skill mismatch and skill deficiency (Silva et al., 2018). This is common, especially during Industry 4.0, where a lack of skills has been identified as a significant barrier to adoption (Adekunle et al., 2020).

By extension, lack of skills and labour shortages affect the construction industry's productivity as an input. Hughes and Thorpe (2014) identified incompetence as one of the critical factors affecting the construction industry. In addition, lack of talent is also a vital factor in barriers to productivity (Rob Elvidge & PlanGrid, 2019). Poor productivity negatively impacts construction projects' profitability (Keogh & Smallwood, 2021). Thus, the poor productivity of the construction industry has a far-reaching effect on the construction industry and its stakeholders. Some of the impacts include a waste of money and time (Al et al.,

DOI: 10.1201/9781003380559-7

2009), leading to a decrease in investments by the client. Consequently, a critical ingredient in improving construction productivity is human or personnel resources and enhancement for enhanced performance.

Human Augmentation

According to neoclassical growth theory, to achieve long-term growth, technology is a critical factor (Hahn, 2010; Sredojević et al., 2016). Technological adoption is thus non-negotiable for the construction industry to achieve the necessary growth. This is imperative due to the labour problems confronting the industry; labour scarcity has, however, been observed to induce technological advancement (Acemoglu, 2009).

Human augmentation refers to technologies for the improvement of a human's ability to do work. It involves adopting technology for healthy persons who want to improve their work rate or to replace a lost part for those with impairment or disability (De Boeck & Vaes, 2021). Human augmentation is all about overcoming limitations and improving human efficiency through technology.

Table 7.1 presents a classification of human augmentation. It is evident that human augmentation covers diverse areas of human life for improvement. It can

Table 7.1 Human augmentation aspects

			Dimensions		
			1	*2*	*3*
			Replicating human ability	*Supplementing human ability*	*Exceeding human ability*
Categories	1	Social augmentation	Facemask that lights up when words are spoken and mimic mouth movement	Virtual reality headset to improve teamwork	A smart necklace that sets off an alarm when/if someone comes too close
	2	Cognitive augmentation	Memory improving Brain–computer interface	Virtual reality headset that stores mental images	Augmented reality headset providing optimised navigation support
	3	Physical augmentation	Leg or arm prosthesis	Exoskeletons that help with load lifting	Jetpack that enables flying
	4	Sensory augmentation	Hearing aids, glasses	Smart headphones that filter out noise	X-ray vision lenses

Source: Adapted from De Boeck & Vaes (2021).

be used to replicate or supplement human activities by making their performance more effective than they would be or to exceed human abilities. This can be adopted in social activities, cognitive enhancements, physically, or to enhance the sensory capabilities of humans. For instance, it involves transforming brain activity for external devices through control for people with severe disabilities (Valeriani et al., 2019). This continues to receive further attention from researchers to improve non-muscle communication through technology (Rezeika et al., 2018). Other aspects include the development of an exoskeleton for walking assistance (He et al., 2019).

These human technologies are embarked upon to enhance human ability and help them achieve beyond their natural capabilities. When adopted, it can be used to improve physical, psychological, and social performance, depending on the usage.

Productivity in the construction industry has been observed to be flat, and all productivity improvement efforts have yielded little or no results in the past 50 years (WEF & BCG, 2016). This narrative requires a change, and adoption of technology is necessary to achieve this growth. This study, therefore, focuses on the adoption of human augmentation technologies in the construction industry i) to identify human augmentation technologies for use in the construction industry and ii) to critically analyse the important attendant issues for their adoption in the construction industry.

The adoption of human enhancement technology in the construction industry was especially required during the Covid-19 pandemic, during which physical human interactions were restricted. The occurrence of this disease impacted many aspects of human living. It caused significant global destabilisation (Schindler et al., 2020) in every facet of life, one of which was touted to have economically extreme impacts (Bartik et al., 2020). Every sector of the economy was affected, and loss was suffered in different magnitudes. For instance, according to Bartik et al. (2020), 43% of small businesses were reported to close due to Covid-19.

Traditionally, the construction industry is a busy industry that employs many hands working concurrently; hence, human contact is inevitable. For instance, several structures were constructed during the pandemic to accommodate victims, despite pandemic lockdown measures, among which are the high-standard isolation units in Shenzen, China (Zhou et al., 2020). Thus, the industry is essential, and its productivity and manpower challenges must be solved for it to be responsive to demands. The next section outlines the study methods, while the subsequent sections dwell on the findings, implications, and conclusion.

Materials and Methods

The study adopted a multi-staged approach to understanding the adoption of human augmentation in the construction industry and analysed the adoption issues. An interpretivist philosophical viewpoint (Aghimien et al., 2021) was adopted. The strategy adopted included three stages similar to Che Ibrahim et al. (2020); it involves the selection of the database, data cleaning, and analysis of collected data. Published articles on human augmentation were sourced, irrespective of the

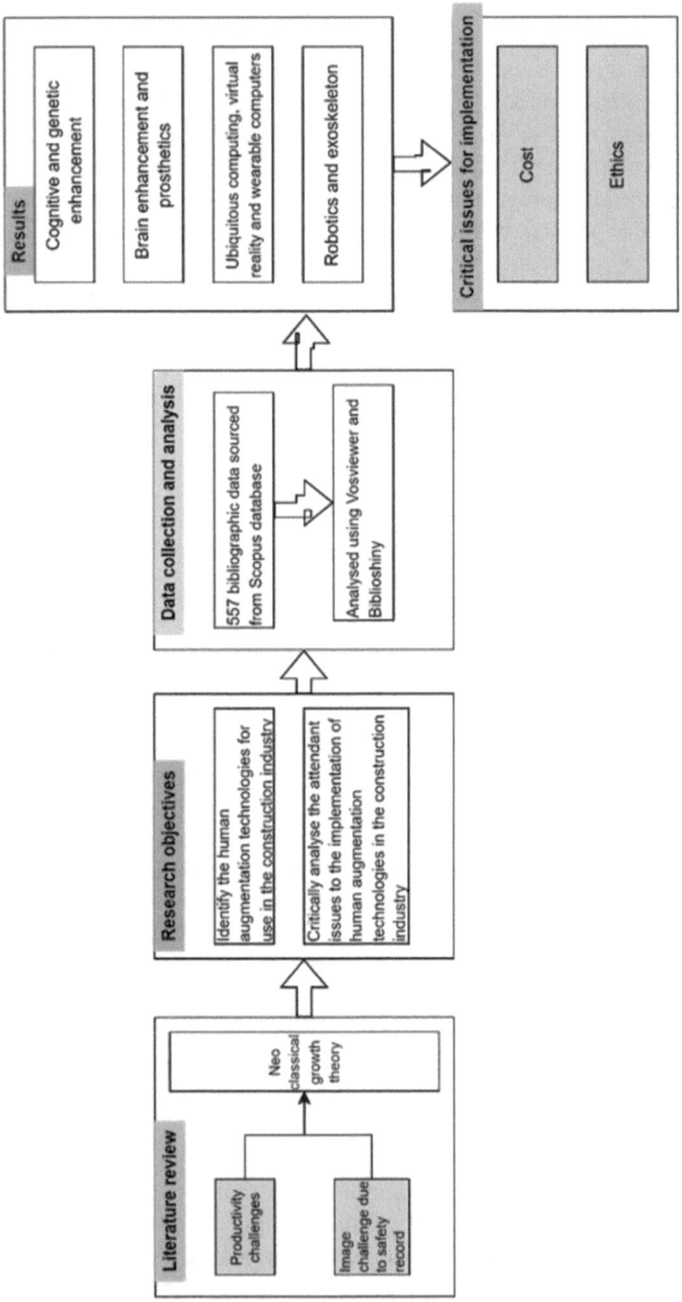

Figure 7.1 Research framework.

research field. This provided insight into the human augmentation research field and helped identify the various technologies. Figure 7.1 presents the research framework for the study.

Database Selection

The database selected for this study is the Scopus database. It was chosen due to due coverage, and it has been adopted by several studies recently for reviews (Adekunle, Aigbavboa, et al., 2021; Aghimien et al., 2021; Che Ibrahim et al., 2020) in different areas in construction digital transformation designer competency for prevention through design and dynamic capabilities for digitisation, among other research focuses. The study adopted the search for studies on human augmentation using the keywords "human augmentation". The retrieved document consists of all document types, including articles, conferences, book chapters, and books. The search covered publications in this research focus up till 2021.

Article Selection

After identifying human augmentation research works, the study examined the data collected, and they were cleaned up appropriately to fit into the study objectives. Firstly, the study identified the background information on the research publications by analysing research trends over the years, citations, and others. Secondly, the study cleaned the collected data to ensure that the articles specifically focus on human augmentation. Thus, through screening, the study removed studies that were not human augmentation tailored. In all, 557 articles were adopted for the analysis.

Analysis

To achieve appropriate data analysis for the collected data, the study examined similar existing review studies (Adekunle et al., 2022a; Adekunle, Ejohwomu, et al., 2021; Aghimien et al., 2019; Janik et al., 2020; Liu et al., 2019); these studies analysed bibliometric data collection through visualisation. They employed this technique to understand the data structure, trends, and patterns. Similarly, this study adopted visualisation to understand the trends and patterns in the collected data. The results were presented using cluster maps. The investigation started by understanding the background information about the articles; later, the clusters were identified and discussed.

Results

This section presents the results and analysis. Various diagrams were adopted in the presentation of the study results.

Publication by Country

Data collected revealed that the USA is the most active and productive in human augmentation research. This is followed, but with a wide gap, by the United Kingdom, Japan, and Canada. Portugal is the least productive; however, it is worth noting that African countries are yet to begin research in this area (Figure 7.2).

These results might not be unconnected to the fact that human augmentation technology adoption is still in its infancy stage. It is not surprising to see that most of the work has been done in developed countries, as they implement new technologies earlier than developing countries (Adekunle, Aigbavboa, et al., 2021).

Although the USA is the most productive, it is noticed that the UK is the most cited of the active human augmentation research countries. However, the USA is the second most cited country. The USA has 304 documents and 849 citations, while the UK has 126 documents but 1333 citations. Japan has 86 documents and 41 citations.

The non-existent research output in this area in African and, by extension, many developing countries provides a gap to be filled by researchers for more productive labour in the construction industry. It is considered a field requiring a lot of attention especially in the ethical and cultural contexts.

Publications per Year

Further analysis of collected data showed that although there were some efforts in the 20th century, the number of publications experienced a significant leap in the 21st century. However, it is still a developing research area, and its application in various fields is still unfolding.

A critical look at the analysis shows that this research area has witnessed steady growth and interest. However, there was a steep decline in the years 2016 and 2020. This is presented in Figure 7.3. These steps might not be unconnected to the diverse challenges this field faces for progressing research on it; this includes privacy and ethical challenges. Therefore, the progress observed is dependent on the success achieved in overcoming these challenges.

Publication by Source

An analysis of the sources of human augmentation publications was carried out. The document publication, citation, and h-index were analysed to achieve this. It was observed that the ACM international conference proceeding series contributed the most documents to this area of research; this is followed by *Nanoethics* and the *Journal of Medical Ethics* in the third position (Figure 7.4).

The focus of the second and third top contributors makes their contributions understandable. For the top contributor, a conference proceeding provides an avenue for faster dissemination of research efforts irrespective of the stage of work. Thus, allowing authors to take advantage of faster publication opportunities as opposed to publishing in many journals with longer turnaround times.

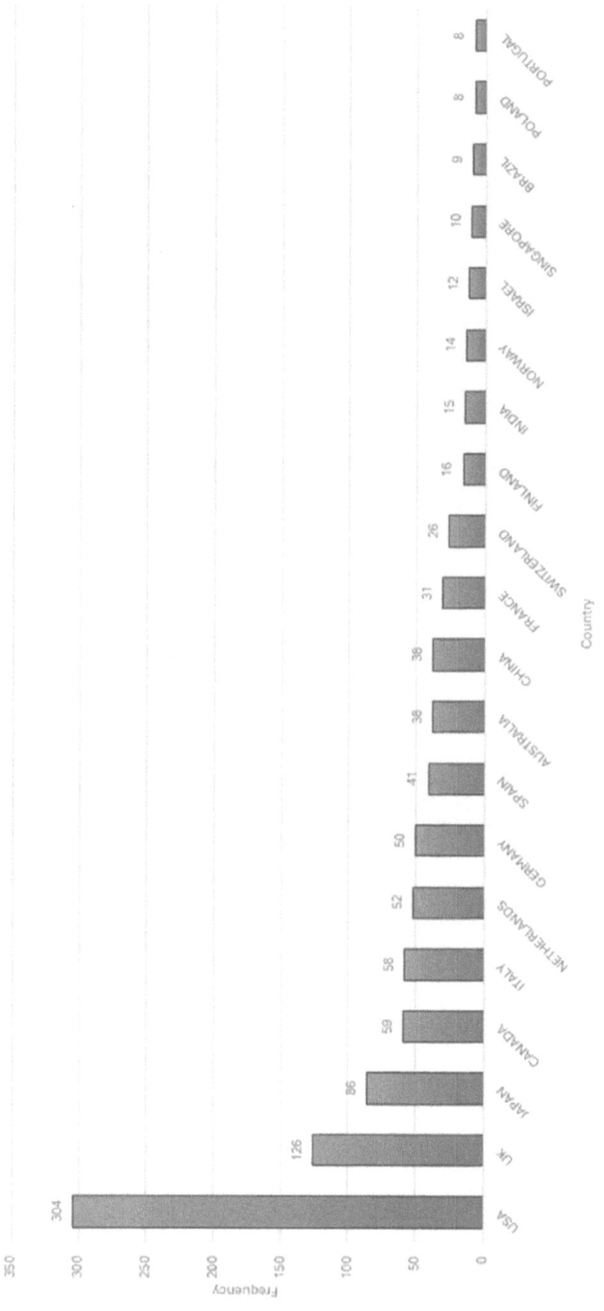

Figure 7.2 Country research trend.

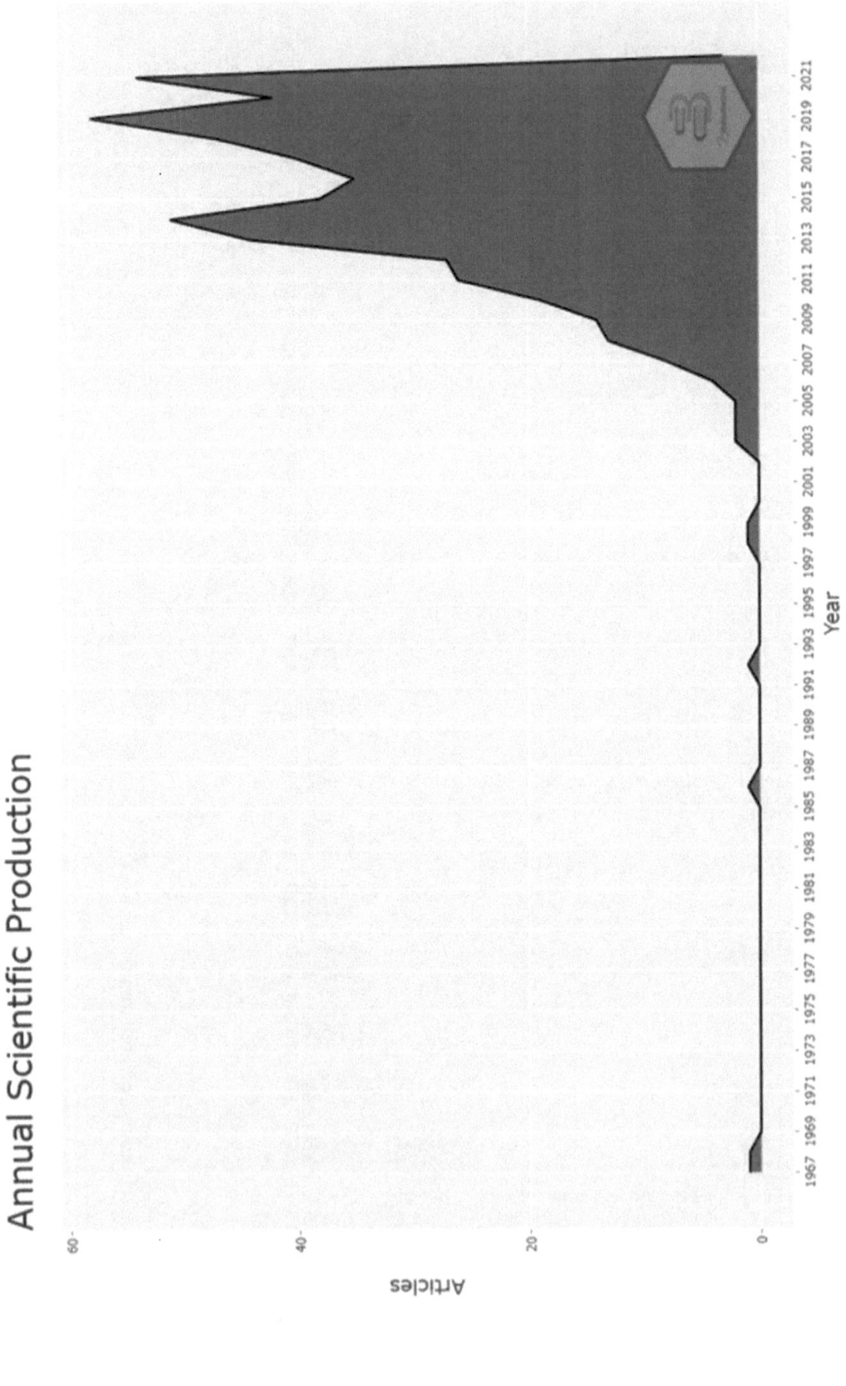

Figure 7.3 Annual publications.

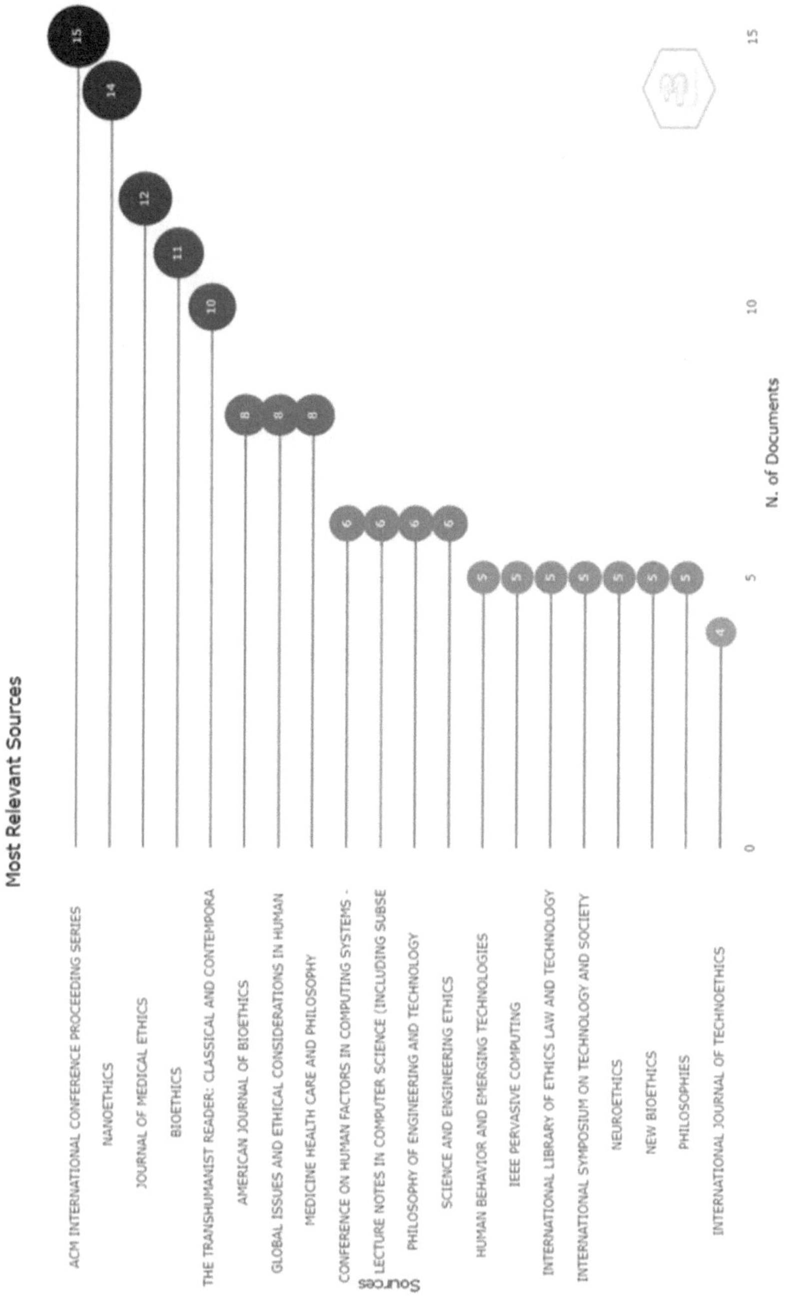

Figure 7.4 Source of publications.

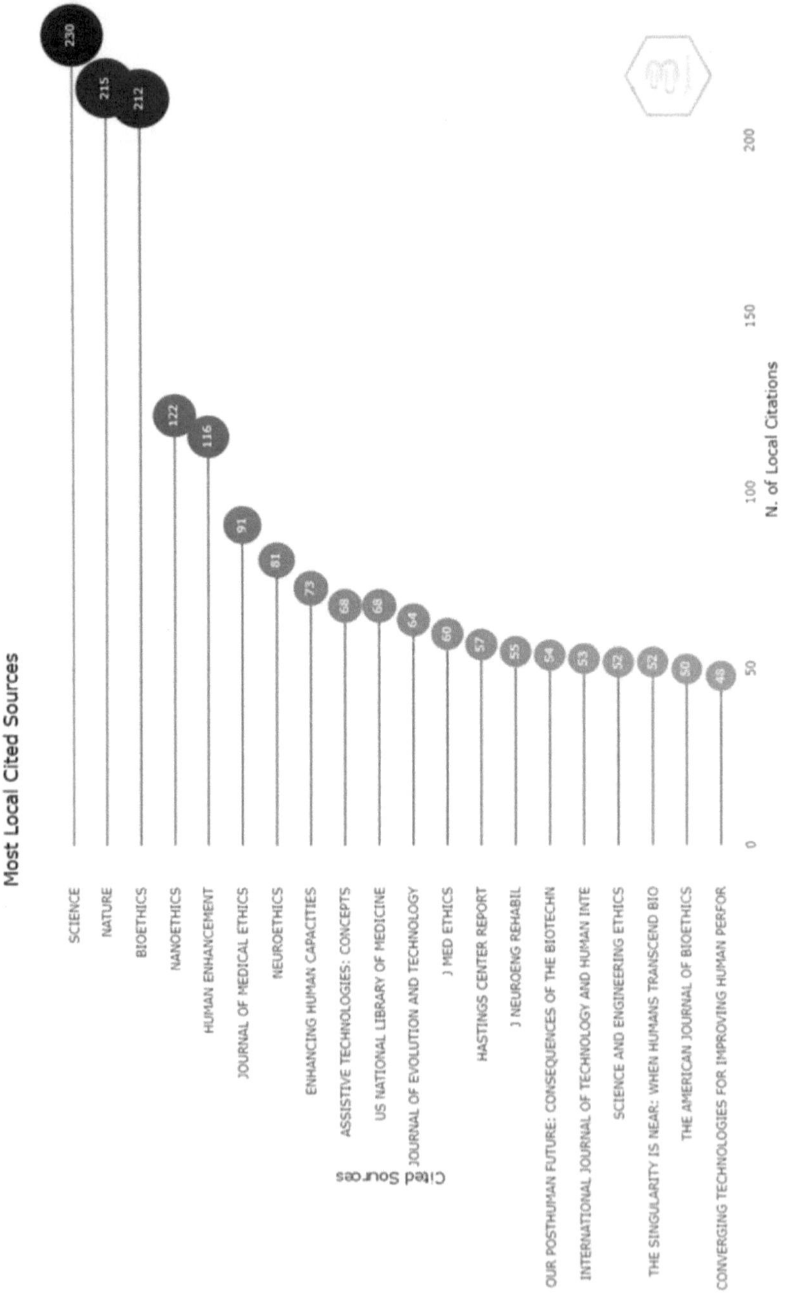

Figure 7.5 Citation counts of sources.

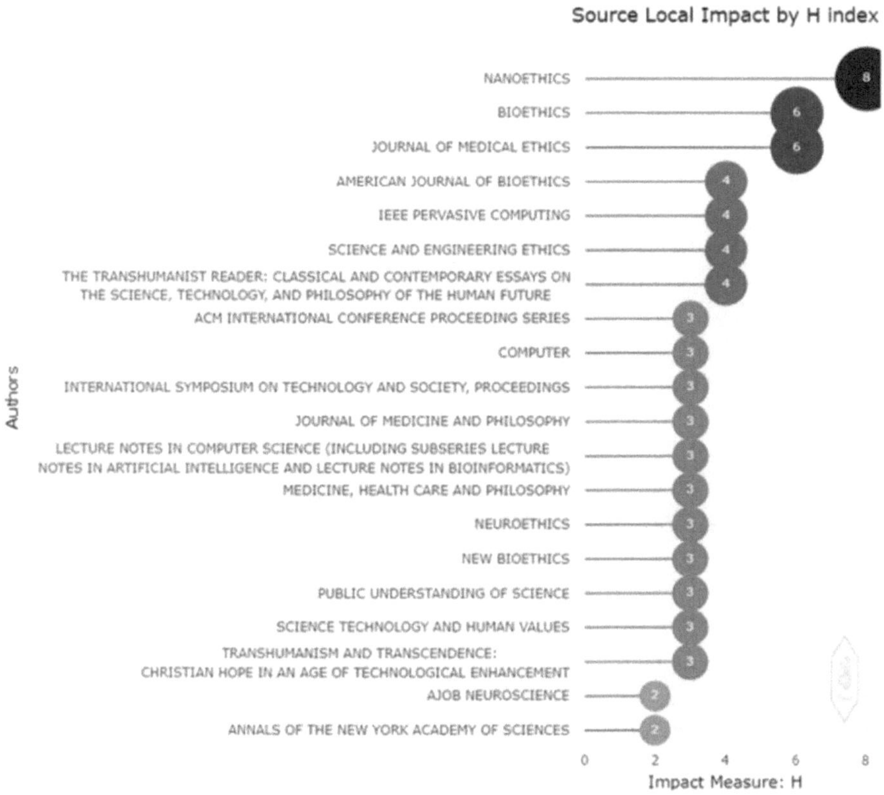

Figure 7.6 Impact of sources.

Figure 7.5 and Figure 7.6 present the citation and impact of each source. A critical study shows that *Science*, *Nature*, and *Bioethics* are the most cited. In addition, *Nanoethics*, *Bioethics*, and the *Journal of Medical Ethics* are the top three sources in terms of impact. *Nanoethics* has 14 documents, 122 citations, and an h-index of 8. *Bioethics* has 11 documents, 212 citations, and an h-index of 6.

Structure of Human Augmentation Research

To understand the trends, structure, and patterns of human augmentation research, two approaches were adopted. Firstly, the conceptual structure was achieved through a factorial analysis. Figure 7.7 presents the factorial analysis of the collected data using the keyword. The keywords were classified into two categories: blue classification and red. The blue classification includes four keywords: These are philosophical aspects, human enhancements, human bodies, and emerging technologies.

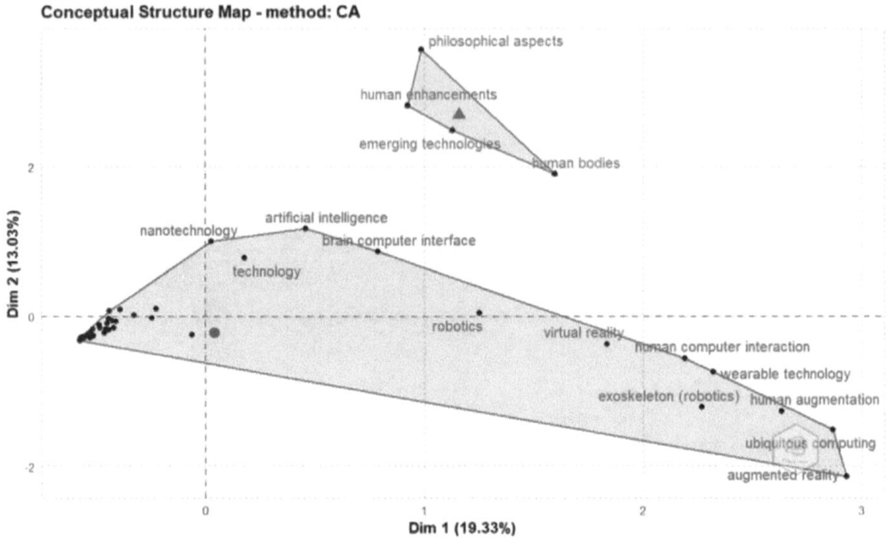

Figure 7.7 Conceptual structure map.

The second classification contained the following keywords: nanotechnology, artificial intelligence, technology, brain–computer interface, robotics, virtual reality, human–computer interaction, wearable technology, exoskeleton (robotics), human augmentation, ubiquitous computing, and augmented reality.

A closer look at these classifications shows that the first classification focused on the philosophy and fundamentals guiding human enhancement. It looks at the interaction between, and ethics among, other emerging issues related to the use of emerging technologies on the human body for enhancement. The second classification contains the various emerging technologies adopted for human enhancement.

The research focus of existing studies was further analysed to understand clustering through the occurrence of keywords using VOSviewer (van Eck & Waltman, 2019). All keywords were employed to achieve the network visualisation. Five clusters were identified, showing the focus of the existing works. The clusters are named accordingly and analysed in depth. Figure 7.8 presents the network visualisation map. The five clusters contain 63 keywords and a total link strength of 816 with 399 links.

Cluster 1: This is represented by the red cluster. It is named *human enhancement through cognitive enhancement and genetic enhancement*. It contains 19 keywords; the cluster contains keywords like gene editing, cognitive enhancement, and human genome. Cognitive enhancement is the intervention to improve the brain's attention, concentration, and processing of information capacity (Glannon, 2020) and to increase emotional intelligence (Eack et al., 2007). Cognitive enhancement thus improves brain function and makes workers more effective in delivering their duties, although this has been a subject of ethical debates and morality by several scholars (Douglas, 2013; Dubljević & Ranisch, 2018).

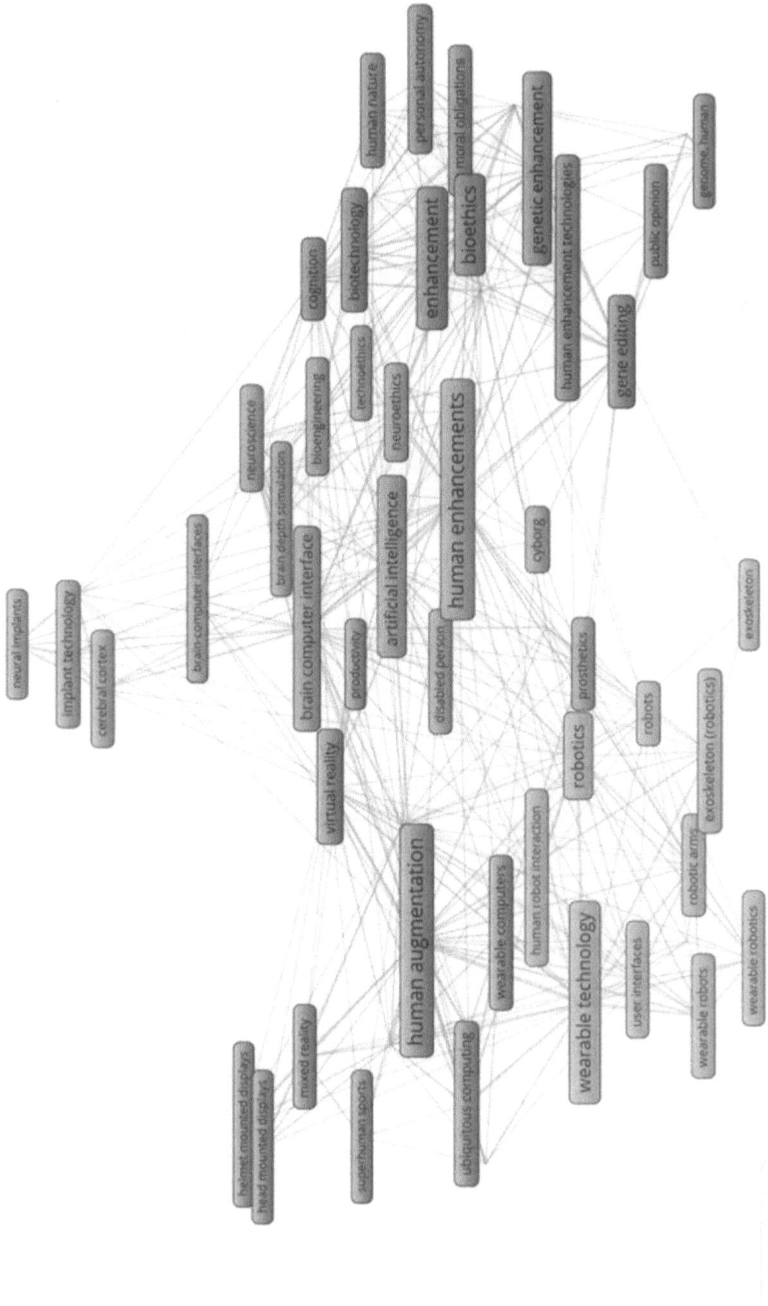

Figure 7.8 Network visualisation.

Cluster 2: The cluster is represented in green and has 15 keywords. It involves *human enhancement through brain enhancements* (brain depth stimulation and brain–computer interfaces) *and prosthetics.* Prosthetics refers to the artificial replacement of missing body parts. Strait (2006) chronicles the evolution of prosthetics and the different materials adopted in developing countries. Similarly, Cummings (1996) identified some of the factors necessary for its adoption. In the 21st century, the technology of prosthetics has advanced. It provides an opportunity for people with disabilities or with missing body parts to live as closely as possible to physically fit people. This has been researched extensively, even for rehabilitation (Iqbal et al., 2010, 2014).

Prosthetics provides an opportunity for the construction industry to accommodate a formerly skilled workforce limited by missing body parts to be employed and function fully. It provides an avenue for skilled labour additions for the construction industry. The presence of prosthetics does not encourage workers to be careless through the reassurance that their parts can always be replaced in the event of a loss through site accidents. However, it provides a lifeline for life-threatening accident victims on site. It thereby provides hope and an opportunity to be reintegrated into the workforce in the event of an accident.

Cluster 3: This cluster contains 13 keywords and is represented by the colour blue. The cluster is focused on *human enhancement through ubiquitous computing* (head-mounted displays, helmet-mounted displays), *wearable computers, and virtual reality.* This cluster is focused on human–computer interaction using ubiquitous computing. This refers to making the everyday object more effective and performing tasks by incorporating computational capability into them. This makes computing possible anywhere without the need to be in an office with a computer.

Advantages of head-mounted displays include the prevention of accidents, and it has been applied in different sectors, including the military, commercial applications, gaming, and firefighting, among others. It can be adopted for safety and productivity in the construction industry when worn by site workers. Through data collected using the head-mounted displays, workers' environment and safety can be analysed and accidents prevented, among other possibilities. This provides workers with a sense of safety while carrying out site activities.

Cluster 4: This is represented by the colour yellow, and it contains 11 items. The cluster is focused on *human augmentation through human–robot interaction* (robotics, exoskeleton, robotic arms, wearable robots). Exoskeletons include wearable systems that enhance the wearer's ability to do things to reduce the physical requirements and strain that would otherwise be required. Exoskeletons refer to technological systems and devices that provide protection, support, enhancement, and sensing (Andres et al., 2018). They can either be relied upon to perform the task in terms of persons with disabilities or used to augment or increase the ability of an able-bodied individual (Rupal et al., 2017).

Exoskeletons are built for different parts of the body and can also be designed for use separately, for instance in chairless chairs (Fox et al., 2020). Exoskeletons have been adopted in the manufacturing process for different processes, including assembling and installation (Fox et al., 2020). Similarly, exoskeletons can be applied in the construction process. The construction process is known to involve a lot of lifting of heavy objects and installations, depending on the project type.

The adoption of exoskeletons in the construction industry will enhance productivity as more can be done by workers in less time, especially with respect to the lifting and placement of heavy objects on site. In addition, it will also help in reducing strain and enhancing safety in site workers. Capitani et al. (2021) presented an exoskeleton model for construction workers in shotcrete projection. Similarly, Cho et al. (2018) conducted a study to demonstrate an exoskeleton's ability to enhance construction workers' safety and health.

Cluster 5: This is the purple cluster containing four items. The cluster is focused on *human augmentation through neural implants* (brain–computer interface); it is considered to be similar to cluster 2. It is focused on enhancing brain–computer interactions through implants. A critical look at the outlined human augmentation technologies reveals that they are mostly adopted during the construction phase on site. They are thus central to achieving productivity in the construction industry.

The identified human augmentation is 4th Industrial Revolution (4IR) inclined and IoT-supportive. It is worth noting that this field is still developing and continuously being improved upon. Presently, mixed reality, wearable computers, robotic arms, wearable robots, and gene editing are areas that are evolving and attracting considerable attention from researchers globally in terms of enhancing human effectiveness. It is essential that researchers in the construction industry pay attention to these areas and be strategic in incorporating them into the construction industry process.

Adoption Issues in the Construction Industry

Adoption of the identified technologies is not without challenge. Adopting these technologies will also enable organisations to achieve dynamic capabilities (Aghimien et al., 2021). Still, the construction industry is able to address the following issues strategically to achieve the adoption of human augmentation technologies to enhance productivity. Although there are different issues and challenges to be considered, including infrastructural requirements and ease of use, among others, the study focused on two challenges – cost and ethics, which are considered critical implementation issues in the construction industry.

This was achieved through a review of existing studies on technology adoption in the construction industry and the implementation of human augmentation. Studies commonly indicate that financial or economic variables are a critical challenge for adoption (Akinradewo et al., 2022; Amuda-Yusuf et al., 2017).

Cost

Cost has always been a critical factor in the adoption of innovations in the construction industry. Bademosi and Issa (2021) classified cost into initial investment, operation, and maintenance. The cost of investment is observed to be huge. However, construction stakeholders must view the cost required from an investment perspective rather than seeing it from a cost perspective. This changes the perception, making stakeholders see from a value perspective.

The investment cost required for the implementation of these technologies is huge, but mostly the ROI is deemed to be more rewarding. Furthermore, the fear for most stakeholders in the industry stems from the commitment costs for maintenance and operation. Most emerging technologies require periodic subscription costs as part of the maintenance and operation costs. The ability of stakeholders to meet this is dependent on many factors, including cash flow and project execution profit margin, among others. Local availability and sourcing might assist in achieving more comprehensive implementation and reducing the cost involved.

Ethics

This is a critical issue central to the adoption and implementation of human augmentation technologies. Most researchers argue about the morality of human augmentation, and many question the "level of humanity" of humans after augmentation. Thus, the argument is centred on the alteration of the human and the promotion of inequality among humans. However, some other authors called for a balance between human augmentation or enhancement and ethical or moral issues (Selgelid, 2014). Cinel et al. (2019) argued for tracking these technologies as they evolve, considering ethical implications rather than pre-empting out of fear.

This debate has been described as a sham debate from a scientific standpoint (Dubljević & Ranisch, 2018). Other factors aside from ethics and morality, based on the geographical location of the industry, include beliefs and cultural inclinations. This is because most construction industries operate within and are responsive to their specific context. Another critical factor is the cultural context and prevailing norms of society, especially as they affect augmenting human strengths. Other foreseen challenges include privacy issues due to the collection of data in real time.

Future Research Opportunities

The adoption of human augmentation technology in the construction industry presents a fertile research ground for researchers and other stakeholders. It offers new and exciting possibilities owing to its rapid growth. Consequently, there are areas of focus and opportunities to be explored by researchers to improve the productivity, efficiency, and safety of the construction industry. This section presents some outlook on these opportunities.

Firstly is the development of exoskeletons that are easy to adopt by site workers for the enhancement of strength and endurance. This is critical to help with the proper lifting of heavy loads and objects on site. Also, this will prevent muscular injuries often experienced by construction workers as a result of lifting, carrying, or putting objects down (Safe Work Australia, 2015).

Secondly, researchers and stakeholders should focus on the adoption of wearable technologies and sensors to achieve safe construction sites. This will help monitor site workers' health and safety. They will detect and alert workers in real time, saving them from accidents and unsafe environments. Other areas include exploring augmented reality (AR) and virtual reality (VR) tools to assist construction workers in their tasks on the sites and developing resilient materials for PPE to achieve safer work environments.

An important aspect of the adoption of technology is human-related factors. This is critical and has been extensively researched from different perspectives. The digital transformation of the construction industry is impossible without the human actor's essential role (Adekunle, Aigbavboa, et al., 2021). There is a need for proper coevolution of the different aspects of technological advancement (Gu et al., 2014).

Resistance to technology adoption incurs due to job security, technical skills, pedagogical review, and skill development, among others (Adekunle et al., 2022b, 2020; Aliu, Aghimien, Aigbavboa, Ebekozien, et al., 2022; Aliu, Aghimien, Aigbavboa, Oke, et al., 2022; Ronie Navon et al., 1993). The human factor is a critical area of focus that requires further research for human augmentation adoption in the construction industry.

Implication and Benefits of the Study

The Covid-19 pandemic impacted every area of life globally, including different sectors of the economy. The pandemic further compounded and exposed the construction industry's longstanding productivity and manpower challenges. It was observed that many construction industries in different countries suffered losses as they could not keep up with demands associated with the pandemic measures and needs. To overcome these challenges, it is imperative that the construction industry position itself strategically. A vital ingredient to achieving this is the adoption of technology and being at par with the industrial revolution (Ebekozien & Aigbavboa, 2021; Ejohwomu et al., 2021).

This chapter identified human augmentation technologies capable of helping the construction industry to overcome these challenges and prepare it for future pandemic occurrences. The findings of this chapter will offer stakeholders the required insight for achieving productivity and safety in the construction industry. The adoption of the study's result also helps to make the construction industry more attractive to prospective workers. The ability to predict accidents and prevent accidents before they happen will reduce accident rates on construction sites. It will also reduce or eliminate the cost burden as a result, thus reducing overall construction project costs.

Adopting these technologies reduces the strain on construction site workers on. It also enhances their productivity, as the technology can be adopted to improve productivity through either supplementing their ability or making it exceed workers' natural capabilities. Moreover, it helps the construction industry achieve the sustainable development goals of the United Nations (UN, 2015), which has become a global target for all sectors. This study helps the construction industry overcome its longstanding slow adoption of technology. Thus, the study's outcomes benefit construction site workers at an individual level as well as at the level of the construction industry and society.

Conclusions

The chapter studied emerging technologies, focusing on human augmentation technology and its adoption in the construction industry to improve productivity. Considering the low productivity, loss of workforce, and unattractive image of the construction industry due to its high accident rate, the study presents human augmentation as a viable solution to solve the problem, through its adoption to increase industry productivity and provide a lifeline in the event of an accident on site. Observing the identified technologies critically, it is evident that they have emerged from the medical field and are thus being incorporated into the construction industry to solve its problems. The construction industry also needs to be ready if a global pandemic strikes in the future.

To achieve the primary objective of the study, a thorough analysis of published works within the research focus was carried out. The analysis of the data retrieved from the Scopus database using keywords was carried out using Biblioshiny from the Rpackage. The analysis reveals a growing interest and output in human augmentation research, although some years experienced a decline in the number of outputs, for instance 2016. This was attributed to various challenges faced by researchers in progressing research in the area. It also identified the critical publication outlets and countries. It was not surprising to observe a dearth of research from developing countries.

The study identified human augmentation in the following dimensions for the construction industry: cognitive and genetic enhancement, brain enhancement and prosthetics, ubiquitous computing, VR and wearable computers, and robotics and exoskeleton. The study presented the adoption of enhancement of the brain to be more effective in communication with computers; this helps workers' work rate and productivity. Also, prosthetics present a solution to workers losing their jobs in the event of life-threatening accidents during site operations. Instead of construction workers suffering potential job loss due to work-related physical disability, they have the opportunity to be reintegrated into the industry workforce through the replacement of missing or damaged limbs. Areas for future research that require more attention as they continue to evolve are also identified. Researchers in the construction industry must track the development of the presented research focus in order to address the morality and ethical issues as it evolves.

The study also analysed implementation issues in terms of human augmentation technology implementation in the construction industry. Despite the array of challenges, the study focused on two critical issues, which are ethics and cost. However, the study posited that the industry could overcome these to achieve the adoption of these technologies and increase productivity. It should, however, be noted that the data for this study was solely sourced from the Scopus database.

Also, analysis was performed using VOSviewer and Biblioshiny. The use of other databases or a combination of more than one database might present a different result. Further studies might assess the practical impact of productivity of each of the identified technologies and provide an elaborated framework for their adoption in the construction industry.

References

Acemoglu, D. (2009). *When Does Labor Scarcity Encourage Innovation?* https://doi.org/10.3386/W14809

Adekunle, S. A., Aigbavboa, C., & Ejohwomu, O. A. (2022a). SCAN TO BIM: A systematic literature review network analysis. *IOP Conference Series: Materials Science and Engineering, 1218*(1), 012057. https://doi.org/10.1088/1757-899x/1218/1/012057

Adekunle, S. A., Aigbavboa, C. O., & Ejohwomu, O. A. (2022b). Understanding the BIM actor role: A study of employer and employee preference and availability in the construction industry. *Engineering, Construction and Architectural Management.* https://doi.org/10.1108/ECAM-08-2021-0714

Adekunle, S. A., Aigbavboa, C. O., & Ejohwomu, O. A. (2020). BIM implementation: Articulating the hurdles in developing countries. *8th International Conference on Innovative Production and Construction (IPC)*, 47–54, The Hong Kong University of Science and Technology.

Adekunle, S. A., Aigbavboa, C. O., Ejohwomu, O., Adekunle, E. A., & Thwala, W. D. (2021). Digital transformation in the construction industry: A bibliometric review. *Journal of Engineering, Design and Technology, 2013*, 1–29. https://doi.org/10.1108/JEDT-08-2021-0442

Adekunle, S. A., Ejohwomu, O., & Aigbavboa, C. O. (2021). Building Information Modelling diffusion research in developing countries: A user meta-model approach. *Buildings, 11*(7), 264. https://doi.org/10.3390/buildings11070264

Aghimien, D., Aigbavboa, C. O., Oke, A. E., Edwards, D., & Roberts, C. J. (2021). Dynamic capabilities for digitalisation in the AECO sector – A scientometric review. *Engineering, Construction and Architectural Management.* https://doi.org/10.1108/ECAM-12-2020-1012

Aghimien, D. O., Aigbavboa, C. O., Oke, A. E., & Thwala, W. D. (2019). Mapping out research focus for robotics and automation research in construction-related studies: A bibliometric approach. *Journal of Engineering, Design and Technology.* https://doi.org/10.1108/JEDT-09-2019-0237

Akinradewo, O. I., Aigbavboa, C. O., Edwards, D. J., & Oke, A. E. (2022). A principal component analysis of barriers to the implementation of blockchain technology in the South African built environment. *Journal of Engineering, Design and Technology.* https://doi.org/10.1108/JEDT-05-2021-0292

Al, S., Tumi, H., Omran, A., & Hamid, A. (2009). Causes of delay in construction industry in Libya. *The International Conference on Economics and Administration, Faculty of Administration and Business, University of Bucharest, Romania ICEA – FAA Bucharest, 14–15th November 2009*, 265–272. http://conference.faa.ro

Aliu, J., Aghimien, D., Aigbavboa, C., Ebekozien, A., Oke, A. E., Adekunle, S. A., Akinradewo, O., & Akinshipe, O. (2022). Developing emotionally competent engineers for the ever-changing built environment. *Engineering Construction and Architectural Management.* https://doi.org/10.1108/ECAM-08-2022-0806

Aliu, J., Aghimien, D., Aigbavboa, C., Oke, A., Ebekozien, A., & Temidayo, O. (2022). Empirical investigation of discipline-specific skills required for the employability of built environment graduates. *International Journal of Construction Education and Research, 19*(13), 1–20. https://doi.org/10.1080/15578771.2022.2159589

Amuda-Yusuf, G., Adebiyi, R. T., Olowa, T. O. O., & Oladapo, I. B. (2017). Barriers to Building Information Modelling adoption in Nigeria. *Journal of Research Information in Civil Engineering, 14*(2), 1574–1591.

Andres, F. R. O., Alberto Lopez-Delis, & Adson Ferreira da Rocha. (2018). Upper and lower extremity exoskeletons. In Jacob Segil (Ed.), *Handbook of Biomechatronics* (pp. 283–317). Elsevier Inc. https://doi.org/10.1016/B978-0-12-812539-7.00011-8

Bademosi, F., & Issa, R. R. A. (2021). Factors influencing adoption and integration of construction robotics and automation technology in the US. *Journal of Construction Engineering Management, 8*(147). https://doi.org/10.1061/(ASCE)CO.1943-7862.0002103

Bartik, A. W., Bertrand, M., Cullen, Z. B., Glaeser, E. L., Luca, M., & Stanton, C. T. (2020). *How Are Small Businesses Adjusting to COVID-19? Early Evidence from a Survey*. National Bureau of Economic Research. www.nber.org/papers/w26989

Capitani, S. L., Bianchi, M., Secciani, N., Pagliai, M., Meli, E., & Ridolfi, A. (2021). Model-based mechanical design of a passive lower-limb exoskeleton for assisting workers in shotcrete projection. *Meccanica, 56*(1), 195–210. https://doi.org/10.1007/S11012-020-01282-3/FIGURES/13

Che Ibrahim, C. K. I., Belayutham, S., Manu, P., & Mahamadu, A. M. (2020). Key attributes of designers' competency for Prevention through Design (PtD) practices in construction: A review. *Engineering Construction and Architectural Management*. https://doi.org/10.1108/ECAM-04-2020-0252

Cho, Y. K., Kim, K., Ma, S., & Ueda, J. (2018). A robotic wearable exoskeleton for construction worker's safety and health. *Construction Research Congress 2018: Safety and Disaster Management – Selected Papers from the Construction Research Congress 2018, April*, 19–28. ASCE. https://doi.org/10.1061/9780784481288.003

Cinel, C., Valeriani, D., & Poli, R. (2019). Neurotechnologies for human cognitive augmentation: Current state of the art and future prospects. *Frontiers in Human Neuroscience, 13*, 13. https://doi.org/10.3389/FNHUM.2019.00013/BIBTEX

Cummings, D. (1996). Prosthetics in the developing world: A review of the literature. *Prosthetics and Orthotics International, 20*(1), 51–60. https://doi.org/10.3109/03093649609164416

De Boeck, M., & Vaes, K. (2021). Structuring human augmentation within product design. *International Conference on Engineering Design, ICED21 16–20 August 2021, Gothenburg, Sweden Structuring*, 16–20. https://doi.org/10.1017/pds.2021.534

Douglas, T. (2013). Enhancement, Biomedical. *International Encyclopedia of Ethics*. Wiley. https://doi.org/10.1002/9781444367072.WBIEE560

Dubljević, V., & Ranisch, R. (2018). Moral enhancement – And the opposite: Making people more or less moral. *ELS*, 1–8. https://doi.org/10.1002/9780470015902.A0027993

Eack, S. M., Hogarty, G. E., Greenwald, D. P., Hogarty, S. S., & Keshavan, M. S. (2007). Cognitive enhancement therapy improves emotional intelligence in early course schizophrenia: Preliminary effects. *Schizophrenia Research, 89*(1–3), 308–311. https://doi.org/10.1016/J.SCHRES.2006.08.018

Ebekozien, A., & Aigbavboa, C. (2021). COVID-19 recovery for the Nigerian construction sites: The role of the fourth industrial revolution technologies. *Sustainable Cities and Society, 69*, 102803. https://doi.org/10.1016/J.SCS.2021.102803

Ejohwomu, O., Adekunle, S. A., Aigbavboa, O. C., & Bukoye, T. (2021). Construction and Fourth Industrial Revolution: Issues and strategies. In E. Adinyira & K. Agyekum (Eds.), *The Construction Industry: Global Trends, Job Burnout and Safety Issues* (First). (pp. 15–33). Nova Science Publishers. https://doi.org/https://doi.org/10.52305/JDFM1229

Fox, S., Aranko Terveystalo, O., Heilala, J., & Vahala, P. (2020). Exoskeletons: Comprehensive, comparative and critical analyses of their potential to improve manufacturing performance. *Journal of Manufacturing Technology Management*, *31*(6), 1261–1280. https://doi.org/10.1108/JMTM-01-2019-0023

Glannon, W. (2020). Neuroethics: Cognitive enhancement. *Oxford Handbook, December* (pp. 1–24). Oxford Handbooks Editorial Board, Oxford Academic. https://doi.org/10.1093/oxfordhb/9780199935314.013.43

Gu, N., Singh, V., & London, K. (2014). BIM ecosystem: The coevolution of products, processes, and people. In K. M. Kensek & D. Noble (Eds.), *Building Information Modeling: BIM in Current and Future Practice* (Issue 1, pp. 197–210). John Wiley & Sons, Inc. https://doi.org/10.1002/9781119174752.ch15

Hahn, F. H. (2010). Neoclassical growth theory. *Economic Growth*, 172–192. https://doi.org/10.1057/9780230280823_24

Haskel, J., & Martin, C. (2001). Technology, wages, and skill shortages: Evidence from UK micro data. *Oxford Economic Papers*, *53*(4), 642–658. https://about.jstor.org/terms

Haupt, T., & Harinarain, N. (2017). The image of the construction industry and its employment attractiveness. *Acta Structilia 23*(2), 79–108. https://doi.org/10.18820/24150487/as23i2.4

He, Y., Li, N., Wang, C., Xia, L., Yong, X., & Wu, X.-yu. (2019). Development of a novel autonomous lower extremity exoskeleton robot for walking assistance. *Frontiers of Information Technology and Electronic Engineering*, *20*(3), 318–329. https://doi.org/10.1631/FITEE.1800561

Hughes, R., & Thorpe, D. (2014). A review of enabling factors in construction industry productivity in an Australian environment. *Construction Innovation*, *14*(2), 210–228. https://doi.org/10.1108/CI-03-2013-0016

Khanal, N., Singh, S.K., Dhungel, S., & Wagley, S., (2020), Impact of COVID-19 on the construction sector of Nepal. *AMC Indian Journal of Civil Engineering*, *3*(2). DOI: 10.17010/ijce/2020/v3i2/156368

Iqbal, J., Khan, H., Tsagarakis, N. G., & Caldwell, D. G. (2014). A novel exoskeleton robotic system for hand rehabilitation – Conceptualization to prototyping. *Biocybernetics and Biomedical Engineering*, *34*(2), 79–89. https://doi.org/10.1016/j.bbe.2014.01.003

Iqbal, J., Tsagarakis, N. G., Fiorilla, A. E., & Caldwell, D. G. (2010). A portable rehabilitation device for the hand. *2010 Annual International Conference of the IEEE Engineering in Medicine and Biology Society, EMBC'10*, 3694–3697. IEEE. https://doi.org/10.1109/IEMBS.2010.5627448

Janik, A., Ryszko, A., & Szafraniec, M. (2020). Scientific landscape of smart and sustainable cities literature: A bibliometric analysis. *Sustainability (Switzerland)*, *12*(3). https://doi.org/10.3390/su12030779

Keogh, M., & Smallwood, J. J. J. (2021). The role of the 4th Industrial Revolution (4IR) in enhancing performance within the construction industry. *IOP Conference Series: Earth and Environmental Science*, *654*, 12021. https://doi.org/10.1088/1755-1315/654/1/012021

Liu, Z., Lu, Y., & Peh, L. C. (2019). A review and scientometric analysis of Global Building Information Modeling (BIM) Research in the Architecture, Engineering and Construction (AEC) industry. *Buildings*, *9*(10), 1–36. https://doi.org/10.3390/buildings9100210

Rezeika, A., Benda, M., Stawicki, P., Gembler, F., Saboor, A., & Volosyak, I. (2018). Brain–computer interface spellers: A review. *Brain Sciences*, *8*(4). https://doi.org/10.3390/BRAINSCI8040057

Rob Elvidge, & PlanGrid. (2019). *Using Digital to Close the Productivity Gap – BIM+*. www.bimplus.co.uk/using-digital-close-productivity-gap/

Ronie Navon, B., Kelly, P. W., & Johnston, D. W. (1993). Human factors in introducing on-site construction automation. *Journal Construction Engineering Management*, 801–812.

Rupal, B. S., Rafique, S., Singla, A., Singla, E., Isaksson, M., & Virk, G. S. (2017). Lower-limb exoskeletons: Research trends and regulatory guidelines in medical and non-medical applications. *International Journal of Advanced Robotic Systems*, *14*(6). https://doi.org/10.1177/1729881417743554

Safe Work Australia. (2015). *Construction Industry Profile – Main Causes of Injury*. www.swa.gov.au

Schindler, S., Jepson, N., & Cui, W. (2020). Covid-19, China and the future of global development. *Research in Globalization*, *2*, 100020. https://doi.org/10.1016/j.resglo.2020.100020

Selgelid, M. J. (2014). Moderate eugenics and human enhancement. *Medicine, Health Care and Philosophy*, *17*(1), 3–12. https://doi.org/10.1007/S11019-013-9485-1

Silva, G. A. S. K., Warnakulasuriya, B. N. F., & Arachchige, B. J. H. (2018). A review of the skill shortage challenge in construction industry in Sri Lanka. *International Journal of Economics, Business and Management Research*, *2*(01), 75–89. www.ijebmr.com

Sredojević, D., Cvetanović, S., & Bošković, G. (2016). Technological changes in economic growth theory: Neoclassical, endogenous, and evolutionary-institutional approach. *Economic Themes*, *54*(2), 177–194. https://doi.org/10.1515/ethemes-2016-0009

Strait, E. (2006). *Prosthetics in Developing Countries*. Prosthetic Resident.

UN. (2015). *Transforming Our World: The 2030 Agenda for Sustainable Development*.

Valeriani, D., Cinel, C., & Poli, R. (2019). Brain–computer interfaces for human augmentation. *Brain Sciences*, *9*(2), 22. https://doi.org/10.3390/BRAINSCI9020022

van Eck, N. J., & Waltman, L. (2019). *Manual for VOSviewer version 1.6.10*. Univeristeit Leiden. www.vosviewer.com/documentation/Manual_VOSviewer_1.6.1.pdf

WEF & BCG. (2016). *Shaping the Future of Construction: A Breakthrough in Mindset and Technology*.

Zhou, Y., Zhang, Z., Wang, B., Ren, G., Qi, H., & Wang, X. (2020). Construction time, cost and testing data of a prefabricated isolation medical unit for COVID-19. *Data in Brief*, *32*. https://doi.org/10.1016/j.dib.2020.106068

8 Rendering the Digital Future

Understanding the Social-Spatial Perception of a Dynamic Data Intensive Environment as an Architect

Jiayi Jin and Mingyu Zhu

Introduction

We seek to comprehend digital technologies, their significance in our lives, and their impact on our individuality and social connections as we move into a new phase in the Industrial Revolution. According to Colomina and Wigley (2016), each innovation of an artefact leads to the development of new practices for handling its implications, eventually leading to the evolution of humanity into entirely new cultures and even species. As a result, inventors and designers (including architects) play a crucial role in society, as they indirectly participate in the reinvention of humanity itself. This evolution began with designing artefacts to compensate for physical limitations and eventually progressed to the desire to extend mental and social capacities, leading to computational and digital modifications of artefacts (Nabian et al., 2013). The first computers aided humans in analysing and recording observations, ultimately resulting in an intricate network of intelligent ubiquitous devices that communicate with humanity and track agencies, fluxes, and movements (Townsend, 2013).

Each action taken by a person in the digital age generates a trace of their existence. Every photo uploaded to any digital service creates a new reality from their memories. While access to our digital lives used to be limited to stationary points, primarily through desktops, the development of digital mobile devices has created a range of new capabilities for generating 'digital pheromones' of human lives (Gershenson, 2013; Townsend, 2013). As a result, these devices have transformed human life into a virtual counterpart, creating a digital data trail accessible for anyone to obtain and analyse (Manovich, 2018). The digital data generated by users has become immensely valuable and financially lucrative for various institutions and service providers (Lupton, 2016), as these services have been able to manufacture, personalise, and optimise to satisfy the ever-changing needs and desires of current society.

Unfortunately, events such as the Facebook–Cambridge Analytica scandal in 2018 have exposed how certain entities have exploited the digital trails of social networks and identities for political and cultural gain (Schnädelbach & Kirk, 2019a). These incidents of 'corrupted' digitally enhanced services have prompted society to become more aware of their digital footprints, leading intergovernmental

DOI: 10.1201/9781003380559-8

institutions to introduce data protection laws that empower individuals to control their personal data (Manovich, 2018). These events also reflect upon the realm of architecture. Advancements in digital artefacts have enabled buildings to become more interactive and customisable to human needs (Colomina & Wigley, 2016; Schnädelbach & Kirk, 2019b). This shift, along with the datafication of society, has led to a transition from assigned locations to more ambiguous spaces, where the built environment and our perception of it are increasingly bound to and transformed by the content and nature of these digital interactions (Baldwin, 2019).

This chapter investigates the social-spatial perception and significance of data-intensive environments through the eyes of architects. The authors notice digital adaptations are currently being overlooked by architects and spatial designers, as the industry still focuses on the conventional architectural design process based on two design dimensions: the physical environment based on the context and the functional requirements based on users. There is also a third dimension, which is the embedded computational and communication capabilities with physical settings that allow personal data to be increasingly utilised by these new services. We hope to demonstrate a genuine desire to comprehend human experience with these ubiquitous computing services of different environments. Through a systematic review focused on the new realm created by big and small data, the social-spatial impacts of metropolitan information flows highlight new ethical and practical challenges raised when digitally enhanced services and systems are integrated into our lives and surroundings.

Big and small data from ubiquitous computing services

Weiser's vision of ubiquitous computing, presented in 1991, has inspired the development of computing technologies integrated with built spaces over the past 32 years (Schnädelbach & Kirk, 2019a). As the ubiquitous world scales and spreads towards architecture and the built environment, data becomes more relevant and nuanced in architectural design than ever before. According to Doyle, Savíc, and Bühlmann (2019), digital data is now a dominant factor in various fields and disciplines, including science, politics, and design, and is gradually being utilised in the spatial and policymaking of the built environment. However, to effectively use data as a design tool, architects and spatial designers must understand how to handle it (Bridle, 2019), considering data is a very complex tool with multiple stakeholders and components.

Unfortunately, there is a widespread misconception that data and information are synonymous when, as a matter of fact, information is entropic (Doyle, Savíc & Bühlmann, 2019). It can be regarded as 'produced knowledge' that has been processed and opinionated using subjective, politically, and culturally charged findings (Milan, 2018; Kneidinger-Müller, 2018). While data in its rawest form is decontextualised, it lacks meaning; it is unprocessed and ciphered (Doyle, Savíc, & Bühlmann, 2019). Over the centuries data became more essential to society for governance, science, and commerce (Kitchin, 2014). Eventually, the philosophy of the Enlightenment being a large contributor to the value of data as more knowledge and information often led to better decisions (Bridle, 2019). Despite how precise the data had become, the observations were still hampered by the human apparatus – the mind – because recording observations manually still relies on the subjectivity and precision of the operator's intentions and capabilities. As mere humans, we will still fail to comprehend the vast scale of information a device could handle.

With the exponential growth and development of digital infrastructure, data has grown enormously over the past few decades. The early 1990s witnessed the introduction of publicly accessible internet, and platform services ushered in a new era of culture and economy (Kitchin, 2014), allowing anyone to create and consume content and data. This gave rise to platforms such as social media and content providers, significantly increasing data production (Ratti & Claudel, 2016). With the digitalisation of the world, sensors and other ubiquitous devices were also introduced to monitor physical objects and events, commonly known as the Internet of Things (IoT). Eventually, the information generated by IoT will surpass the data produced by humans (Townsend, 2013).

Mankind has reached a crucial turning point in history as the overall production of data has doubled each year, surpassing the entire history of data production (Ratti & Claudel, 2016; Carpo, 2017). This abundance of data has led to the classification of methods and quantities of data production before the 1990s as 'small' data, and current methods and sizes of data as 'big data' (Kitchin, 2014). However, the definition of small data does not solely rely on the volume of collected information, as it can also comprise large datasets, such as nationwide censuses (Kitchin, 2014). Small data is intentionally produced by individuals and small organisations and contains fewer data points than big data, making it more distinguishable, identifiable, and contextualised (Kitchin, 2014; Lupton, 2016, 2021).

Despite the advantages of big data's velocity, variety, flexibility, and resolution, small data remains relevant in research and data science. By leveraging big data's capability of generating an objective overview of large and nuanced informational patterns, sampling techniques of small data can perform more precise and refined searches for specific answers (Schnädelbach & Kirk, 2019b). Big data's precise and consistent indexing allows different sets to be easily merged and linked to each other. However, small data can lack the relational aspects of big datasets. Nevertheless, small data need not be discarded in the field of research and data science, as it can still provide valuable insights when used in conjunction with big data techniques.

Personal data acts as a bridge between small and big data, enabling individual data to contribute to large-scale observations at a national or international level (Lupton, 2016). While personal data previously referred only to information that individuals owned, today, any digital device or platform can generate personal data that can be accessed by providers or institutions. According to the European Union's General Data Protection Regulation (GDPR) of 2018, personal data includes any information that can be directly associated with the identification of a natural person, known as a 'data subject' (Schnädelbach et al., 2019). Personal data can be classified into three main categories: static, dynamic, and derived (Schnädelbach & Kirk, 2019a). The first category (static) encompasses data related to both offline and online identities, assets, and physical, physiological, financial, legal, social, cultural, and relational information. The second category (dynamic) encompasses historical and real-time patterns, traces, and flows of records and resources. Lastly, derived data refers to profiles, routines, and trends of the subject that have been analysed and processed. The third category may have gone through many derivations, analyses, and aggregations, travelling through various physical and digital locations (Milan, 2018; Schnädelbach & Kirk, 2019b). Over time, the data can become a living reflection of the natural person and is sometimes referred to as 'lively data' (Milan, 2018).

Privacy of information used to be defined as hidden or withdrawn from the public sphere by spatial or behavioural dimensions. But the increasing number of mobile and ubiquitous devices has blurred the spatial constraints of privacy (Lupton, 2016, 2021; Schnädelbach & Kirk, 2019b). The GDPR in 2016 established legal privacy obligations for any form of personal data that can relate or identify a natural person/data subject, including data acquired from ubiquitous devices in the built environment. It also regulates the more sensitive 'special categories' of personal data, such as information related to health, gender, sexual orientation, ethnicity, culture, and politics (Schnädelbach & Kirk, 2019a), and provides six legal bases for assessing personal data compliance, most of which include governmental or public interests (except for contractual performances and the data subject's consent). However, given the GDPR's increased focus on protecting rights and providing control to individuals over their personal data, alternative approaches of justification by the individual are now required for analysing personal data, leaving various services to explore alternative approaches, as obtaining an individual's consent can be impractical in certain public situations and urban environments (Schnädelbach & Kirk, 2019b).

The datafication of society

The computational or digital realm of artefacts encompasses everything from early computers to a network of devices that are now integrated into everyday objects, as described by Ratti and Claudel (2016). These devices are equipped with sensors, processors, and other sophisticated technological components and connected to a vast wired and wireless infrastructure that allows physical objects to be rendered in digital form (Howard, 2017). The term 'datafication' was introduced in a 2013 review of 'big data' processes across business and the social sciences (Mayer-Schönberger & Cukier, 2013, chapter 5; Mejias & Couldry, 2019). The process of datafication enables increasingly intelligent algorithms to perform a variety of intellectual tasks for humans (Ratti & Claudel, 2016). As these artefacts become more advanced, they also have a more precise and nuanced impact on humanity (Colomina & Wigley, 2016). Early iterations of digital enhancements were more easily perceptible, such as the increased mobility and flexibility of mobile devices and wireless networks. However, newer developments have had a more subtle impact, such as the displacement of human linguistic and cognitive abilities using algorithms to interpret signals and sounds (Carpo, 2017). Overall, the datafication of society is a complex and evolving phenomenon with significant implications for our lives and society.

The emergence of digital technology has also transformed ordinary objects and functions and enabled greater mobility and flexibility for individuals. Mobile devices have revolutionised the way we live and work, freeing us from the constraints of fixed schedules and meeting points (Townsend, 2013). This newfound mobility has been compared to the nomadic lifestyle of our ancestors and has enabled us to perform daily tasks from virtually anywhere, at any time of the day. As a result, distances and time have become less of a hindrance, leading to new approaches to living. While it was expected that this increased mobility, in combination with traditional modes of transportation such as cars, radios, and landlines, would lead to the demise of urban areas (Mitchell, 1999), the opposite has occurred. Digital technology has enhanced the vitality of cities, providing greater connectivity and mobility for their inhabitants. This has allowed individuals to navigate the complexities of urban life and access a variety of physical and digital services, resulting in a hybrid digital-physical urban fabric. Overall, the introduction of digital technology has had a significant impact on our daily lives and transformed the way we interact with our built environment.

The rise of mobile devices and internet connectivity has facilitated the emergence of 'platformisation' as a replacement for physical socio-cultural transactions among actors. Although the initial versions of these platforms were often primitive (such as forums), they laid the foundation for a neoliberal social and cultural medium that encouraged user-generated content and decentralised power, influence, and practice (Towsend, 2013; Manovich, 2018; Schnädelbach & Kirk, 2019b). Alaimo and Kallinikos describe the numerous and significant changes to social life that come with datafication as 'computed sociality' (2017, p. 177). Similarly, Van Dijck (2013, p. 5) uses the term 'platformed sociality' to describe this transformation. Unlike previous cultural mediums such as music, television, and movies, which were dominated by Western (American) culture and promoted a global homogeneous perspective, the decentralisation of socio-cultural mediums has led to the diversification and expansion of communities and subcultures by removing geographic barriers (Manovich, 2018). These platforms serve as a medium for like-minded individuals to connect and express their opinions publicly, even on politically charged issues, thus facilitating the formation of public opinion (Citton, 2017, pp. 28–31; Colomina & Wigley, 2016; Manovich, 2018).

As the algorithms and interfaces of digital platforms advanced, neoliberal structures began to capture the world's population (Manovich, 2018). Encouraging users to shift their daily routines and interactions to digital technology has led to the emergence of a data-rich era (Milan, 2018), so data has become highly valuable and profitable, especially from information on people's behaviours, habits, and preferences (Lupton, 2016, p. 44). In consequence, there is a new form of capitalism/consumerism called 'prosumption', where individuals interact with digital artefacts by both consuming and creating digital content. Free services have emerged that analyse digital traces to determine user behaviours and preferences, leading to the datafication of behaviourism (Kitchin & Dodge, 2014; Lupton, 2016; Manovich, 2018). Our spatial and social activities are tracked and logged. Facebook, tweets, Uber calls, TripAdvisor reviews, and check-ins are becoming the natural activities of daily living (Ratti & Claudel, 2016, p. 51), and some Chinese counterparts like Baidu, Weibo, DiDi, and Wechat to smaller players across what can be called the 'social quantification sector' (Couldry & Mejias, 2018, 2019). These services are expanding to include interactions with physical objects and bodies, promoting new activities that are linked to everyday lives. Data is often circulated and analysed to predict social relationships and individual behaviour, leading to the concept of lively data, where traces become a hybrid of the physical user and digital footprint (Lupton, 2016; Lupton et al., 2018).

Social-spatial impacts of metropolitan information flows

As data increasingly reflects human life, analysing the philosophical, cultural, and social implications of self-tracking movements becomes intriguing. Individuals who engage in self-monitoring practices have fully embraced the datafication of their lives, using lively data to measure aspects of everyday life and embodiment (Kneidinger-Müller, 2018). These individuals often have an extended array of (bio) sensor devices (for example, Apple Watch) to accurately track any aspect of human or non-human activity. Algorithms and other digital artefacts transform this data into culturally readable symbols, allowing members to share their patterns with their community through various platforms and forums (Lupton & Pink, 2018). According to practitioners of the movement, transparency and data exchange motivate members to perform better and foster stronger bonds within the community. The success of developing a better self through numbers draws strength from the theory of social translucence (ST) by Erickson and Kellogg (2000), which suggests that motivating desired behaviour requires more than the exchange of one's behavioural data with their social network.

The concepts of biopolitics and selfhood, as discussed by Foucault (1975) and Lupton (2016), relate to the increasing presence of neoliberal structures that involve personal data and social control in everyday life. The idea of selfhood is centred around self-reflection for the purpose of personal growth, achievement, health, and well-being. With the increasing digitisation of society, this behaviour is further encouraged, allowing individuals to share every aspect of their lives in real time. As a result, the pursuit of an improved self-image expands beyond physical appearance to include socio-cultural appearance, with individuals seeking to become the perfect citizen or inhabitants of the world. This self-development process empowers individuals with self-knowledge and enhances their socio-cultural positioning in a fluid world, aligning with the neoliberal political agenda of soft power over hard power on the public (Cusset, 2018, p. 19; Lupton, 2016).

The prevalence of digital technologies in society has made it impossible to avoid interactions with them, regardless of whether individuals embrace them or not (Howard, 2017). Integrating digital technologies in both public and private spheres has enabled objects to communicate and interact autonomously with humans, going beyond traditional human–object connections (Lupton, 2016; Kitchin & Dodge, 2014). This confirms actor–network theory (ANT), which acknowledges non-human actors as active participants in social networks alongside humans. The digital apparatus can be adequate whole new social dynamics and ambiences in the public and private spheres, through the various medium outputs for interactions between human and non-human actors. The dystopian TV show *Black Mirror* aired an episode called 'Nosedive',[1] which depicted a society where individuals rate each other on a 5-star scale using their cell phones after every social interaction and everyone's average score is publicly accessible. In this world, any behaviour or opinion deemed unpopular results in lower ratings and eventual social ruin.

While it is true that invasive scoring mechanisms are not yet prevalent throughout a significant portion of the European population with GDPR implemented for data protection and privacy, this does not mean that they will not be in the future. In fact, the reality of the situation may be even more insidious than what is portrayed in fiction (Kayser-Bril, 2019). For example, algorithms are being used by intelligence services in France to identify abnormal online behaviour based on search data, while in Mannheim, Germany, surveillance cameras will notify the police when their algorithm detects specific behaviours from individuals. In Spain, an algorithm assesses the eligibility of tenants for subsidised electricity prices based on their income and rent data (Kayser-Bril, 2019). These examples contribute to the idea that notions of social materialism of the digitally enhanced non-human actors go beyond mere interaction and communication, as they affect the selfhood and embodiment of the individual and society (Lupton, 2016; Latour, 2007).

Presently, a diverse array of algorithms has been incorporated into numerous social, cultural, and political establishments to observe populations impartially and comprehensively. Depending on their consequences, these algorithms have the potential to mould the societal and cultural outlooks of entire populations, resulting in the concept being coined as the 'culture machine' (Finn, 2019). Unlike human operatives, culture machines can operate without alternative motives, depending on their structure and implications (Finn, 2019; Rouvroy, 2020). As a result, the position of the translator of digital data into information and implications holds significant power and can perform like a parasite, translating acquired knowledge into its own biases (Doyle, Savíc, & Bühlmann, 2019). In its most sophisticated form, the culture machines can incorporate even the most correlated and microscopic anomalies into its observations to provide detailed and optimised services, information, or policies (Rouvroy, 2020; Finn, 2019; Doyle, Savíc, & Bühlmann, 2019). This algorithmic governmentality, while analysing an accurate and detailed observation of lively data by everyone, is shaping a more homogenous culture through objectiveness (Rouvroy, 2020). The influence of culture machines may not be associated with a specific cultural or socio-political context and can assess entire aggregated population-wide datasets, unbound by geographical size and place, to define its services or policies (Manovich, 2018).

Rendering the digital future as architects

In an era characterised by the ubiquitous presence of computing and artificial intelligence, the relationship between the physical and digital realms has become increasingly intertwined (Kitchin & Dodge, 2014). Architects are keenly interested in this trend since the earliest days of digital technology, as physical structures are replaced or enhanced by digital artefacts and services (Schnädelbach & Kirk, 2019a). This has gradually given rise to a more hybrid-built environment and society. The impact of digital artefacts on the behaviour and desires of society will ultimately manifest indirectly in the physical world, akin to the transformative effects of automobiles on urban spaces (Ratti & Claudel, 2016). As a result, interconnected digital apparatuses are rethinking not only facilities but also the entire urban fabric such as workplaces and libraries, leading to new hybrid services and buildings like pop-up stores and drone ports (Colomina & Wigley, 2016). These services also generate vast amounts of data, which can be harnessed to create custom and optimised services that satisfy the ever-changing needs and desires of contemporary society. This sets the stage for the development of living and dynamic structures like adaptive architecture, participatory or mass-customised buildings, and augmented reality with the physical environments (Ratti & Claudel, 2016; Towsend, 2013).

In the realm of the digital and physical world, two primary categories of digital manifestation can be distinguished: digitalised structures and digital hybrid structures (Kitchin & Dodge, 2014). The former refers to a digital manifestation where the physical object's production and function are influenced by digital processes but do not rely on their digital counterpart to function. It is rare in contemporary Western society to find any physical object that are treated by some form of digital process (Colomina & Wigley, 2016; Kitchin & Dodge, 2014). In the field of architecture, for instance, the use of different digital drawing software has become a typical practice in the process of designing a building. On the other hand, the latter category—digital hybrid structures—denotes a dyadic relationship between the physical and the digital where one cannot exist or function without the other.

The integration of digital enhancements with the physical built environment has led to the emergence of adaptive design, also known as robotic architecture, generative architecture, and interactive architecture (Schnädelbach & Kirk, 2019a). With actuators, the physical structure can adapt to changing conditions by responding to inputs from sensors and manual modifications made by occupants. This type of manifestation has become increasingly prevalent in recent years with the rise of ubiquitous devices in the built environment (Schnädelbach & Kirk, 2019b). Architects have been exploring the intersection of architecture and cybernetics since the 1960s, with pioneers like Leon Groisser and Nicholas Negroponte applying their knowledge of architecture to the fields of cybernetics and computer science (Ratti, 2015). These architects drew inspiration from Le Corbusier's concept of 'a house is a Machine for Living in', which sought to optimise architecture in terms of not only structural engineering and design but also mass production and social functions (Ratti, 2015).

One of the most well-known research projects in this field was the Architecture Machine Group at MITs School of Architecture and Planning, which was founded by the architects. Their objective was to overcome the barrier between humans and machines and achieve 'Humanism through Intelligent Machines', as described in Negroponte's book, *The Architecture Machine* (Negroponte, 1975, p. 151). As part of their research, the group created artefacts such as the SEEK device, which was equipped with multiple attachments, including magnets, photocells, and markers. The device can be computer-controlled to position these attachments in three dimensions and serves as a peripheral device for students' experiments in developing sensors and effectors that interact with the physical built environment. The machine successfully accomplished tasks such as recognising hand sketches, comprehending human speech, following commands, and executing basic design rules. However, the machine mastered many experiments but failed at the most important one: it could not learn. Negroponte mentioned, 'the fantasies of an intelligent and responsive physical environment are too easily limited by the gap between the technology of making things and the science of understanding them' (Negroponte, 1975); however, this gap is slowly disappearing with the increasing technicity of adaptive environments (Schnädelbach & Kirk, 2019b).

The advent of the digital revolution in architecture held the potential for novel shapes and intricate dematerialisation. In the 1990s, Bernard Tchumi's Paperless Studio at GSAPP, Columbia University, emerged as a trailblazer for a computer-centric future in architectural education, offering freedom from the cumbersome burden of physical documents. Architectural firms like Frank Gehry's studio also started devising fully digital techniques for construction. Information could create and depict, enabling a building to be constructed directly from the digital blueprint to its ultimate physical form (Fankhänel & Lepik, 2020).

Advancements in digital technology have revolutionised the field of architecture, resulting in buildings that are increasingly interactive and adaptable to the needs and preferences of their occupants (Schnädelbach & Kirk, 2019a). This trend, coupled with the datafication of modern society, has led to a shift from traditional, fixed locations to more fluid, ambiguous spaces (Colomina & Wigley, 2016). However, to achieve this dynamic environment, it is necessary to utilise personal data from the users, which can lead to a more transparent and fluid lifestyle (Lupton, 2016). While personalised services and fluid structures can enhance the human experience, they can also impact an individual's sense of identity and place, and inappropriate design choices may result in feelings of alienation and discomfort (Carpo, 2017; Schnädelbach & Kirk, 2019b). Architects, however, frequently tend to prioritise aesthetic, technical, financial, and sustainable considerations over the social and spatial implications of interactive architecture design and digital artefact adaptation.

To fully comprehend the significance of interactive elements in building design, architects could draw on decades of research in the field of human–computer interface (HCI) and specifically human–building interaction (HBI) as sources of inspiration, to understand, incorporate, or enhance user-centric design and planning in their creative workflows (Krukar, Dalton & Hölscher, 2016). As HCI/HBI research examines human behaviour, cognitive processes, and operational routines when physical and digital content coexists in the new media space, it is essential that future architects adapt their designs to integrate with metropolitan information flows, creating spaces that are connected and responsive to both the individual end-users and the larger urban ecosystem in envisioning and shaping the future of living.

In recent years, the concept of digital twins (DT) as one example of urban ecosystems has been gaining attention across academia, industry, and government as the solution for connecting the spatial world and its counterpart data version. Originating in the aerospace and manufacturing industries, the concept of DT had been applied in modelling and simulating the dynamic of different assets/systems to mimic their real-world behaviour (Liu et al., 2021). Mechanical engineers use digital twins to create highly sophisticated mock-ups and test their performance and optimise their designs. However, the system faced by architects is far more complicated. Unlike a closed manufacturing workshop, cities are composed of fully open, ever-changing, and interconnected objects and systems. As a result, developing an 'Urban Digital Twin' (Dembski et al., 2020; Ferré-Bigorra, 2022) demands a sophisticated and customised approach based on the local spatial and social context. Although Urban Digital Twin applications are currently being utilised for building energy benchmarking, emergency response, road traffic control, and crowd monitoring, these applications are predominantly limited to the construction and facility management of the built environment. The task of leveraging Urban Digital Twins to provide practical design and planning solutions remains a challenging one.

Towards the future: a new social-spatial perspective

According to sociology, the perception of social-spatial dynamics is based on four temporal scales that are constructed over time (Alavi et al., 2019), as shown in Figure 8.1. This perception can lead to different interpretations of similar events happening in different locations, which can ultimately alter the perception of the proceeding space (Kitchin & Dodge, 2014). These constructs are also essential for interactive artefacts that utilise lively data and reshape everyday life. The first two temporal scales (minutes–hours and days–weeks), known as 'micro mutations', involve calibrating the present needs and preferences of humans during direct interaction (Alavi et al., 2019). The third temporal scale (months–years), called 'meso developments', involves reconfiguring a personalised interaction by analysing

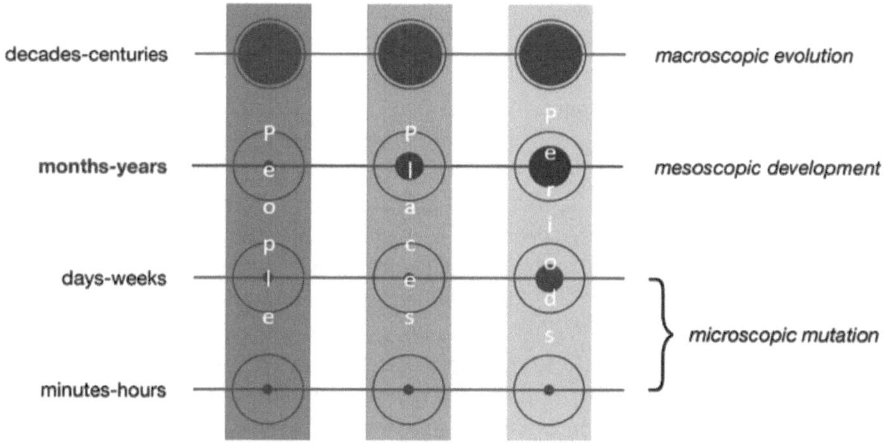

Figure 8.1 Four temporal scales of adaptation concerning people, place, and periods, as outlined by Alavi et al. (2019).

previous encounters, while the fourth temporal scale (decades–centuries), also called 'macro evolutions', involves forming a profile or memory of the encountered actors (Alavi et al., 2019).

Regarding the spatial scale, micro mutations are associated with the specific interaction of the actors, whereas meso and macro developments aggregate data from multiple encounters and actors (Alavi et al., 2019). Although the digital apparatus can record these interactions in fine detail in terms of time and location, the human perception of the exact location and moment of the interaction becomes harder to define (Schnädelbach & Kirk, 2019b). This notion of fractured ecologies could cause a feeling of estrangement in certain locations, as humans will not entirely perceive the full reality of the physical construct. However, the digitally enhanced artefact recognises the human actors and their behavioural routines and slowly develops for better affordances with cognitive signals that identify the capabilities of interactions with each object or actor (Kitchin & Dodge, 2014; Lupton, 2016; Finn, 2019).

This affordance can also extend to social-spatial notions, such as identifying the required procedures and conducts in a particular environment (Kitchin & Dodge, 2014; Lupton, 2016). Passive digital engagements encompass structural, mechanical, and forensic components or analogue cognitions. Conversely, proactive digital entities can engage in personalised interactions with the human social network through auditory, visual, health-related, vibrational, and olfactory signals (Schnädelbach & Kirk, 2019a). Although the first encounters may require a range of cognitive signals and interactions to perform the procedures, the forthcoming procedures could perform a smooth interaction, thereby stimulating self-reflection and a more profound connection (Lupton, 2016). Ultimately, these proactive procedures can transform the perception of interactions and the associated social-spatial notions.

The introduction of new interactive elements in the built environment is changing how humans perceive space. This has been discussed in multiple architectural books like *Building Dynamics: Exploring Architecture of change* (Kolarevic & Parlac, 2015), *Interactive Architecture: Adaptive World* (Fox, 2016), *Architectural Robotics* by Green (2016), *Architecture and Interaction* (Dalton et al., 2016) and *Data, Architecture and the Experience of Place* (Karandinou, 2019). This fluidity can lead to feelings of estrangement and disorientation in urban or architectural settings (Carpo, 2017, p. 104; Doyle, Savíc, Bühlmann, 2019). However, by implementing clear orientation points, open visual perception, and interactive elements that indicate directions, occupants can more easily orient themselves (Doyle, Savíc, Bühlmann, 2019; Schnädelbach & Kirk, 2019a; Karandinou, 2019). Apart from the HCI design, these also offer opportunities for architects to create new social experiences and atmospheres for occupants to engage with the digital setting; by using interactivity to tell a story and encourage engagement with the environment, architects can guide occupants through their journeys, which will enable the overall narrative of interactions to be approachable and recognisable for the occupants. To achieve this, the occupants' mobile devices could connect to the environment and direct the narration, allowing them to orientate, connect, and anchor themselves in the fluidity of the space.

Dynamic data reinstates the empowerment of the actors in the flow of liquid society and can stimulate emotional connections, restores, and visualises the truth about origins. Forensic Architecture, a multidisciplinary practice led by the architect Eyal Weizman, offers a great illustration of this. The practice assists civil society groups in confronting state crimes and human rights violations worldwide by providing evidence. They accomplish this by visualising events and scenarios using a diverse array of digital mapping and modelling technologies, and by relying on an unconventional combination of human and non-human testimony to accurately reconstruct and map what happened to people, buildings, and their surrounding urban settings (Weizman, 2014, p. 151). For example, throughout and after the Israeli attack on Gaza, the media repeatedly featured a new and previously unfamiliar weapon. These weapons were shown in images depicting explosions in the air, which released smoking fragments that fell onto densely populated areas of the Gaza strip. These digitally reconstructed materials alongside other depictions of the widespread destruction of buildings and infrastructure were among the most defining images of the 'Cast Lead' conflict (Weizman, 2014, p. 337). By locating these images with the 3D model, Forensic Architecture can utilise the 3D model with embedded big and small data, allowing them to identify the bombs based on their shadows and the shape of their cloud formations. This technique enables them to determine the precise time and location of each image and construct a coherent account of the events of that day.

Similar to Forensic Architecture, a pilot DT of a maritime port is proposed by the authors. This project entails a virtual replica of the port's operations process that runs parallel to the real port and displays the energy consumption of each piece of equipment and facility. The DT also considers various factors, such as renewable energy sources, incoming vessels, shore and offshore power, equipment handling efficiency, and connecting inland smart logistics schedule, to predict future energy demands of the smart port. By anticipating changes in the system and predicting possible future behaviours, the DT can comprehend the equipment's dynamic, interdependency, and performance fluctuations over time. In the event of a disruptive occurrence affecting normal operation, such as energy cuts, terrorist attacks, or mechanical failures, the port operator can assess the impact and take real-time control actions to mitigate the negative impacts via this DT platform, and inform other stakeholders, such as shipping liners and cargo receivers, of the situation.

The structure of the proposed DT is composed of three stages, as detailed in Figure 8.2. In the first stage, models are developed to represent the applicable

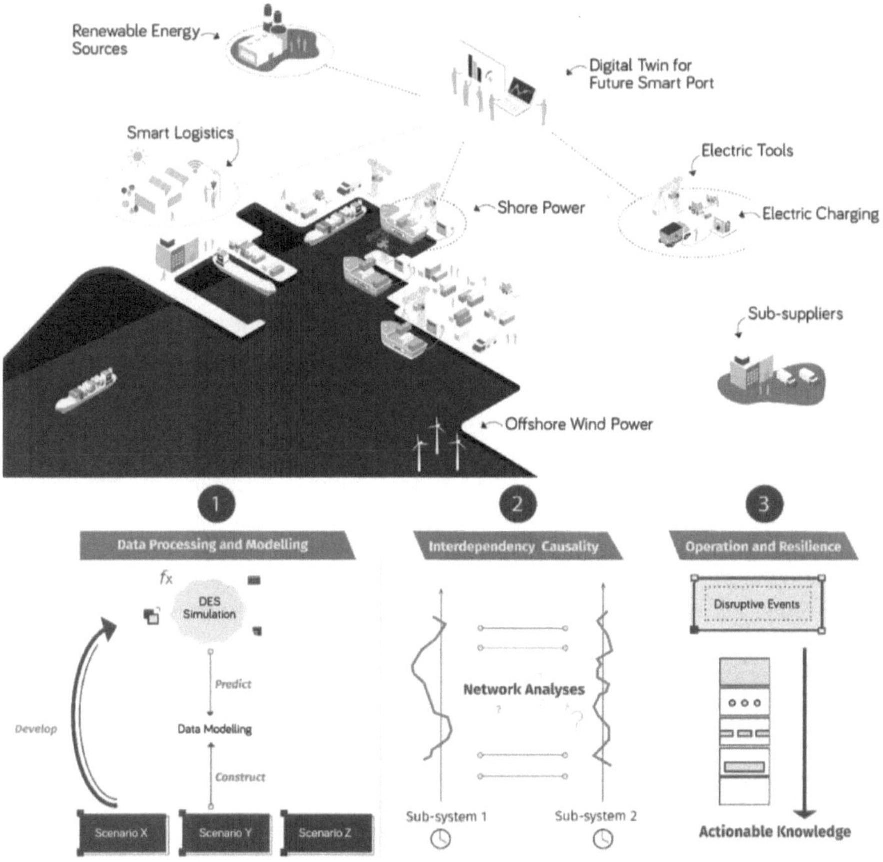

Figure 8.2 The framework of DT, proposed by the authors.

energy consumption of each piece of equipment. The input of these models includes the dynamic parameters and static parameters. A discrete event simulation is developed to generate dynamic input to the energy model. In this way, the future energy demand can be obtained as several time-series data series. Such demand is set as the baseline for normal port operation. In the second stage, network analysis is utilised to depict the interdependency between equipment in the port. The time-series data generated in the previous stage are treated as physical quantities, such as energy demand, that instrument each node of the complex network. The result can be used to infer the topology of the network. In the port operation, the operator can first use the observed actual energy usage to compare with the baseline scenario and identify the irregularity of any equipment, then assess how such irregularity will propagate and impact the rest of the system. Finally, with the insights gained from the previous stages, the DT provides pre-event estimations, identifies irregularities during an event, and proposes recovery strategies.

This twinning system covers all four temporal scales (minutes–hours, days–weeks, months–years, and decade–centuries), and the sensing and simulation of this smart port has the potential to be integrated into local community management across diverse social infrastructure systems; this includes safety and security, commerce, health, education, and governance. An interconnected urban ecosystem will thus be formed to bridge the gap between the port and surrounding residents and communities, with a shared identity and purpose that extends beyond physical boundaries.

Conclusion

The proliferation of digital data has enabled the creation of an immense amount of information, and the development of sophisticated algorithms has made it possible to process and analyse this data to identify patterns and trends. As a result, digital data has become a powerful tool for providing customised and liberated services and policies. It can be used to detect even the most nuanced unintentional patterns, allowing for the provision of services that are tailored to an individual's specific needs and preferences. This has led to a fundamental shift in the way that people live their lives. Digital data has enabled people to have greater control and to make decisions about their social status and personal relationships. It has also enabled the development of new technologies and services that are designed to visualise the historical event, empower the people, provide pre-event estimations, identify irregularities, and propose recovery strategies.

While this can be beneficial in many ways, there are also risks associated with the increasing reliance on algorithms and digital devices. One concern is that these technologies may reinforce existing biases and inequalities in society, and new forms of discrimination and exclusion could be created like the mentioned scoring mechanisms. Additionally, there is a risk that the power of jurisdiction over an entire population may shift from human actors to a closed algorithmic system, leading to a loss of individual autonomy and control. Therefore, it is important to strike a balance between the benefits of digital data and the potential risks associated with its use. This can be achieved through the development of robust ethical and regulatory frameworks that ensure that personal data are fully protected, and algorithms and digital devices are used in ways that are fair, transparent, and accountable.

The concept of adaptive environments has been gaining traction in the field of architecture in recent years, as the shown examples from the Architecture Machine Group to Bernard Tchumi's Paperless Studio; from Forensic Architecture's hybrid physical and virtual images to our digital twin modelling of the Port of Tyne. Architects have an important role to play in shaping the experiences and interactions of occupants with a dynamic data-intensive environment. Beyond the functional design of human–computer interfaces, we can use interactivity as a tool for storytelling and engagement, to create new social experiences and atmospheres that stimulate the senses and encourage users to interact with their surroundings in new ways.

As cities and buildings become more interconnected through the use of technology, architects need to be able to navigate and comprehend the complex data flows that take place within and around the built environment. This will allow them to engage more fully with each contextual setting and take advantage of the transformative power of adaptive environments that lies in their ability to transform buildings and their inhabitants into culture machines. By stimulating the creation of communities between occupants and surrounding residents, adaptive environments can foster a sense of shared identity and purpose that transcends the boundaries of physical space.

Fundings

The author(s) disclosed receipt of the following financial support for the research: AHRC Design Exchange Partnerships research fund under the project: *'Participatory neighbourhood planning for the "Brown to Green" transition in Tyneside'* (Grant Ref: AH/Y000250/1) led by Jiayi Jin.

Note

1 In the episode 'Nosedive', every individual lives in a version of America where every tiny interaction is ranked by the people involved on an app that syncs with augmented-reality contact lenses. This allows others to see the people with a digital layer of their scores immediately popped out, meaning that reality has morphed into a pastel-coloured nightmare of aggressive cheeriness as all citizens attempt to out-nice each other and bump up their ratings.

References

Alaimo, C., & Kallinikos, J. (2017). Computing the everyday: Social media as data platforms. *The Information Society*, 33(4), 175–191. doi:10.1080/01972243.2017.1318327

Alavi, H. S., Verma, H., Mlynar, J., & Lalanne, D. (2019). On the temporality of adaptive built environments. In: Schnädelbach, H., & Kirk, D. (Eds.) *People, Personal Data and the Built Environment*. Springer Series in Adaptive Environments. Springer, Cham. https://doi.org/10.1007/978-3-319-70875-1_2

Baldwin, P. J. (2019). Narrative Interactions. In: Schnädelbach, H., & Kirk, D. (Eds.) *People, Personal Data and the Built Environment*. Springer Series in Adaptive Environments. Springer. https://doi.org/10.1007/978-3-319-70875-1_7

Bridle, J. (2019). *New Dark Age: Technology and the End of the Future (Reprint ed.)*. Verso.

Carpo, M. (2017). *The Second Digital Turn: Design Beyond Intelligence* (pp. 36–37). The MIT Press.

Colomina, B., & Wigley, M. (2016). *Are We Human? Notes on an Archaeology of Design*. Lars Müller Publishers.

Couldry, N., & Mejias, U. A. (2018). Data colonialism: Rethinking big data's relation to the contemporary subject. *Television & New Media*, 20(4). doi:10.1177/1527476418796632

Couldry, N., & Mejias, U. A. (2019). *The Costs of Connection: How Data is Colonizing Human Life and Appropriating It for Capitalism*. Stanford University Press.

Cusset, F. (2018). *How the World Swung to the Right: Fifty Years of Counterrevolutions* (Vol. 25, p. 19). MIT Press.

Dalton, N. S., Schnädelbach, H., Wiberg, M., & Varoudis, T. (2016). *Architecture and Interaction*. Springer. doi: 10.1007/978-3-319-30028-3

Dembski, F., Wössner, U., Letzgus, M., Ruddat, M., & Yamu, C. (2020). Urban digital twins for smart cities and citizens: The case study of Herrenberg, Germany. *Sustainability*, 12(6), 2307.

Doyle, M. R., Savić, S., & Bühlmann, V. (2019). *Ghosts of Transparency: Shadows Cast and Shadows Cast Out* (Vol. 13). Birkhäuser.

Erickson, T., & Kellogg, W. A. (2000). Social translucence: An approach to designing systems that support social processes. *ACM Transactions on Computer–Human Interaction (TOCHI)*, 7(1), 59–83.

Fankhänel, T., & Lepik, A. (Eds.). (2020). *The Architecture Machine: The Role of Computers in Architecture*. Birkhäuser.

Ferré-Bigorra, J., Casals, M., & Gangolells, M. (2022). The adoption of urban digital twins. *Cities*, 131, 103905.

Finn, E. (2019). The black box of the present: Time in the age of algorithms. *Social Research: An International Quarterly*, 86(2), 557–579.

Foucault, M. (1975). *Discipline and Punish* (Sheridan, A., Tr.). Gallimard.

Fox, M. (Ed.). (2016). *Interactive Architecture: Adaptive World*. Chronicle Books.

Gershenson, C. (2013). Living in living cities. *Artificial Life*, 19(3–4), 401–420.

Green, K. E. (2016). *Architectural Robotics: Ecosystems of Bits, Bytes, and Biology*. MIT Press.

Howard, J. (2017). Radical technologies: The design of everyday life. *TLS: Times Literary Supplement*, 5980, 35–36.

Karandinou, A. (2019). *Data, Architecture and the Experience of Place*. Routledge.

Kayser-Bril, N. (2019). *Personal Scoring in the EU: Not Quite Black Mirror Yet, at Least if You're Rich*. Retrieved from https://algorithmwatch.org/en/personal-scoring-in-the-eu-not-quite-black-mirror-yet-at-least-if-youre-rich/

Kitchin, R. (2014). *The Data Revolution: Big Data, Open Data, Data Infrastructures and Their Consequences*. Sage.

Kitchin, R., & Dodge, M. (2014). *Code/Space: Software and Everyday Life.* MIT Press.

Kneidinger-Müller, B. (2018). Self-tracking data as digital traces of identity: A theoretical analysis of contextual factors of self-observation practices. *International Journal of Communication,* 12, 629–646.

Kolarevic, B., & Parlac, V. (2015). *Building Dynamics: Exploring Architecture of Change.* Routledge.

Krukar, J., Dalton, R. C., & Hölscher, C. (2016). Applying HCI methods and concepts to architectural design (or why architects could use HCI even if they don't know it). *Architecture and Interaction: Human Computer Interaction in Space and Place* In: Dalton, N., Schnadelbach, H., Wiberg, M., & Varoudis, T. (Eds.) (pp. 17–35). Springer.

Latour, B. (2007). *Reassembling the Social: An Introduction to Actor-Network-Theory.* Oxford University Press.

Liu, M., Fang, S., Dong, H., & Xu, C. (2021). Review of digital twin about concepts, technologies, and industrial applications. *Journal of Manufacturing Systems,* 58, 346–361.

Lupton, D. (2016). *The Quantified Self.* John Wiley & Sons.

Lupton, D. (2021). Self-tracking. In: Kennerly M., Frederick S., & Abel J. (Eds.). *Information* (pp. 187–198). Columbia University Press.

Lupton, D., Pink, S., Heyes LaBond, C., & Sumartojo, S. (2018). Digital traces in context: Personal data contexts, data sense, and self-tracking cycling. *International Journal of Communications Special Issue on Digital Traces in Context,* Vol 12, 647–665

Manovich, L. (2018). Can we think without categories? *Digital Culture & Society,* 4(1), 17–28.

Mayer-Schönberger, V., & Cukier, K. (2013). *Big Data: A Revolution That Will Transform How We Live, Work and Think.* John Murray.

Mejias, U. A., & Couldry, N. (2019). Datafication. Internet Policy Review, 8(4). https://doi.org/10.14763/2019.4.1428

Milan, S. (2018). Political agency, digital traces, and bottom-up data practices. *International Journal of Communication,* 12, 507–527.

Mitchell, W. J. (1999). *E-topia: Urban Life, Jim – But Not as We Know It.* MIT Press.

Nabian, N., Offenhuber, D., Vanky, A., & Ratti, C. (2013). Data dimension: Accessing urban data and making it accessible. *Proceedings of the Institution of Civil Engineers – Urban Design and Planning,* 166(1), 60–75.

Negroponte, N. (1975). The architecture machine. *Computer-Aided Design,* 7(3), 190–195.

Negroponte, N. (2021). Humanism through intelligent machines. *The Architecture Machine.* MIT Press. Retrieved from https://doi.org/10.1162/a8667414.177f4cfd

Ratti, C., & Claudel, M. (2016). *The City of Tomorrow: Sensors, Networks, Hackers, and the Future of Urban Life* (pp. 52–55). Yale University Press.

Rouvroy, A. (2020). Algorithmic governmentality and the death of politics. *Green European Journal,* 27. www.greeneuropeanjournal.eu/algorithmic-governmentality-and-the-death-of-politics/

Schnädelbach, H., & Kirk, D. (2019a). Introduction. In: Schnädelbach, H., & Kirk, D. (Eds.) *People, Personal Data and the Built Environment.* Springer Series in Adaptive Environments. Springer, Cham. https://doi.org/10.1007/978-3-319-70875-1_1

Schnädelbach, H., & Kirk, D. (Eds.). (2019b). *People, Personal Data and the Built Environment.* Springer.
Townsend, A. M. (2013). *Smart Cities: Big Data, Civic Hackers, and the Quest for a New Utopia*. WW Norton & Company.
van Dijck, J. (2013). *The Culture of Connectivity*. Oxford University Press.
Weizman, E. (2014). *Forensis: The Architecture of Public Truth.* Sternberg.

9 Digital Technologies and Alternative Social Imaginaries

Producing Architectural Counternarratives in Eco-communities

Nadia Bertolino

Introduction

This chapter discusses to what extent digital technologies enable practices of radical sustainability in eco-communities and how these can eventually contribute to the production of architectural counternarratives. In 2005, Pierre Lévy, in his essay "Collective Intelligence, A Civilisation," reclaimed the importance of collective imagination to produce reality. In his view, cyberspace is a poetic figure, "a placeless place" that reveals "the infinite openness which is the essence of man" (p. 197). Drawing on Lévy, I assume that the process of cultural evolution brought in by the rapid development of digital technologies creates closer connections between members of the same community, opening up their mental space and enabling new forms of communication. According to de Kerckhove, under the influence of ICT tools people develop the ability to quickly search through reality and learn through cooperation, yet they also lose the habit of linear thinking and begin to think in a quick and careless, even chaotic, way, as if they are clicking links on a website without any plan (2001).

This chapter investigates how digital technologies apply to one specific type of community, which has widely been investigated over the past few years: so-called eco-communities. A geographical focus is kept on three English case studies: Hockerton housing project in Southwell, Landmatters co-operative in Devon, and Old Hall Community in Colchester. All these respond to the definition of "eco-communities" given by the Global Ecovillage Network: "an ecovillage is an intentional, traditional or urban community that is consciously designed through locally owned, participatory processes in all four dimensions of sustainability (social, culture, ecology, and economy) to regenerate their social and natural environments."

These three case studies have been chosen for their architectural settings, socio-political agenda, and radical approach towards a non-exploitive growth model. What makes the reflection here even more challenging is that such communities flourished in the 1960s, echoing the back-to-the-land movement, to claim a greater degree of autonomy and inclusion than was found in the predominant high-tech post-war society (Martell, 2023). Hence, this chapter will question if and how

DOI: 10.1201/9781003380559-9

a collective cultural shift took place, from an intentionally low-tech paradigm towards the integration of digital technologies into everyday community activities and types of social interaction. This is seen as a virtual layer that overlaps, and to some extent complements, that of the physical space that is adapted and inhabited by the community, reiterating the real/virtual dichotomy theorised by Maldonado (1994).

Maldonado (1994) described a virtual model as a creative and cognitive strategy, through which it is possible to learn more about the real world and visualise an infinite range of alternative scenarios for its future development. Thus, to fully understand how the digital realm entwines the physical substrate of the eco-community and contributes to the production of new collective imaginaries, it is required firstly to define what the key features of such communities are and why referring to these experiences is extremely relevant within architectural discourse today. It is sensible to say that eco-communities today form a world-wide social movement, only partially coinciding with the Global Ecovillage Network (GEN), promoting anti-capitalist responses to today's social, economic, and environmental challenges.

Castells (2012) investigated the linkages between the rise and diffusion of networked social movements in relation to the rapid development of digital technologies. As Castells wrote: "Humans create meaning by interacting with their natural and social environment, by networking their neural networks with the networks of nature and with social networks. This networking is operated by the act of communication" (Castells, 2012, p. 6), intended as the process of sharing meaning through the exchange of information. Thus, a priori focus is given to a reflection about understanding values and limitations of digital technologies to establish or strengthen networks across communities with similar agendas. It is argued that the role played by the implementation of digital technologies in eco-communities can be confined to one specific strand. Instead, through literature and a community-based assessment, a few hypotheses on possible applications leading to diversified forms of digital social innovation are ventured.

Building on previous research, presented at the seventeenth annual conference of the Architectural Humanities Research Association (Pratt Institute, October 2022), an ad-hoc questionnaire is developed to investigate the relationships that community members established with digital technologies, type, and main reasons for usage. This provided the basis for a structured reflection on the potentials and limitations of such tools in terms of enhancing inclusion, improving local and trans-local networks, and determining new forms of interactions supported by the physical and virtual spaces produced and inhabited by the community. Finally, the chapter discusses the potential to establish an architectural counternarrative to define eco-communities, based on de Kerckhove principles of *connecting architecture* (2001): this draws on a process of collective imagination, emerging from the overlap of the real and the virtual layers, to create new forms of spatial and social interactions.

Integrating the concept of radical ecologies in the definition of "eco-community"

This section introduces the notion of radical ecology, as the basis to understand and frame sustainable communities, reflecting on the relevance and timeliness of investigating such settlements within the current architectural debate. While there is extensive literature providing definitions of eco-communities, it is here proposed to embed in such a definition the concept of "radical sustainability." This allows the introduction of the three case studies and clarification of the analytical categories for reflecting on the role played by digital innovation technologies to enhance specific aspects of everyday life within eco-communities.

In 2021, because of the pandemic, communal life has been considered a viable solution to tackle the new issues imposed by the lockdowns and the constraints in the use of shared spaces. The *Guardian* has reported that communes have been "inundated by new applicants driven by the Extinction Rebellion movement and the glimpse that lockdown has offered of simpler, less consumption driven, lifestyles" (Guardian, 17 Jan. 2021). These communities are "characterized by close relationships, shared common values, and can be found throughout the world in varying degrees of community integration" (Lulavy, 2019). Interestingly, the lockdown appears to have shone a new light on the alternatives these communities can offer to de-commodify received and typically unquestioned beliefs about human nature and environmental agency, economy, and the material geographies of life. As Helen Jarvis has noted, this can be seen as a recognition that mainstream contemporary society and the lifestyles of the past are permanently broken (Jarvis, 2017).

In Britain, most of these communities flourished in the 1960s and 1970s, echoing the back-to-the-land movement and claiming a greater degree of autonomy and inclusion than could be found in the increasingly high-tech post-war society (Martell, 2023). The generation of "diggers" (named after the seventeenth-century English communards) challenged social rules and constructs, opting out of "the grab-game of straight society," as Richard Neville wrote in *Oz* magazine (1968) about the first London digger commune, the Love Commune in Full Flowers. In *The World Turned Upside Down*, Christopher Hill highlighted how the digger communities "might have given the counterculture an economic basis" (1991, p. 78), as they proposed a social system in which the abolition of exploitation would enable the creation of a society that would "transcend the property system" (1991, p. 83).

John Gurney defined the diggers as

> a movement that had set out to declare the earth a common treasury and to call for an end to all private property and buying and selling, an episode rooted in local experience, and one that reflected tensions and conflicts that had long affected the community.
>
> (Gurney, 2019, p. 153)

This resonates with the experience of some contemporary eco-communities, which are "a mixture of the practically minded and utopianisms" (Howard, 2021), and emphasises the close link between the political agenda of such communities and their collective commitment to enabling practices of environmental care.

What makes the study of eco-communities particularly timely and relevant within today's architectural debate is that these push the concept of sustainability to a more radical level, translating into action Ward's idea of *planetary housekeeping* (1976). This was a rather disruptive idea when she first conceived it in the 1970s, as it highlighted that sustainability could only be achieved through a collective effort to practise environmental care and promote social equity, seen by Ward as two faces of the same coin. Similarly, Adger (2000) described social and environmental resilience as two interdependent aspects and defined resilience as the ability of communities to maintain themselves in the face of external disturbances because of social, political, and environmental changes.

All this interweaves with key issues in the current architectural debate: there is an increasingly urgent demand to conceive other ways of doing architecture in the face of the current climate emergency and pandemic, avoiding further forms of environmental and social depletion in favour of integration, circularity, durability, and social resilience (Space Caviar, 2021).

According to the Global Ecovillage Network, "an ecovillage is an intentional, traditional or urban community that is consciously designed through locally owned, participatory processes in all four dimensions of sustainability (social, culture, ecology, and economy) to regenerate their social and natural environments" (Marantz, 2019). Dawson reports that "ecovillages—local communities which aim to minimise their ecological impact while maximising human wellbeing and happiness—are found all over the world" (Dawson, 2015). Intentional communities are generally groups of families and individuals who have chosen to live together, drawing on a shared environmental agenda and a desire to live an egalitarian, co-operative life "to enhance their shared values or for some other mutually agreed-upon purpose" (Sargent, 2012: 55). Such communities stand in opposition to centralised political systems and social constructs and produce inclusive and environmentally resilient ways of living. In fact, the "global ecological crisis along with social alienation has coalesced the community impulse in the form of eco-villages" (Liftin, 2014).

Eco-communities adopt a sustainable and self-reliant approach. They aim to strike a balance between economic development and social wellbeing while preserving the natural environment and working towards avoiding extinction of resources for future use. Recycling, reusing, self-construction using locally sourced material, and growing their own food are just some of the methods to which they adhere. Efficient management of resources that are common to all members of the community is a significant measure of the quality of life: "The physical and social materiality of eco-communities are purposefully built to shape the commons" (Pickerill, 2015). Although these communities are not always extroverted in nature, they can bring about drastic changes in society, As Sager noted, some of them trigger the need for social change while acting within their boundaries (Sager, 2018).

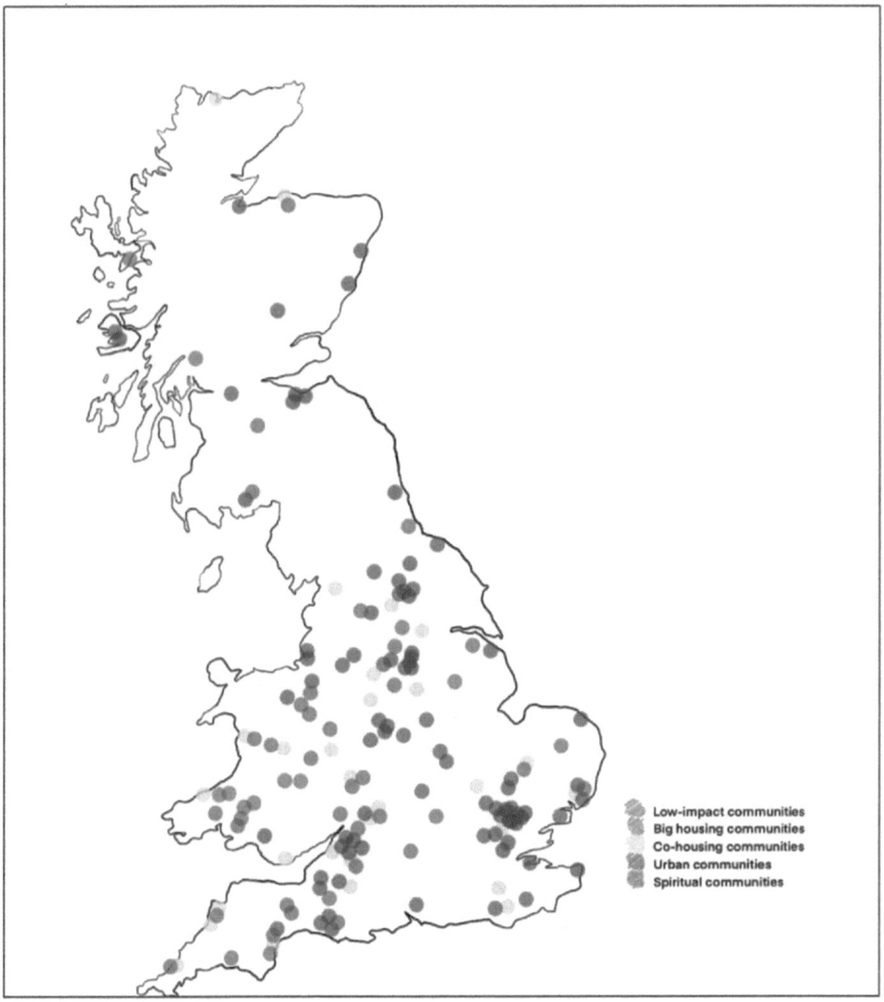

Figure 9.1 Territorial distribution of intentional communities across the UK.

In the United Kingdom, intentional communities are spread across the country with an increased distribution in the South, as shown in Figure 9.1. According to the collective "Diggers and Dreamers." these can be of different natures: Cohousing Community, Big House Commune, Urban Communal, Spiritual Community, Low-Impact Community. For the extent of this discussion, I focused on the last category, as they offer a wide range of more radical alternatives to mainstream collective living.

In fact, these are often off-grid and have an ambition to reach a self-sufficient status through the drastic reduction of energy and goods consumption and the production of clean energy through renewable sources. Low-impact communities are settled in the countryside, most of them in southwest England, and benefit from large open areas where many collective activities take place: growing crops, breeding animals, mending, upcycling small items, and socialising.

In the next section, I describe in depth three of these communities, discuss spatial arrangements in response to their social and environmental agendas, and, eventually, reflect upon the role of IT technologies in enabling further and more radical forms of radical ecologies.

The Old Hall Community, Hockerton Housing Project, and Landmatters Co-operative: architectural settings, social values and environmental agendas

Three low-impact communities in the United Kingdom (The Old Hall Community, Hockerton Housing Project, and Landmatters Co-operative) are considered an integral part of this research for a comprehensive understanding of the alternative lifestyles they have adopted and the relevance of the architectural counternarrative they produce. Despite having a striking similarity in their agendas, community principles, and the environmentally friendly techniques that they adhere to, they differ significantly in the architectural setting within the community. The type of built structures that the members occupy is very different, resulting in varying living conditions within each community. This makes it interesting to investigate and understand how the differences in living conditions can affect the community's approach. Geographically, all three communities are situated in rural locations.

The Old Hall Community, located in Colchester, is the oldest, constituted in 1974. Hockerton Housing Project in Southwell and Landmatters Co-operative in Devon were more recently formed, in 1995 and 2002 respectively. The three communities have been selected for their agendas, the different spatial settings they offer, and the use of digital technologies by their members. All share interesting architectural arrangements. The Old Hall occupies a Grade 1 listed building, while Hockerton Housing Project, in Nottinghamshire, is built on a 25-acre site and is "the UK's first earth-sheltered, self-sufficient, ecological housing development. The residents of the five houses generate their own clean energy, harvest their water, and recycle waste materials, causing no pollution or emissions of carbon dioxide" (Harridge, 2006). Landmatters Co-operative offers a totally different set-up, with its residents occupying temporary dwellings in symbiosis with the surrounding natural environment.

The Old Hall is a large community of about 60 people who share an old friary and 65 acres of farmland. They share the workload to be nearly self-sufficient in food and energy. Their water is heated in part by a large wood-fired biomass boiler they call the dragon. This is supplemented by a ground source heat pump, currently in need of repair. Oil-fired and gas-fired boilers are used as back-up. They have a 30kw array of 120 solar PV panels to supplement electricity usage and water comes from a borehole with mains water back-up. Because of their communal commitment, members find that they can only work about three days a week outside the community to bring in enough money to pay the bills, while spending the remaining days working for and within the community. This work–life balance brings with it a quality, stress-free outside existence that benefits one's health. They all help and support each other. In the summer months they enjoy help from friends from all over the world who turn up for two weeks at a time.

The Hockerton housing project is a smaller community compared to Old Hall; it has around 15 members, with adults required to contribute approximately 300 hours annually towards communal activities. They are paid equally, where applicable. It is a sustainable community that focuses on low-energy and environmentally friendly living. The project was initiated in the 1990s by a group of five households aiming to create an ecologically responsible way of living. The Hockerton Housing Project encompasses several noteworthy characteristics. First, sustainable design is a foundational element, with houses constructed using environmentally friendly techniques and materials. These dwellings boast high insulation levels and leverage passive solar design principles for optimal energy efficiency.

Another prominent feature is the commitment to renewable energy. The community generates its own sustainable power through a variety of methods, such as solar panels, wind turbines, and biomass boilers. The houses are meticulously designed to minimise energy consumption, promoting self-sufficiency within the community. Water conservation is a key focus of the project, incorporating rainwater harvesting systems to reduce overall water consumption. Additionally, wastewater undergoes on-site treatment using reed bed filtration systems. Furthermore, the Hockerton Housing Project emphasises sustainable land use. The development includes communal gardens and allotments, providing residents with spaces to cultivate their own food using organic methods.

This holistic approach underscores the project's dedication to ecological principles and community self-sufficiency. The project also emphasises wildlife conservation and biodiversity enhancement. Brenda and Robert Vale, Hockerton architects, ensured the houses had "high thermal mass and high levels of insulation which efficiently eliminated the need for heating systems" (Diggers and Dreamers, 2023).

The entire development represents a significant step towards sustainability. Its members grow and raise their own food: fruits, vegetables, honey, eggs, fish, and some meat. This has minimised the number of food packaging products on site. Food waste is composted, while other materials are re-used to the greatest extent possible before being recycled. From the preliminary stages, this project was intended to be exemplary of a sustainable lifestyle and has succeeded in living up to its principles.

Landmatters co-operative differs from The Old Hall and Hockerton housing project in drawing on the concept of permaculture, though its residents hope to contribute profoundly to confronting the current environmental and ecological crisis. The project focuses on land-based activities, ecological stewardship, and co-operative living. The Landmatters Co-operative incorporates distinctive attributes. Firstly, it prioritises land stewardship, committing to sustainable land use and management. Co-operative members collaborate to nurture and revitalise the land, applying permaculture principles and practices.

Additionally, Landmatters operates as a co-operative living community, where decisions are collectively made, and responsibilities are shared among members. The co-operative aims to cultivate an inclusive and supportive community ethos, emphasising cooperation, equality, and mutual respect. Another feature is the project's focus on low-impact infrastructure, advocating for sustainable building practices. Structures on the land are intentionally designed to leave a minimal ecological footprint, utilising natural materials and renewable energy sources. Furthermore, Landmatters actively engages in education and workshops, offering programmes and internships to disseminate knowledge about sustainable living, permaculture, and land-based skills. The goal is to inspire and empower others to establish regenerative communities, aligning with the co-operative's broader mission.

Figure 9.2 provides a comprehensive comparative analysis, highlighting the different spatial settings of these communities, alongside data regarding their

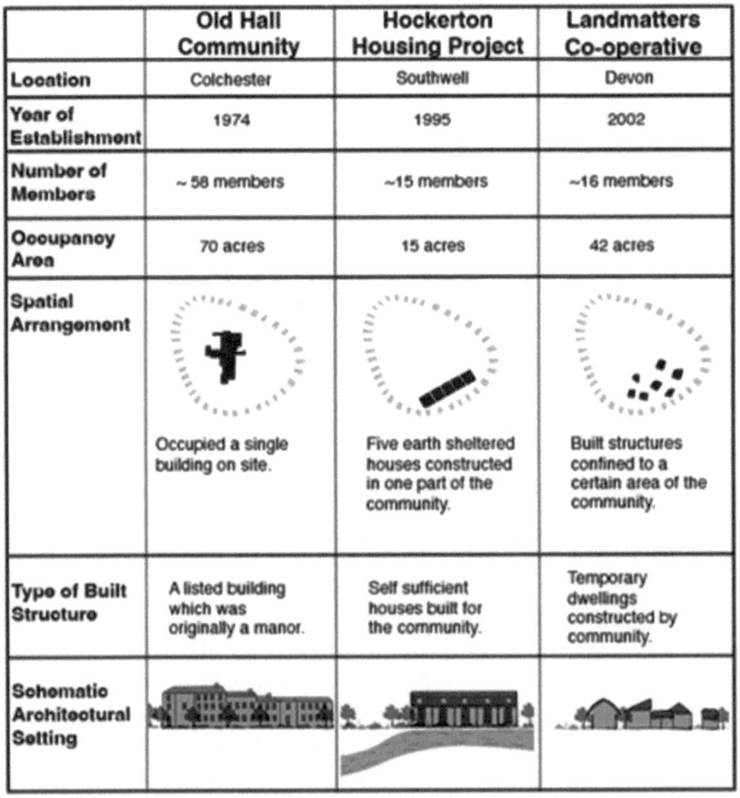

	Old Hall Community	Hockerton Housing Project	Landmatters Co-operative
Location	Colchester	Southwell	Devon
Year of Establishment	1974	1995	2002
Number of Members	~ 58 members	~15 members	~16 members
Occupancy Area	70 acres	15 acres	42 acres
Spatial Arrangement	Occupied a single building on site.	Five earth sheltered houses constructed in one part of the community.	Built structures confined to a certain area of the community.
Type of Built Structure	A listed building which was originally a manor.	Self sufficient houses built for the community.	Temporary dwellings constructed by community.
Schematic Architectural Setting			

Figure 9.2 Case studies: a comparative analysis.

year of foundation, size, area occupied, and typology of buildings. This overview is relevant to introduce the role played using digital technologies to foster their idea of "sustainability" and create opportunities for territorial networking through enhanced connectivity. Next, the data gathered through an online survey circulated across members of the three communities is discussed and reflections are made upon the key findings, trying to identify shared trajectories, divergent opinions, and reasons behind some scepticism towards incorporating such technologies within the routine administration of their communities.

All three communities stressed the significance of the development of individual lives without compromising the communal aspects. As members of those communities, the individuals accommodate one another's needs and extend support wherever required. The members have adapted to incorporating the strategies of the community into their everyday lives, thus constructing a well-defined social structure within each community. All three communities strive to make their eco-friendly approaches the very definition of their community ethos.

Deliberate efforts are made by community members, whether making personal choices or acting on behalf of the community, to be responsible and reduce their carbon footprints. Although these communities are in different parts of the United Kingdom, their sites have certain similarities: they are rural and surrounded by green areas. Unlike urban settings, the calm and serene surroundings influence the aura within each community. They differ significantly in their membership policies.

The Old Hall Community has elaborate policies that allow partakers to become associated with the community in the most suitable manner. However, none has impenetrable boundaries; they deal with visitors and volunteers and have thus successfully established social connections through their porous boundaries. The well-defined social and communal aspects are crucial factors shaping these communities.

The Findhorn Ecovillage is literally home to many people, and many others are associated in different ways. Physically, Findhorn is spread across a large area and has many more sustainably built structures than the other two. They have expanded into education and other activities, making it a genuinely large-scale community. Hockerton and Landmatters, on the other hand, have fewer people living within the communities, so they operate on a smaller scale. They include vast areas of grasslands, woodlands, and other cultivable areas that form essential parts of the communities.

Despite the disparities in the three communities, they have remarkably similar underlying principles. Most of the food they consume is grown or raised within the community. When there is a need to source surplus food, the community members make a conscious effort to restrict it to food that is in season to reduce their carbon footprints. They follow efficient waste management systems that ensure minimal impact on the environment. Construction is generally carried out with naturally sourced materials. Hockerton is known for the innovative construction techniques used on the five houses in the community. These earth-sheltered houses help reduce the community's carbon footprint by eliminating the need for heating systems. The buildings in Landmatters co-operative are built using largely naturally sourced materials.

A community-based assessment on the role of digital technologies to enhance the notion of radical ecologies

To comprehend how these communities deal with and embed the use of digital technologies into their everyday life, a concise questionnaire was sent to the communities, first gathering information about the research participants' profile (age, gender, years of residency within their respective communities) and, secondly, exploring the key issues below:

A) Which internet-connected devices do you use most frequently?
B) On average, how many hours per day do you spend on a mobile phone or computer device?
C) Which of the following best describes what you use the internet for? (Multiple choice question: connect with community members and share news or communications; personal use i.e. social media; personal administration and or casual browsing; work; personal and community decisions; personal entertainment; other)
D) How often do you use social media (or internet, more broadly) to engage with members of other eco-communities?
E) How often do you attend online events (such as networking events, public debates on topics of interest for your community etc.)?

Overall, 18 responses were submitted, evenly distributed from across the three communities and with no significant divergences in terms of profile and responses based on the community the respondents belong to. Most respondents fall into the 30–49 age band (40%), with the 28% being 50–69 and 28% being over 70. Interestingly, only one respondent was under 30. Most respondents identify as male (50%), while 45% identify as female and the remaining 5% as non-binary. Approximately 40% of respondents have been living in their respective communities for five to ten years, and 33% for longer than ten years, which suggests they have a comprehensive knowledge of community life and type of interactions across time.

Table 9.1 and Table 9.2 shows data collected and the analysis, with reference to the questions mentioned earlier. It is noted that most respondents rely on individual devices to interact with "the digital" world, with some using their laptop too, but not spending longer than two hours per day on these.

This fact, which might read as irrelevant at first glance, highlights a rather important aspect of eco-community life, that is, the amount of time spent in communal activities and material labour within the community buildings and outdoor spaces. The "digital" component of human interactions is reduced to the very minimum, with some community members having opted to not own a mobile or laptop

Table 9.1 Responses to the questionnaire, showing which internet-connected devices are used more frequently by community members

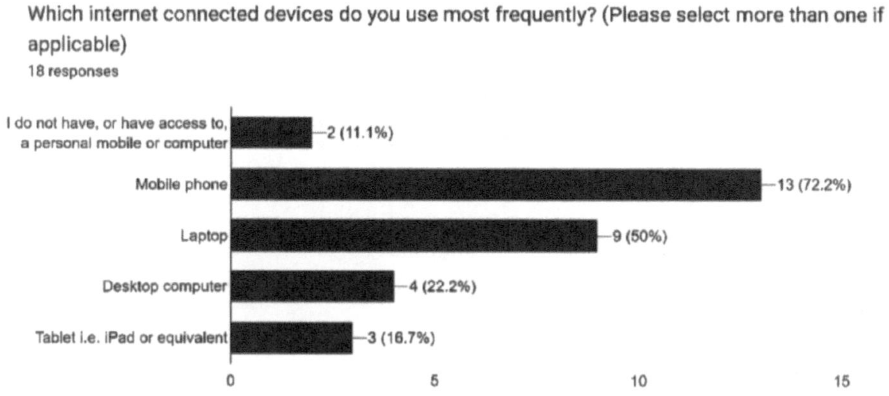

Table 9.2 Responses to the questionnaire, showing how many hours per day community members spend on their connected devices

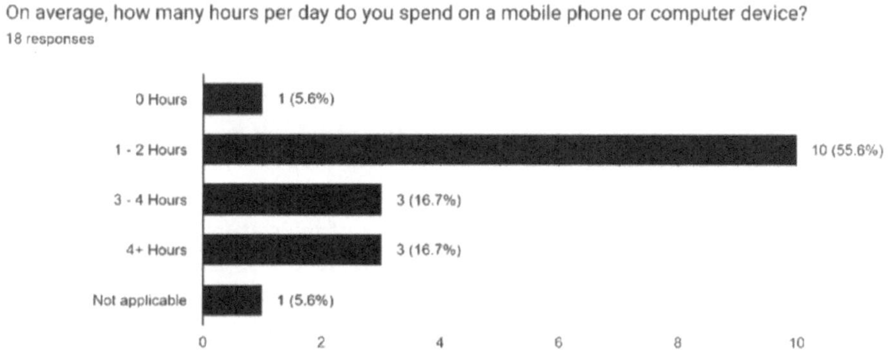

at all. This is extremely consistent with Luke Martell's analysis of sustainable communities established in the 1960s and 1970s, described as "a backlash against hi-tech postwar societies" (2023, p. 87). Analysing more in depth the types of usage that the respondents engage in with internet facilities, it can be noted that approximately 45% of them make personal use of internet (social media, private administration, and causal browsing), with only 11% using the internet for connecting with other community members and sharing news and communications or to make community decisions.

This suggests that there might be an undiscovered potential in digital technologies, as well as in the most advanced AI-based technologies, to support such communities' development and the establishment of larger territorial networks, as I will explain further in the following section. As shown in Table 9.3, more than 60% of respondents do not ever use social media, or the internet more broadly, to establish a connection with members of other eco-communities, while 33%

Table 9.3 Responses to the online questionnaire, showing how often community members use social media to engage with members of other eco-communities

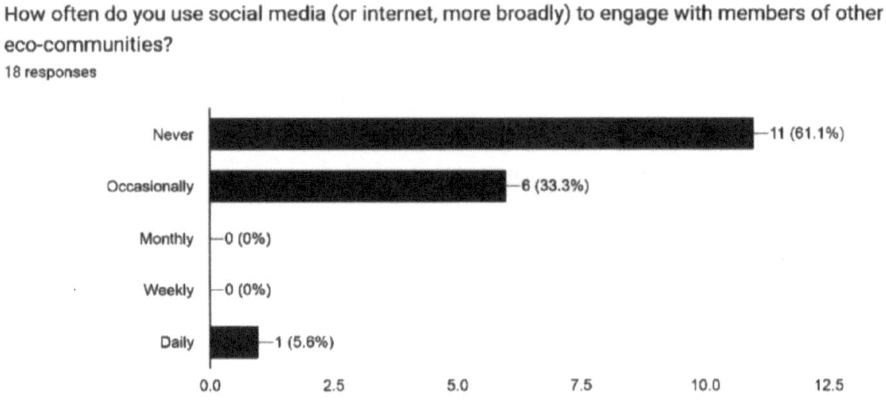

of respondents do this only occasionally. Again, the potential for creating digital networks could be vast and transformative, yet is underestimated by the community members (at least, by those who took the questionnaire). In the following section, drawing on literature and the analysis of further case studies, AI aims to demonstrate the substantial contribution that digital technologies can have for the creation of virtual connections, seen as the natural extension of the actual, material world we live in.

In fact, approximately 67% of respondents do not attend online events, such as networking events or public debates, in topics of interest for their community, while 28% do so only occasionally and one respondent does so weekly. One may argue this is due to a poor internet connection speed within the community boundaries; however, this is not confirmed by data collected (with more than half of respondents having declared very strong and good connection speed).

Simultaneous analysis of responses revealed that constructive social structures established at communal levels heavily rely on everyday "in-person" relationships, whether these take place for personal reasons or for decision-making and tasks delivery: the *diggers* component of Chris Coates' phrase "diggers and dreamers," after which is named his eco-communities' repository. Further informal and unstructured conversations with Old Hall, Hockerton, and Landmatters members emphasised the methods adopted and practised by these communities to address issues of ecological crisis, yet none of these mentioned the potential aid that could come from introducing digital technologies into their everyday ecological practices.

On the potential applications of digital technologies to enable practices of radical sustainability

In the twenty-first century, digital technologies have created the conditions for the automation of all social interactions, making collective will almost inoperative.

In his 1995 book *Out of Control: The New Biology of Machines, Social Systems, and the Economic World*, Kevin Kelly presaged as much, arguing that then-nascent digital networks would eventually create a global mind to which sub-global minds (individual or collective or institutional) would have to submit. It is peculiar that this foresight, which perfectly describes what happened in mainstream society, does not apply to all those social settings that do not conform or propose alternative community models, such as the eco-communities discussed in this chapter.

The potential of digital networks for eco-communities can be diverse and transformative. Nathan Schneider in *Everything for Everyone: The Radical Tradition That Is Shaping the Next Economy* (2018) explores how digital networks can be harnessed to build more equitable and sustainable economies, and he demonstrates how this applies to eco-communities. Drawing on Rushkoff's *Team Human* (2019), it notes the importance of human cooperation and community-building in the digital age, including how eco-communities can leverage technology for positive change. This position is supported by Zuckerman in *Rewire: Digital Cosmopolitans in the Age of Connection* (2013), which explores how digital networks can foster global connections and facilitate collaboration among eco-communities world-wide. This resonates with the concept of digital cosmopolitanism, which refers to the idea that individuals can develop a broad understanding of global issues and cultures through their online interactions.

Zuckerman examines how the internet has the potential to broaden our perspectives and connect us to a diverse range of people and ideas. He argues that the internet can be a powerful tool for fostering empathy, understanding, and cross-cultural communication. However, Zuckerman also acknowledges the challenges that come with this increased connectivity, such as the proliferation of echo chambers and filter bubbles, which can lead to polarisation and reinforce existing biases.

Nicole Starosielski in *The Undersea Network* explores the undersea cables that power the internet and their environmental implications, which can be relevant to discussions on ecology and the digital connectivity of self-sufficient communities. However, the increase in virtuality and dematerialisation that is at the core of Starosielski's position might imply a radical rethink of eco-communities' agendas and ethos, to embed aspects that are not currently relevant within their social and financial models. I will return to this aspect, discussing why and how there is a need to reframe and expand the conventional field of operation of architectural agency and include a dematerialised component for a comprehensive definition of a spatial counternarrative.

It is very relevant at this point to refer to the enlightening work of Sasha Costanza-Chock in *Design Justice: Community-Led Practices to Build the Worlds We Need* (2020), where she discusses how digital networks and design practices can be employed to foster ecological principles in sustainable communities and address consequently environmental justice issues at a larger scale. Here she advocates for the inclusion and empowerment of marginalised communities in the design of digital technologies, policies, and systems that directly affect their lives. The core idea of design justice, which according to Costanza-Chock can be achieved also through the implementation of IT-based iterations, is that the design process should not be dominated by a select group of individuals, corporations, or institutions. Instead, it should be democratised, allowing the people who are most impacted by the design outcomes to be active participants in shaping them.

In fact, digital technologies can have a significant impact on social justice efforts, both positively and negatively. On the one hand, they can empower marginalised communities and facilitate social justice initiatives. The internet and digital platforms have democratised access to information, making knowledge and resources more accessible to people around the world. Online activism and awareness campaigns on social media can quickly gain global attention and spark conversations about pressing social issues. Digital technologies also facilitate organising and networking, enabling social justice movements to connect across geographical boundaries.

Big data and analytics can be used to understand social issues, track progress, and develop evidence-based policy solutions, informing more effective strategies for social justice organisations. Additionally, digital technologies have opened opportunities for online education, reducing barriers to education for marginalised communities. However, there are also concerns about how such technologies can perpetuate social injustices.

The digital divide, where certain communities lack access to digital technologies and the internet due to affordability, infrastructure limitations, or digital illiteracy, remains a major challenge (Castells, van Dijk, Markham, and others). Social media platforms can become spaces for harassment, hate speech, and cyberbullying, disproportionately affecting marginalised groups. Data privacy and surveillance can compromise the security and privacy of vulnerable communities.

Algorithms used in digital technologies can perpetuate bias and discrimination, leading to unfair outcomes for marginalised individuals in various aspects of life. It must be added to this the inherent sceptical attitude of members of low-tech communities towards such topics, as demonstrated by the responses to the questionnaire, which is only partially resolved with a shift of the average age of community members. This point is made very clear by danah boyd, whose work explores issues related to privacy, digital inequality, and youth engagement in digital spaces (2014).

However, some previous experiences can pave the way for a more structured integration of digital tools within eco-communities. For instance, Pia Mancini's "Open Collective," a platform that facilitates transparent fundraising for community projects, might be of great inspiration. Her work focuses on using digital networks to empower local communities and promote sustainability. Similarly, one may refer to the ambitious project "Civic Square," an initiative that sees the neighbourhood as the most important unit of radical societal transformation. It is legitimate to think that a similar approach might work well to sustain and support communities such as those described here.

Digital technologies and alternative social imaginaries: towards an architectural counternarrative

In the preceding sections, it was discussed how the communities of Old Hall, Hockerton, and Landmatters share similar ideologies regarding social values, political ethos, ecological approaches (including internal food cultivation, recycling, reusing), and communal responsibilities. However, each community possesses distinct practices addressing environmental challenges. For instance, members of Old Hall prioritise material reuse over recycling, actively exploring a broader range of recyclable materials within the community and advocating for local council support. In the Hockerton housing project, residents emphasise consuming primarily community-grown food to reduce nonessential packaging. Any surplus food is chosen to be in season, reflecting a conscious effort to minimise the community's carbon footprint. At Landmatters, members strictly adhere to permaculture principles, integrating eco-friendly techniques into their daily lives, closely aligning themselves with nature.

All three communities raise awareness about their multifaceted sustainability approaches by sharing knowledge with visitors through websites, interviews, and reports in national and international magazines and newspapers. While it might be unrealistic to implement such low-impact practices throughout contemporary society, disseminating and practising eco-friendly ideologies could indeed address a portion of today's environmental challenges through relatively simple practices and tools.

The question arises whether these practices can be replicated, scaled up, and adapted to different geographical contexts while maintaining the same effectiveness and inclusiveness. It's important to acknowledge that Adger (2000) views social and environmental resilience as interconnected. He defines resilience as communities' ability to withstand external disturbances amid social, political, and environmental changes. From this perspective, lessons drawn from eco-communities could foster local resilience by enhancing "adaptive capacities, reorganization, renewal, self-organization, and learning" (Bohle et al., 2009), facilitated by social actors leading spatial production processes.

Simultaneously, the investigation highlights that the shared socio-spatial configurations of these communities significantly drive the adoption of eco-friendly and sustainable techniques, directly linked to their implementation in their respective locations. I argue that architects can act as mediators by studying these spatial arrangements thoroughly, replicating, extending, and adapting environmental strategies to address environmental concerns in other spaces.

The examination of the three eco-communities reveals their intentional lack of closed boundaries, suggesting the potential for interactions beyond their geographical borders. These permeable boundaries allow for the inclusion of new members, visitors, and volunteers based on community policies, facilitating the broad dissemination and sharing of knowledge and experiences regarding ecological practices and related spatial narratives.

Considering this, our reflection proposes a significant dichotomy, contrasting the absence of physical borders with the lack of virtual and digital connections between the community and its wider local and hyper-local territorial context. This work, recent literature, and case studies lead us to examine the role of digital technologies in sustainable communities from a distinct perspective, seeking potential applications to establish a lasting social and environmental impact on hyper-local contexts and questioning the role of spatial practitioners in these processes.

As Maldonado suggested, can we consider the virtual world as an extension of the real one, forming an architectural counternarrative that blends the virtual and real in a balanced manner? If so, how can digital technologies be introduced into the ethos and agenda of communities originally conceived as intentionally low-tech, without compromising their inherent nature? This spatial counternarrative resonates with the challenging coexistence of humans and machines depicted in Isaac Asimov's "Robot Visions" and the "Three Laws of Robotics," a set of ethical guidelines programmed into robots to ensure their behaviour aligns with human values and safety.

It is commonly agreed that digital networks can connect people from all over the world who share a common interest in environmental sustainability and eco-friendly living. However, responses to the questionnaire discussed here revealed that such potential is widely unexpressed in the context of eco-communities. Drawing on some recent literature and data collected, it is legitimate to outline a field of practice in which digital technologies, and in particular digital networks, can be implemented to empower and enhance eco-communities.

The utilisation of digital networks is instrumental in fostering a culture of eco-consciousness and sustainable living within communities. These networks serve as platforms for individuals to share valuable information, resources, and educational materials related to sustainable practices, renewable energy, and waste reduction (Smith et al., 2020). The use of IT technologies facilitates efficient communication and collaboration among members of low-impact communities. Online platforms, social media, and digital tools play a crucial role in enabling information exchange, coordinating activities, and organising communities in a wider and more horizontal manner (Jones & Brown, 2019).

Moreover, digital networks empower eco-communities to collaboratively address environmental challenges through initiatives such as developing sustainable products, organising clean-up events, or supporting green technology startups (Robinson & Green, 2021). The collective action of community members is amplified through digital collaboration. In terms of advocacy, digital networks prove to be powerful tools for raising awareness about environmental issues and influencing policy changes at various levels (Miller & Johnson, 2018). These platforms provide access to educational resources, support skill development through online courses, and mobilise a larger audience through campaigns and social media, urging decision-makers to adopt environmentally responsible actions.

Resource sharing and the promotion of circular economy principles are facilitated through digital networks (Brown & White, 2017). Members can exchange goods, share tools, and reduce overall consumption by reusing and repurposing items within eco-communities. Additionally, digital networks offer avenues for accessing environmentally friendly products and services (Jones et al., 2019). Online marketplaces dedicated to eco-friendly products support sustainable businesses and encourage conscious consumerism. IT technologies enable remote work, aligning with community values, and online platforms can serve as a space for crowdfunding and financial support for eco-friendly projects.

Geographical barriers are eradicated through digital networks, allowing individuals from different parts of the world to connect and collaborate on common eco-friendly goals (Smith & Green, 2020). This fosters cross-cultural exchange and the sharing of diverse perspectives and solutions. Digital networks also facilitate the organisation of virtual events, workshops, and webinars on sustainable practices, promoting continuous learning and skill development. Finally, in terms of monitoring and data collection, digital networks can be leveraged to gather and analyse environmental data (Robinson et al., 2019). IT tools assist in collecting data on energy usage, resource consumption, and environmental impact, guiding decision-making for greater community sustainability.

Certainly, there are challenges that eco-communities might face in introducing digital technologies into their agendas, as these require radical rethinking of the lifestyle and everyday habits of community members. Overreliance on technology could undermine the community's goal of self-sufficiency and low-impact living, and a delicate balance must be struck between utilising technology and maintaining the community's core values as described at the beginning of the chapter. Many IT technologies require electricity, which might conflict with the low-energy or off-the-grid principles of some sustainable communities.

The energy used to power IT infrastructure can counteract the community's efforts to reduce consumption, and thus a balance must be achieved. Introducing IT technologies assumes access to reliable internet and digital devices: the digital divide, where some community members might lack access due to different abilities or different age, could create inequalities within the community. Furthermore, integrating IT technologies into the community governance might raise concerns about data privacy and surveillance: sustainable communities often value privacy and autonomy, and adopting technology could pose risks to these principles.

Extremely relevant to this point is the assessment of the cultural and social impact of embedding IT technologies into the eco-community agenda. In fact, it might change social dynamics and interactions within the community: face-to-face interactions, which are proven to be essential and the largely favourite medium of communication, might diminish in favour of online communication, potentially affecting the community's cohesion.

In a recent article for *E-Flux* journal, Franco "Bifo" Berardi speculates on the transition from a status of social disorder to order, not yet completed, when the introduction of the digital began to challenge established norms (2023): "The change is indeed prepared for"; however, he concludes: it is not yet in its place. While Berardi's focus has been more on the cultural and social dimensions of digital culture and IT technologies, he has highlighted how capitalism's drive for endless growth is incompatible with ecological limits. He critiques the idea that technology alone can solve environmental problems, arguing for a deeper transformation of values and economic systems.

What is extremely relevant to the argument made here is that Berardi has explored the potential of digital technologies to foster alternative imaginaries and forms of social organisation and resistance. He has discussed the possibilities of digital activism, networked cultures, and collective intelligence as tools for imagining new futures and challenging dominant narratives (2009). This applies particularly well to the discussion around the potentials and challenges of digital technologies in eco-communities: to achieve digital social justice, it is crucial to address these and involve work towards a more equitable and inclusive digital environment.

Byung-Chul Han in "Psychopolitics: Neoliberalism and New Technologies of Power" highlights how contemporary society is characterised by a form of power that he calls "psychopolitics," which is a product of the neoliberal, capitalist era. Han builds on and critiques Foucault's notion of biopolitics, suggesting that in the current era, power operates through the exploitation of individuals' psyches by unseen global forces of capital, leading to a situation where even our unconscious fears and desires are bought and sold in the digital marketplace and the political arena. Han also provides an original theory of Big Data and a lucid phenomenology of emotion, offering surprising alternatives to the current situation and thus contributing to defining the set of social counternarratives discussed here (Han, 2017).

It follows that spatial practitioners, policymakers, and technologists should consider the social implications of technology and involve community members in conceiving, shaping, and testing digital tools and platforms that meet their needs and develop technologies that are consistent with the ethos and agenda of low-impact communities. Bridging the digital divide through policy measures and infrastructural investments is vital.

Conclusion: radical ecology as an interpretive spatial category

Based on the data collected in the three eco-communities discussed here, the potential of introducing digital networks into eco-communities lies in their ability to amplify collective efforts, foster collaboration, and leverage technology for positive environmental and social impact, sharing practices of what I formerly defined as "radical ecology." By harnessing the power of these networks, eco-communities can accelerate the transition towards a more sustainable and resilient future, where social, environmental, and economic aspects intertwine.

Aiming to outline alternative digital imaginaries is a way forward to encourage creative thinking about how technology can be harnessed for positive ends: this involves considering ethical and sustainable uses of technology, exploring new forms of digital interaction, or reimagining the relationships between humans and machines. As suggested by Donna Haraway, this implies the redefinition of boundaries and intersections of technology, biology, and culture towards a hybrid coexistence (1991) to challenge conventional relations and propose alternative ways of thinking about human–machine relationships.

Embedded within the fabric of contemporary architectural discourse, there is a discernible exploration of radical ecological concepts. Orr, echoing through the narratives of architectural practices, underscores the imperative for individuals engaged in design and planning within the architectural field to cultivate a profound understanding of ecological principles (Orr, 1991). Within the spatial design context, this resonates as a call for a re-evaluation, pushing for a more nuanced understanding of the integration of ecological considerations in both educational and professional spheres. A similar position has been taken by Richard Register, who passionately advocates for the reconstruction of human landscapes based on decentralised systems (Register, 2006).

As noted in the three eco-communities, and discussed in Lister's work, the importance of fostering social-ecological complexity within human-made landscapes is essential. In fact, her ethnographic research intricately intertwines human experiences with the environment (Lister, 2008) and provides an interpretive tool to assess the spatial unfolding of sustainable human behaviours, individually and collectively. Lister's contributions enrich the overarching dialogue on integrating ecological principles into community design, presenting a nuanced perspective within the ethnographic architectural research milieu, in line with the thematic underpinnings of radical ecology.

Moreover, as an interpretive analytical category, radical ecology unfurls as a powerful lens for architectural researchers to critically engage with and unravel established frameworks in the spatial discourse. It beckons scholars as well as practitioners to scrutinise the socio-political dimensions of ecological issues through immersive field studies and a deep reading of certain contexts. By incorporating radical ecology as an interpretive category, architectural research can reveal the intricate relationships between human activities and the environment within the lived experiences of diverse communities and their everyday geographies. This interpretive lens prompts a re-evaluation of spatial practices, urging a more holistic and transformative approach in community-based architectural research and design.

References

Adger, W. N., Social and Ecological Resilience: Are They Related? *Progress in Human Geography*, 24(3), pp. 347–364, 2000.

Berardi, F. "Bifo," *The Soul at Work: From Alienation to Autonomy*. Los Angeles: Semiotext(e), 2009.

Berardi, F. "Bifo," The Completion, *E-Flux Journal*, 137, 2023. www.e-flux.com/journal/137/544269/the-completion/

Bohle, H. G., Etzold, B., and Keck M., Resilience as Agency. *IHDP-Update 2 (2009)*: 8–13, 2009.

boyd, d., *It's Complicated: The Social Lives of Networked Teens*. New Haven, Connecticut: Yale University Press, 2014.

Brown, C., & White, L., Circular Economy Practices in Digital Networks: A Case Study of Resource Sharing in Eco-Communities, *Sustainable Technology Research*, 10(2), pp. 201–218, 2017.

Castells, M., *Networks of Outrage and Hope: Social Movements in the Internet Age*. Cambridge, UK: Polity, 2012.

Costanza-Chock, S., *Design Justice: Community-Led Practices to Build the Worlds We Need*. Cambridge, Massachusetts: MIT Press, 2020.

Dawson, J., *Ecovillages: New Frontiers for Sustainability*. Cambridge: UIT Cambridge, 2015.

de Kerckhove, D., *The Architecture of Intelligence*. Basel-Boston-Berlin: Birkhäuser, 2001.

Han, B.C., *Psychopolitics: Neoliberalism and New Technologies of Power*. London: Verso Books, 2017.

Haraway, D., *Simians, Cyborgs, and Women: The Reinvention of Nature*. London: Routledge, 1991.

Harridge-March, S. Can the building of trust overcome consumer perceived risk online? *Marketing Intelligence & Planning*, 24(7), pp. 746–761, 2026. https://doi.org/10.1108/02634500610711897

Hill, C., *The World Turned Upside Down: Radical Ideas During the English Revolution*. London: Penguin, 1991.

Jarvis, H., Sharing, Togetherness and Intentional Degrowth, *Progress in Human Geography*, 43(2), pp. 256–275, 2017.

Jones, P., et al., Eco-Friendly Product Access in Online Marketplaces: A Study of Sustainable Business Models, *Journal of Green Commerce*, 14(4), pp. 567–582, 2019.

Jones, A., & Brown, K., The Role of Online Platforms in Facilitating Horizontal Communication in Eco-Communities, *Environmental Communication*, 36(4), pp. 321–338, 2019.

Lévy, P., Collective Intelligence, A Civilization: Towards a Method of Positive Interpretation, *International Journal of Politics, Culture, and Society*, 18(3/4), 189–198, 2005.

Lister, N. M., *The Ecosystem Approach: Complexity, Uncertainty, and Managing for Sustainability*. New York: Columbia University Press, 2008.

Litfin, K. T., *Ecovillages: Lessons for Sustainable Community*. Cambridge: Polity Press, 2014.

Lulavy, M., Intentional Communities: Sustainable for You and the Environment. *Environmental Journalism*, 2019. https://envhumanities.sites.gettysburg.edu/environmental-journalism/intentional-communities-sustainable-for-you-and-the-environment/

Maldonado, T., *Lo Real Y Lo Virtual*. Barcelona: Gedisa Editorial, Colección Multimedia, 1994.

Marantz, Y., *What Is an Ecovillage?* Global Ecovillage Network, 2019. https://ecovillage.org/what-is-an-ecovillage/

Martell, L., *Alternative Societies: For a Pluralist Socialism*. Bristol: Bristol University Press, 2023.

Miller, R., & Johnson, T., Advocacy in the Digital Age: Harnessing Online Platforms for Environmental Policy Change, *Journal of Environmental Advocacy*, 42(1), pp. 89–104, 2018.

Neville, R., The Love Commune in Full Flowers. *Oz Magazine*, p. 9, 1968.

Orr, D., *Ecological Literacy: Education and the Transition to a Postmodern World*. New York: State University of New York Press, pp. 1–22, 1991.

Pickerill, J. M., Building the Commons in Eco-Communities. In S. Kirwan, L. Dawney, & J. Brigstocke (Eds.), *Space, Power and the Commons*. London: Routledge, 2015.

Register, R., *Ecocities: Rebuilding Cities in Balance with Nature*. Gabriola Island: New Society Publishers, 2006.

Robinson, M., et al., Data-Driven Decision-Making in Sustainable Communities: The Role of Digital Networks, *Environmental Data Analysis*, 22(3), pp. 245–260, 2019.

Robinson, M., & Green, S., Collaborative Initiatives for Environmental Solutions: Insights from Digital Networks, *Sustainable Communities Journal*, 18(3), pp. 45–62, 2021.
Rushkoff, D., *Team Human*. London-New York: Norton & Company Publisher, 2019.
Sargent, L. T., Theorizing Intentional Community in the 21st Century. In E. Ben-Rafael, Y. Oved, & M. Topel (Eds.), *The Communal Idea in the 21st Century*. Leiden: Brill, pp. 53–72, 2012.
Schneider, N., *Everything for Everyone: The Radical Tradition That Is Shaping the Next Economy*. New York: Bold Type Books, 2018.
Smith, J., et al., Digital Networks and Sustainable Living: A Community Perspective, *Journal of Sustainable Development*, 25(2), pp. 123–140, 2020.
Smith, M., & Green, A., Breaking Down Geographical Barriers: Global Collaboration in Eco-Friendly Initiatives, *International Journal of Environmental Connections*, 8(1), pp. 76–91, 2020.
Space Caviar, *Non-Extractive Architecture. On Designing without Depletion*. Berlin: Sternberg Press, 2021.
Starosielski, N., *The Undersea Network*. Durham and London: Duke University Press, 2015.
Ward, B., *The Home of Man*. New York: WW Norton & Company Inc., 1976.
Zuckerman, E., *Rewire: Digital Cosmopolitans in the Age of Connection*. New York: Norton & Company, 2013.

10 Measuring Walkability through Data-led Analysis and Spatial Modelling

Jennifer Wiles, Seraphim Alvanides, and Ayse Ozbil Torun corresponding author

Introduction

Recent decades have seen a significant decline in the number of children walking to school globally. In Germany, for example, there was a significant decrease in the number of children walking to school between 1990 and 2010 (Shaw et al. 2013). Similarly, considerable declines in walking to school, with corresponding increases in car use, have been observed in Norway, Denmark, and Finland (Fyhri et al. 2011). This trend is echoed in many other medium and high-income countries, throughout Europe and globally (Buehler et al. 2020; Carver et al. 2019; Smith et al. 2020). In 2019, 47% of primary school children in England usually walked to school (Department for Transport 2020). The rate has fluctuated around the 50% mark for the last 15 years, but in the 1970s, rates were significantly higher, at over 70% (Living Streets 2016). As a result, the UK government released its first Cycling and Walking Investment Strategy, which includes a target to 'increase the percentage of children aged 5 to 10 who usually walk to school from 49% in 2014 to 55% in 2025' (Department for Transport 2017).

Active school travel (AST) offers a promising opportunity to increase daily physical activity (PA) among children. Additionally, increasing the rates of walking to school (and reducing car journeys) has the potential to positively impact UK and international priorities, including air quality and pollution (Ahern et al. 2017), obesity (Lorenc et al. 2008), mental health (Sport England 2018), and congestion (National Institute for Health and Care Excellence 2012). As city populations continue to grow, these issues are of particular concern in more urban areas. Hence, it is important to better understand the factors affecting AST behaviours.

A wide range of factors affect the choice to walk to school, and individual behaviours are the result of a complex interaction between these factors (Mitra 2013). Despite the breadth of literature on the school journey, a key research limitation in the UK, and particularly in England, is a lack of data on school travel modes, especially at finer-grained geographical scales. By investigating the associations of street network design and school social characteristics with rates of walking to primary schools in England, this study aims to provide insight into incorporating spatial and other datasets into AST studies. The focus is on the extent to which urban knowledge can be assembled and explored through digitalisation to

DOI: 10.1201/9781003380559-10

provide evidence-based guidance for informing spatial policies and interventions. It introduces a data-led methodology, including spatial modelling, that might be used and replicated to initiate a modal shift in schools, such as identifying schools with the greatest potential for increased walking. In addition to contributing to the literature, this study seeks to build on existing evidence, to move forward the academic understanding of how street network design influences rates of walking to school, alongside other factors such as socio-economic status. It can also help to inform policy relating to school allocation, planning, and urban design, for example, by offering recommendations about which street network layouts are most supportive of walking to school.

Correlates of active school travel

There has been increasing research interest in children's mobility, across a variety of fields including medical, social science, and engineering (Rojas Lopez and Wong 2017). Distance is generally agreed to be a key influencer on choice of mode (Easton and Ferrari 2015; Ahern et al. 2017). However, an extensive review of academic articles relating to active travel (walking and cycling) to school found that the reasons for choosing to use a particular mode are complex and affected by a variety of environmental, social, economic, and geographical factors (Rojas Lopez and Wong 2017). Several detailed systematic reviews have been undertaken by others, including Pont et al. (2009), Mitra (2013), Rojas Lopez and Wong (2017), and Ikeda et al. (2018), and in this section we summarise the important findings relating to the aim of our study.

Studies that looked at the impact of 'urbanness' on rates of walking/active travel generally found a positive relationship, resulting from higher population densities and therefore shorter average distances to school (Curtis et al. 2015; Murtagh et al. 2016; Ginja et al. 2018). The broad distinction between urban and rural is a useful indicator of whether children are likely to walk to school, but there are other factors involved. A growing body of research has demonstrated that urban form characteristics (such as density and land-use) of children's neighbourhoods can have both positive and negative impacts on their walking rates. For instance, Lin and Chang (2010) found that higher density correlated with higher rates of walking to school. While Kerr et al. (2006) and Pont et al. (2009) did find a positive relationship between land-use mix and rates of walking/active travel to school, Su et al. (2013) and Broberg and Sarjala (2015) found the opposite. Despite a wealth of studies on the relationships between density, land-use, and walking rates, little is known about the relationship between street network configuration and walk-to-school rates.

Several studies have identified a positive association between active travel to school and street connectivity, as a well-connected street network provides direct routes to school from a variety of residential locations (Helbich et al. 2016; Ozbil et al. 2016; Rojas Lopez and Wong 2017; Ikeda et al. 2018; Leung and Loo 2020). On the other hand, some others found that better connected streets resulted in lower rates of walking to school because these streets were often more heavily trafficked and required more crossings (Timperio et al. 2006; Su et al. 2013). The fact that standard measures of connectivity, such as the total street length per area (Ewing et al. 2004; Helbich et al. 2016) and a pedestrian network area, or 'Pedshed' (Giles-Corti et al. 2011), identify the density of streets within an area but fall short in evaluating the structural qualities of these streets is a significant underlying reason for these inconsistent results.

Structural qualities are related to both the density and alignment of streets that have the potential to influence movement between locations. Syntactic measures, as defined by space syntax methodology (Hillier and Hanson 1984), are useful in capturing the various structural qualities of street segments that make up a street at different scales (e.g., local versus global). Prior studies have revealed that syntactic connectivity measures may be more effective than standard measures of connectivity in measuring walkability at different scales (e.g., city, neighbourhood, street) (Ortegon-Sanchez et al. 2021; Zhao et al. 2022). While their application in studies of school travel to date is limited, Ozbil et al. (2021) demonstrated that several space syntax metrics corresponded to children's AST. Hence, the space syntax methodology offers a useful tool to provide policymakers and practitioners with data-led, evidence-based spatial models and maps, identifying specific streets to invest in traffic reduction measures (e.g., installing new pavements and walkways, pedestrianisation schemes), which in turn would encourage a shift to more AST.

An additionally important consideration for understanding children's AST are socio-economic factors. The school neighbourhood location has been found to influence active travel rates (Jarvis and Alvanides 2008), with schools located in more deprived areas experiencing higher active travel rates (Cutumisu et al. 2014; Ikeda et al. 2018). The socio-economic profile of an area is captured by the aggregate social and economic conditions of households, which influence walking/active travel in many ways. Eligibility for free school meals is a household-specific indicator that can be used as a proxy for deprivation (Easton and Ferrari 2015). According to Su et al. (2013), schools with a larger number of children receiving school meal subsidies had a higher proportion of children walking to school. This was despite greater environmental and safety concerns in these lower-income communities, suggesting that socio-economic status is one of the more important factors in determining school travel mode. The causal relationships involving income and deprivation are complex. For example, it might be argued that lower-income families are also more likely to live in dense, urban areas, as found by Giuliano and Narayan (2003) for the USA and would therefore have a shorter distance to travel to school. On the other hand, larger European cities with gentrified central neighbourhoods may have displaced less affluent households to outer city suburbs (Atkinson 2004; Brade et al. 2009) with reduced school provision, resulting in longer distances to school.

While there is considerable evidence and current research activity around the socio-economic influences on school travel behaviour, the impact of urban form, in general, and street network design, in particular, is less well understood. Overall, some studies have found that urban form explains very little modal variation (e.g., Easton and Ferrari 2015), whereas others have found it to be critically important, even overcoming the influence of distance in a recent urban study (Helbich et al. 2016). These mixed findings highlight the complexity of relationships between different aspects of urban form and their relative influence on whether children walk to school. Hence, further clarity is required about how specific urban form attributes impact AST.

This study focuses on a specific aspect of the urban form, namely the street network design around schools, to address this gap in research. Using a robust data-led approach and integrating important socio-economic attributes and street connectivity measures (both standard and syntactic), this study aims to statistically determine (i) the extent to which street network connectivity is associated with AST, and (ii) the relative significance of standard and syntactic measures in explaining active school travel behaviour.

Methods and data

The conceptual framework of the study is presented in Figure 10.1, bringing together a model proposed by McMillan (2005) with a wider range of variables recognised by Mitra (2013) and elements of the behavioural model used by Ginja et al. (2018). The framework illustrates the significance of the key influencing variables on walking to school and shows how these variables are all filtered by the potential impact they may have on parental perceptions of walking to school, ultimately shaping the AST behaviour of their children. The experience of walking to school then generates a feedback loop, which interacts with the parents' perceptions of other factors, to decide if they will walk again. Walking to school may become a habit and part of the daily routine, which will not require active decision-making every time. Families may change their mode of travel depending on variations in

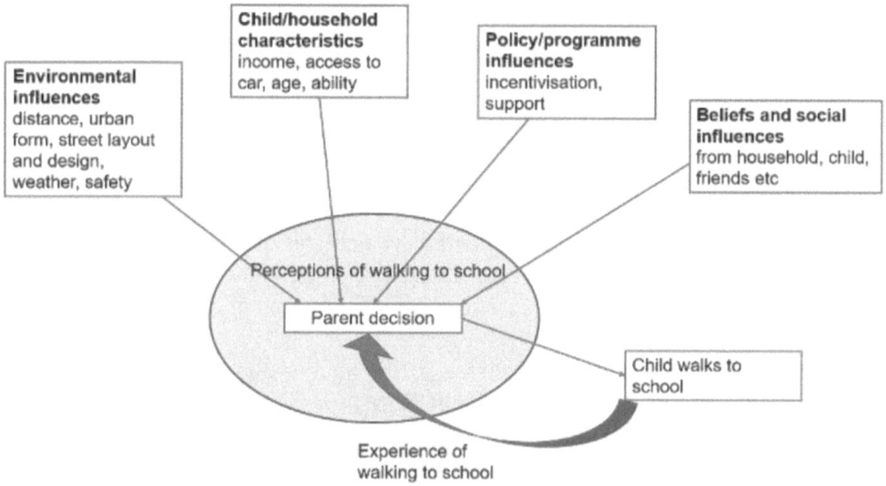

Figure 10.1 Conceptual framework of this study.

the influencing factors, which could be regular (such as walking every Wednesday when the parent has their day off work), occasional (e.g., taking the car if there is heavy rain), or transitional (e.g., moving house). If the school becomes closer to where people live, this change in return can indicate a shift from driving to walking.

Aside from the methodological advances offered by this study, it seeks to add to the emerging research field exploring the association between syntactic street connectivity measures and walking to school. This research develops statistical models to estimate AST rates in two UK settings using syntactic connectivity measures derived through spatial modelling combined with a new school travel data set provided by Living Streets. This combination of new approaches and new data offers a novel perspective on how street network design may interact with other factors to influence rates of walking to school in the UK.

Study area selection and data sources

This study was made possible by Living Streets providing secure access to data collected through its online 'Travel Tracker' (TT) tool. Living Streets, the UK charity for everyday walking, runs WOW, a year-round walk-to-school challenge for primary schools, where students record their mode of travel each day using an online TT and receive badge incentives for regular walking (Living Streets 2023). Living Streets' TT data presents an opportunity to conduct a robust study into how different factors might influence school travel mode choice. To date, limited research has been undertaken using this relatively new data source.

The analysis focused on two large study areas in England: the local authority of County Durham and the combined authority of Greater Manchester (GM). These were selected as two contrasting locations, which would enable interesting comparisons of findings. County Durham is classed as 'largely rural' (Department for Environment Food and Rural Affairs 2014), including a few hub towns and a small city (~65,500 inhabitants). On the other hand, all ten local authorities within Greater Manchester are classed as 'urban with a major conurbation' (ibid.). GM is about half the geographical size of Durham but has over five times the population (see Table 10.1). These study areas were also selected due to both having reasonably high numbers of schools taking part in WOW (Living Streets 2018), and as a result, TT mode of travel data for sufficient numbers of schools to undertake meaningful analysis.

The research method consists of quantitative analysis of publicly available secondary data sources, with access also granted to unpublished Living Streets TT data (Living Streets 2018). Table 10.2 lists the key features of the data sources that were used in this research.

The number of students was included in the analysis as a measure of school size. The percentage of Free School Meals (FSM) for every school and the IDACI (Income Deprivation Affecting Children Index) score of the area where a school is located were used as proxy measures of levels of deprivation/socio-economic status. 'Urban Rural' Code was selected as a measure of 'urbanness'. Additional geospatial data was required to undertake the street network analyses for Durham

Table 10.1 Key statistics for County Durham and Greater Manchester

Statistic	*Data source*	*County Durham*	*Greater Manchester*
Urban–rural character	Department for Environment Food & Rural Affairs (2016)	Largely rural (rural including hub towns)	Urban with a major conurbation
Geographical area	Office for National Statistics (2011)	2,230 km^2	1,277 km^2
Resident population (mid-year 2017)	Office for National Statistics (2018)	523,700	2,798,800
Population density	Calculated	235 people/km^2	2,192 people/km^2
Number of schools taking part in WOW	Living Streets (2018)	31	114

Table 10.2 Data sets, sources, and key features

Data set	*Year*	*Source*	*Key features*
WOW TT data	2017–18	Living Streets (2018)	• School-level mode data for travel to school (student self-reported using online tool) • Mode options: bike, bus, car, 'hop off', other, 'park and stride', rail, scooter, and walk
Number of students, percentage of students with Free School Meals (percentage FSM)	2018	Department for Education (2018)	• School data held by the GIAS service
Income Deprivation Affecting Children Index (IDACI)	2015	Department for Communities and Local Government (2015)	• Measures the proportion of children aged 0 to 15 living in income deprived families • Score used for school postcode
Urban Rural Code	2011	Department for Environment Food & Rural Affairs (2018) data cited in Department for Education (2018)	• Data based on the Rural–Urban Classification of the school Census Output

and GM. To create a 'walkable street network', open-source Ordnance Survey Open Roads (2018) data was used. However, we discovered that this dataset had limitations because it only includes roadways. Since many school-walking routes include, or have the potential to include, off-road trails, Ordnance Survey MasterMap Highways Network-Paths (2018), an open-source dataset that covers the urban path network, was merged with the previous road network data to develop a complete street network. Roads without footways (e.g., motorways, A roads) were then removed to obtain a 'walkable street network'. Finally, Ordnance Survey 1:50,000 Raster (2018) tiles were used to provide a background and reference point for the street network, and Ordnance Survey Boundary-Line™ (2018) data was used to map the Durham and GM area borders.

Outcome variable: walk-to-school rates

School-level data for the academic year 2017–18, consisting of school name, postcode, baseline, and total recorded trips by mode, were filtered and analysed. Schools that did not meet the research or validity criteria, such as schools with no or low numbers of TT recordings, were removed from the dataset. A threshold of at least 1000 trips recorded, which is equivalent to at least 5 students recording every school day at 190 days/year (Long 2023) or 26 students recording once a week, was used in further analysis. This ensured that schools that recorded data less frequently or those that only participated for part of the year remain included, but schools that only recorded a few journeys were excluded to avoid skewness.

The TT dataset was subsequently joined with school data held by the 'Get Information About Schools' (GIAS) service and any duplicate schools were removed. This resulted in a database that included TT data and GIAS data for each of the WOW schools. Indices of Deprivation 2015 – the most relevant data available for the study period – for the school locations were obtained using the 'IMD by postcode' tool (Department for Communities and Local Government 2015) and added to the school dataset. Data on percentage FSM was unavailable for five of the GM schools, and therefore analyses including this variable could only be undertaken for 109 (of the 114) GM schools.

Street network connectivity

To investigate the relationship between street network design and walk-to-school rates, standard and syntactic street connectivity measures were calculated around the sample schools in the study areas using the open-source QGIS. Based on findings of previous studies and conventional wisdom among planners regarding an acceptable walking distance (Timperio et al. 2006; Davis and Huxford 2009; Mitchell and Bendixson 2015), 800m circular buffers were added around each school location, reflecting half a mile or approximately a 10-minute walk.

Within each school buffer, the road, path, and total (road plus path) lengths were estimated to represent standard connectivity measures. These measures represent the availability of streets within walking distance of each school, and they serve as important input for analysis as well as comparators for the structural measures offered by syntactic methods. We utilised segment-based and angular segment analysis of the walking network using Depthmap (Turner and Friedrich 2010). Connectivity measures the number of spaces (street segments) intersecting each space within the network; hence, it is a local measure. Segment Angular Integration measures 'how accessible each space is from all the others within the radius using the least angle measure of distance', while Segment Angular Choice measures 'the extent to which a node is located in between the paths connecting all pairs of origins and destinations' (Hillier and Iida 2005). Therefore, Integration is related to the destination selection and Choice is related to the route selection elements of a trip (Hillier et al. 2012). Both measures were computed for a range of metric and angular radii, including global (radius:n, evaluating all the segments of the network starting from the midpoint of each segment) and local (radius:3, evaluating up to three 'steps' away from each segment within the network, and metric radius 800m, measuring angular turns calculated within an 800m metric radius of each segment in the network). An 800m radius was selected as it relates to the school buffer size. Figure 10.2 visualises these measures computed around a school environment in Durham, UK.

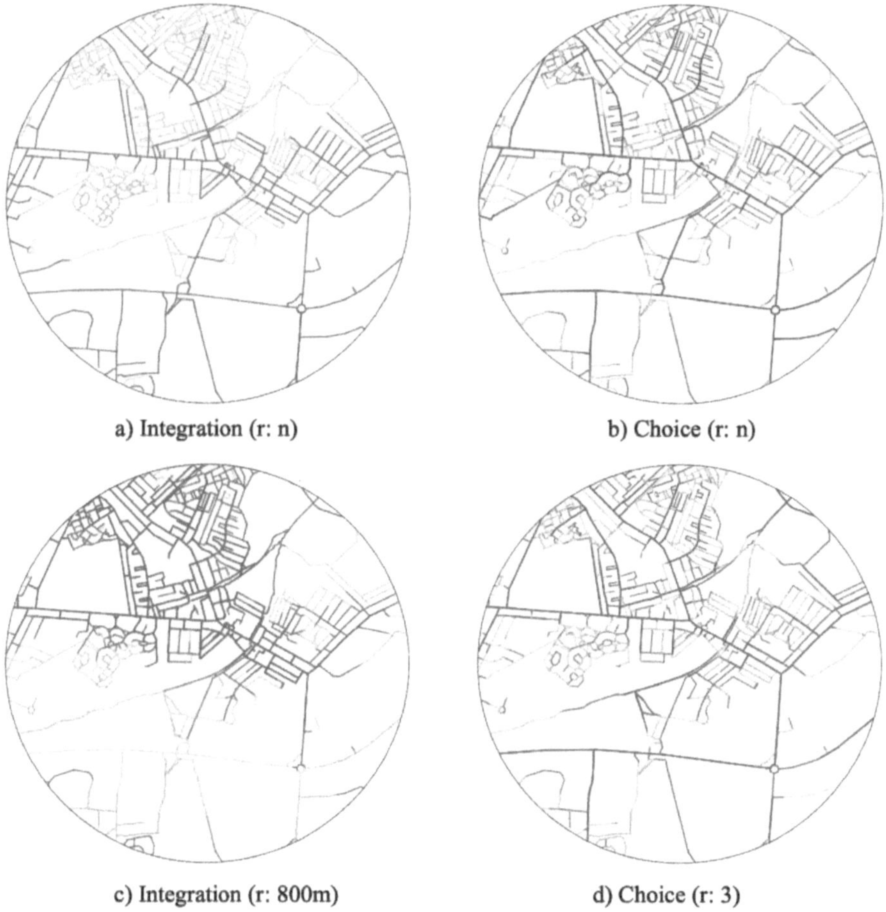

Figure 10.2 Street network around a school environment in Durham (UK) analysed using (a) global Integration (radius: n), (b) global Choice (radius: n), (c) Integration (800m radius), and (d) local Choice (radius: 3). The colours range from red, indicating relatively higher values, to blue, relatively lower values.

Analysis

Bivariate correlations were initially undertaken between each of the independent variables (socio-economic and street connectivity measures) to identify multicollinearity among the variables. Multivariate regression models were then developed to explain the variation in walking rates based on these socio-economic and street network measures. Variables with high correlations (> 0.7) – a common rule-of-thumb validated by Dormann et al. (2013) – were not included simultaneously in the same model to avoid the invalidation of the results. Instead, several models were developed considering these multicollinear measures separately.

Results

Preliminary analysis

Figure 10.3 shows the percentage mode share for trips for the studied Durham and GM schools. There is a higher proportion of walking journeys in GM (57.2%) compared to Durham (45.7%), which is parallel to the higher proportion of car journeys and 'park and stride' (where the journey is partly driven and partly walked) journeys in Durham, all other modes being roughly equal and in very low numbers.

Table 10.3 demonstrates that the only significant positive association for Durham in the predicted direction was between the percentage walking and the percentage FSM ($\beta = .574$, $p < .01$). In GM significant positive correlations were found between percentage walking and percentage FSM ($\beta = .557$, $p < .01$) and IDACI score ($\beta = .521$, $p < .01$), both in the expected direction. In addition, there were weaker positive correlations between walking and AHAH index score ($\beta = .279$, $p < .01$) and to a lesser degree with the number of students ($\beta = .213$, $p < .01$). This difference between GM and Durham could be partly due to GM being a much larger data set (n = 109–114) than Durham (n = 31), thus resulting in stronger relationships. An urban–rural comparison could not be conducted for GM, as all but one of the schools fall into the 'Urban: Major conurbation' category. A finer-grained analysis is therefore required to better understand the impact of the urban environment on walking rates in GM schools. It is also notable that in both study areas, the gender composition does not correlate with walking percentages, at least at the aggregate school level.

No significant correlations were found between the walk-to-school rates for the full Durham dataset and the standard and syntactic street connectivity measures. However, Table 10.3 shows that when Durham schools are split into two sub-categories, urban (n = 15) and rural (n = 16), significant correlations emerge between walking rates and several socio-economic and street connectivity measures only for the rural subset of 16 schools.

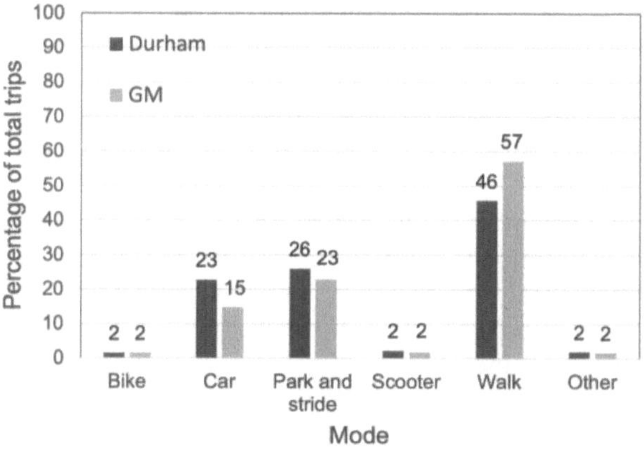

Figure 10.3 Percentage mode share for total trips across 31 Durham schools (N = 36,638) and 114 GM schools (N = 2,326,603). Only modes with trips above 1% are shown.

Table 10.3 Pearson's correlations between independent variables and walk-to-school rates for Durham and GM schools

Variable	*β (number of schools)*		
	Durham		*Greater Manchester*
	Urban (N=15)	*Rural (N=16)*	*(N=109)*
IDACI Score	0.090	0.568*	**0.521****
Percentage FSM	0.492	**0.807****	**0.557****
Path length	0.152	0.398	**0.382****
Road length	−0.044	0.547*	**0.450****
Total length	0.035	0.531*	**0.508****
Connectivity	0.025	0.558*	**0.545****
Integration (r:n)	0.132	0.505*	0.204*
Choice (r:n)	0.303	−0.212	0.124
Integration (r:3)	0.168	0.264	**0.343****
Choice (r:3)	0.218	0.178	**0.257****
Integration (r:800m)	−0.090	0.410	**0.531****
Choice (r:800m)	−0.210	0.410	**0.446****

*p < 0.05; **p < 0.01

Multiple regression analysis was therefore only conducted for this subset to explain variation in AST rates. The percentage of students walking to school correlates significantly with all standard and local syntactic street connectivity measures in GM schools. Correlations with global syntactic street measures were less pronounced, with only global Integration (r:n) appearing to be a significant correlate.

Overall, these results are useful for understanding the strength and direction of the covariation in the outcome variable (walk-mode share) and the independent variables (street connectivity measures). The independent effects of these features are analysed in multivariate regression models in the following section.

Linear models

Multivariate statistics were employed, as has been typical in studies of travel demand, to investigate the connection between street network design and walking to school. Multiple regression analysis allows for the simultaneous inclusion of several independent variables to derive at causal explanations while introducing control variables. Thus, it provides the researcher with definitive statements about the observed relationship while revealing the relative effect sizes of each independent variable.

The empirical models developed here are based on the hypothesis that environments that are connected to support various forms of walking also support AST behaviour by motivating the decision to walk for the school journey. To assess the comparative relevance of street network measures in terms of their relative impact on AST rates while adjusting for socio-economic status, the focus of this study is not so much on improving predictability by developing the best model.

Tables 10.4 and 10.5 summarise the results of regression models for Durham and Greater Manchester respectively. The IDACI score variable was entered into the regression first to allow for the evaluation of socio-economic status in context relative to other variables affecting AST behaviour. Street connectivity measures were then added into the model respectively, first the standard connectivity measure followed by the syntactic measures, to demonstrate the effect of adding each to the model.

Models for rural Durham AST rates (Tables 10.4a and 10.4b) were found to estimate between 27% and 63% of the variation in walk-to-school rates. Table 4a shows models using IDACI score as an indicator of socio-economic status. By

Table 10.4a Multivariate regression models, including IDACI score, Road length and global syntactic measures to explain variation in walking rates between rural schools in Durham

	β	t	$Std\,\beta$	β	t	$Std\,\beta$	β	t	$Std\,\beta$
constant	*35.89*	*6.25*		*31.3*	*4.23*		*−0.53*	*−0.04*	
IDACI Score	65.73*	2.58*	0.57*	42.09	1.2	0.36	30.15	1.09	0.26
Road length				0.00	0.99	0.30	0.00	1.53	0.36
Integration (r:n)							**0.00****	**3.17****	**0.56****
Choice (r:n)							0.00	−1.19	−0.21
N	16			16			16		
R^2	0.32*			0.37*			0.68*		
Adjusted R^2	0.27*			0.27*			0.56*		

*$p < 0.05$; **$p < 0.01$

Table 10.4b Multivariate regression models, including percentage FSM, total length, and global syntactic measures to explain variation in walking rates between rural schools in Durham

	β	*t*	*Std β*	*β*	*t*	*Std β*	*β*	*t*	*Std β*	*β*	*t*	*Std β*
constant	*32.63*	*8.76*		*31.96*	*6.72*		*16.55*	*1.54*		*57.71*	*0.72*	
Percent FSM	**1.10****	**5.12****	**0.81****	**1.06****	**3.73****	**0.78****	**0.95***	**2.86***	**0.70***	**0.91****	**3.14****	**0.67****
Total length				0.00	0.24	0.05	0.00	0.36	0.07	0.00	0.05	0.01
Integration (r:n)							0.00	1.35	0.25	0.00	1.43	0.25
Choice (r:n)							0.00	0.26	0.05			
N	16			16			16			16		
R^2	0.65**			0.65**			0.72**			0.73**		
Adjusted R^2	**0.63****			**0.60****			**0.62****			**0.63****		

*p < 0.05; **p < 0.01

itself, IDACI score was found to explain just over a quarter of the variation in walk-to-school rates (adj $R^2 = .27$, $p < .05$). When standard connectivity measures (road length as reported or path length and/or total length, both not reported) were added to the model, the results were not impacted significantly. However, when global syntactic measures were included, the predictive power of the model increased considerably, explaining approximately half of the variation in walking rates (adj $R^2 = .56$, $p < .05$). In fact, global Integration proved to be as significant as IDACI score in estimating the variation in AST rates (std $\beta = .56$, $p < .01$).

When percentage FSM was used as a proxy of socio-economic status (Table 4b), it was found to be twice as powerful as IDACI score in explaining the variation in walk-to-school rates (adj $R^2 = .63$, $p < .01$) in Durham. Percentage FSM was significant across all models, which produced a higher predictive power than previous models using IDACI score. Local connectivity measures were not significant in any of the models; therefore, these are not reported here.

Multivariate regression models for GM walk-to-school rates are shown in Tables 10.5a and 10.5b. Models were found to explain between 27% and 40% of the variation in walk-to-school rates. IDACI score was found to explain just over a quarter of the variation in walk-to-school rates in GM (adj $R^2 = .27$, $p < .01$). In contrast to the previous models developed for Durham, when the standard street connectivity measure (i.e., Path length) was added to the model, its explanatory power increased, albeit marginally (adj R^2 change = 2–3%, $p < .01$).

On the other hand, the inclusion of syntactic connectivity measures Connectivity and local Integration (radius 800m) resulted in a moderate increase of 6–8% ($p < .01$) in the explanatory power of the models. In fact, local street integration appeared to be as significantly associated with walk-to-school rates in GM as socio-economic status (std $\beta = .34$, $p < .01$). Global syntactic measures and Choice, on the contrary, did not appear to be significant variables in any models, and hence they are not reported here.

These results demonstrate that the density of streets over a given area, as measured by the standard connectivity metric Path length, is significantly associated with the choice to walk to school. However, more refined syntactic measures that consider the spatial structure of the urban grid were stronger correlates of AST rates. The standardised coefficient (std β) for syntactic measures (global Integration and local Integration in the case of Durham and GM, respectively) is positive and statistically significant in all models. This suggests that even after controlling for street density and socio-economic status, increased street network integration around the schools acts as a significant inducement to students' propensity to walk to school.

Discussion

In this study, linear models were developed to investigate the extent to which street network connectivity is associated with school walk-mode shares while controlling for socio-economic status. Durham and Greater Manchester (GM) were selected as contrasting study areas for exploring this association. As expected, the more

Table 10.5a Multivariate regression models, including IDACI score, Path length, and local syntactic measures to explain variation in walking rates between schools in GM

	B	*t*	*Std β*	*β*	*t*	*Std β*	*β*	*t*	*Std β*	*β*	*t*	*Std β*
constant	*45.82*	*24.13*		*41.44*	*15.29*		*−28.40*	*−0.89*		*34.32*	*10.42*	
IDACI Score	**44.10****	**6.46****	**0.52****	**37.06****	**5.00****	**0.44****	**24.48****	**2.96****	**0.29****	**28.30****	**3.77****	**0.33****
Path length				0.00*	2.23*	0.20*	0.00	1.31	0.12	0.00	0.45	0.04
Connectivity							26.03*	2.02*	0.28*			
Integration (r:3)							0.01	0.48	0.05			
Integration (r:800m)										**0.06****	**3.49****	**0.34****
N	114			114			114			114		
R^2	0.27**			0.30**			0.37**			0.37**		
Adjusted R^2	**0.27****			**0.29****			**0.35****			**0.36****		

*p < 0.05; **p < 0.01

Table 10.5b Multivariate regression models, including IDACI score, Path length, and local syntactic measures to explain variation in walking rates between schools in GM

	β	*t*	*Std β*	*β*	*t*	*Std β*	*β*	*t*	*Std β*	*β*	*t*	*Std β*
constant	*43.44*	*21.23*		*39.16*	*14.22*		*−33.38*	*−1.08*		*32.16*	*9.78*	
Percent FSM	**0.67****	**6.94****	**0.56****	**0.57****	**5.40****	**0.47****	**0.38****	**3.35****	**0.32****	**0.45****	**4.25****	**0.37****
Path length				0.00*	2.27*	0.20*	0.00	1.43	0.13	0.00	0.53	0.05
Connectivity							27.11*	2.18*	0.29*			
Integration (r:3)							0.01	0.47	0.05			
Integration (r:800m)										**0.06****	**3.52****	**0.34****
N	109			109			109			109		
R^2	0.31**			0.34**			0.41**			0.41**		
Adjusted R^2	**0.30****			**0.33****			**0.39****			**0.40****		

$*p < 0.05$; $**p < 0.01$

densely populated, urban city-region of GM had higher mean walk-to-school rates (56%) than the more rural, sparsely populated county of Durham (47%).

Overall, the analyses presented here demonstrate that street network design around schools is significantly related to children's choice to walk to/from school, as measured through the percentage of walk-to-school trips. The study's findings provide specific support for two primary conclusions that have important implications for urban planning and urban design decisions targeted at increasing active school commuting and decreasing automobile dependence among children, as discussed below.

Socio-economic status as a determinant of AST behaviour

In GM significant correlations were observed between percentage walking and several socio-economic variables, such as percentage FSM and IDACI score. Only percentage FSM was strongly linked with walking rates in Durham. Percentage FSM is also a more representative measure of the school population, since it relates to individual students, whereas the IDACI score used here is based on the average deprivation experienced by children in the area where the school is located. It may be that, if student home postcodes were known, the mean IDACI score for their home locations would provide comparable results to percentage FSM. When the Durham school dataset was separated into urban and rural subgroups, however, AST rates at rural schools showed significant associations with socio-economic status, as measured by both percentage FSM and IDACI score. Further investigation of these relationships using multiple regression analysis revealed that socio-economic status alone explained nearly two-thirds of the variation in walk-to-school rates in Durham and one-third of the variation in AST rates in GM.

Street network design as a significant correlate of AST behaviour

The results also indicated that local street connectivity around schools (within 800m buffers) can play as significant a role as socio-economic factors in explaining the variation in school walk-mode shares. Street connectivity was measured using standard planning measures (path, road, and total street length) and more refined road-segment measures derived from space syntax (Connectivity, Integration, and Choice). Increased density of streets within school areas is associated with increased AST rates to a certain extent, only in the case of GM. However, the linear models developed also underscored the significance of the spatial structure of street networks, particularly the way in which streets are aligned to create a directional hierarchy.

Directional accessibility, measured through local and global Integration, appeared to be as important as metric accessibility, measured through the standard measures of Road and Path length, in affecting the proportion of students walking to school. Importantly, local standard and syntactic street connectivity measures were far more influential in explaining the variation in walk-to-school rates in the GM models than global syntactic measures, implying that the local street network around the school is critical in densely populated urban areas like GM. These findings support the findings of limited studies around the relationship between the local street network design and AST behaviour (Michail et al. 2022).

Limitations and strengths

Notwithstanding the contributions of this research, the limitations of this study should be acknowledged. One concern is that due to the limited available data, certain potentially important variables were not expressly considered in the analyses. For example, it can be challenging to distinguish between the effects of self-selection and street network design as parents with a positive attitude towards active travel will choose to reside in neighbourhoods supportive of these preferences.

Self-selection has been found to be strongly correlated with travel behaviour in previous studies (Kitamura et al. 1997; Krizek 2003). Through the inclusion of socio-economic indicators that are expected to affect preferences, the study partially controlled individual and household preferences. Future research that incorporates children/parental surveys could reach more conclusive results regarding the potential causal link between local street network layout and AST behaviour.

Another constraining factor is the lack of data on children's home addresses and walking routes to school. It is possible that, due to the largely rural nature of County Durham, many students travel long distances to school, regardless of whether the school is located in an urban or rural area. Also, the relatively small size of the urban and rural subsets in this study may explain the lack of more significant findings.

Additionally, due to the cross-sectional nature of the data used in this study, causal correlations could not be established. Hence, these findings are not meant to be used to draw generalisable conclusions, but rather to contribute to the future development of school environments supportive of AST. Extending the case studies would be a logical next step to determine whether these findings are replicated in other locations. Despite these limitations, the findings presented here support the argument that environments are not isotropic. Some are more supportive of walking due to an increased density of interfaces generated by connectivity patterns and the local spatial structure of street networks.

Conclusions

Aside from theoretical development, this research has further practical implications. The findings corroborate the argument that well-structured and connected street networks promote AST behaviour, particularly in primary schools in England, though they may be extended to other contexts. These findings suggest that planning policy should have an emphasis on siting new schools in densely populated locations where distances to school are walkable, and where street layouts have been designed to create a dense, well-connected, and walkable network.

The same principles should be applied when building new neighbourhood developments. In existing neighbourhoods, towns, and cities, the walkable street network around schools could be densified for example by installing footways on streets that lack them (Ewing et al. 2004) or by opening gated alleyways between rows of terraced houses, while ensuring that these are safe walking spaces. Hence, if the UK government is to fulfil its target of 'increasing the percentage of children aged 5 to 10 who usually walk to school from 49% in 2014 to 55% in 2025' (Department for Transport 2017), national policy regarding street network design around schools should be revised considering this evidence.

The study's findings are best understood in the context of the larger body of research on the syntactic analysis of urban environments, since they may have an impact on street classification systems, particularly as they inform the design of street sections. Given the association between street connectivity and walk-to-school rates, for example, widening the sidewalks on more prominent streets is expected to improve safety and accessibility, thereby maximising the walkability potential of an area's street network. In addition, street management approaches, such as reducing traffic speeds and preventing 'pavement parking', should be a higher priority in locations that have denser street intersections and a clearer internal hierarchy of access based on directional distances.

The findings of this study can also be used to inform the activities of organisations seeking to make the best use of resources to enable more children to walk to school. For schools with a well-connected local street network but lower-than-expected walking rates, for example, low-quality street-level design (e.g., narrow sidewalks, lack of green streets) may be the key barrier to walking, which can be alleviated through modifications to the physical environment. In school locations that have a well-connected street network and high-quality street-level design, however, promoting initiatives such as 'park and stride' may be a preferable strategy to support active commuting to school.

Increased rates of walking to school (and decreased car travel) have the potential to contribute to international priorities such as air quality and pollution, PA, obesity, and mental health. Continued attention to this research field is critical as our population expands and the world becomes more urban. Linear models developed revealed that in addition to socio-economic indicators, standard and syntactic street connectivity measures had a considerable influence on walk-mode shares in both urban and rural areas. Applying a data-led approach and spatial modelling, this study extends the current evidence base by examining the associations of street network design and school social characteristics with rates of walking to primary schools in urban and rural areas of England.

Overall, from a theoretical viewpoint, the data-driven methodological framework provided by this study could be replicated by others to initiate a model shift in schools, such as by identifying schools with the greatest potential for increasing walking. It is recommended that standard measures that evaluate the density of street networks and syntactic measures that assess the structural qualities of street networks are considered in AST studies. From a policy standpoint, the study's findings should be recognised and used by local stakeholders in similar locations globally to inform their decision-making in creating child-friendly environments that promote children to walk to school.

References

Ahern, Sara M, Bronia Arnott, Tim Chatterton, Audrey de Nazelle, Ian Kellar, and Rosemary RC McEachan. 'Understanding parents' school travel choices: A qualitative study using the Theoretical Domains Framework'. *Journal of Transport & Health* 4: 278–293, 2017.

Atkinson, Rowland. 'The evidence on the impact of gentrification: New lessons for the urban renaissance?' *European Journal of Housing Policy* 4, no. 1: 107–131, 2004.

Brade, Isolde, Günter Herfert, and Karin Wiest. 'Recent trends and future prospects of socio-spatial differentiation in urban regions of Central and Eastern Europe: A lull before the storm?' *Cities* 26, no. 5: 233–244, 2009.

Broberg, Anna, and Satu Sarjala. 'School travel mode choice and the characteristics of the urban built environment: The case of Helsinki, Finland'. *Transport Policy* 37: 1–10, 2015.

Buehler, Ralph, John Pucher, and Adrian Bauman. 'Physical activity from walking and cycling for daily travel in the United States, 2001–2017: Demographic, socioeconomic, and geographic variation'. *Journal of Transport & Health* 16: 100811, 2020.

Carver, Alison, Alison Barr, Ankur Singh, Hannah Badland, Suzanne Mavoa, and Rebecca Bentley. 'How are the built environment and household travel characteristics associated with children's active transport in Melbourne, Australia?' *Journal of Transport & Health* 12: 115–129, 2019.

Curtis, Carey, Courtney Babb, and Doina Olaru. 'Built environment and children's travel to school'. *Transport Policy* 42: 21–33, 2015.

Cutumisu, Nicoleta, Ariane Bélanger-Gravel, Marilie Laferté, François Lagarde, Jean-Frédéric Lemay, and Lise Gauvin. 'Influence of area deprivation and perceived neighbourhood safety on active transport to school among urban Quebec preadolescents'. *Canadian Journal of Public Health* 105: e376–e382, 2014.

Davis, Colin, and Robert Huxford. 'Briefing: Manual for Streets'. In *Proceedings of the Institution of Civil Engineers – Municipal Engineer*, vol. 162, no. 3, pp. 129–131. Thomas Telford Ltd, 2009.

Department for Communities and Local Government. *The English Indices of Deprivation 2015*, 30 April 2015. www.gov.uk/government/statistics/english-indices-of-deprivation-2015

Department for Environment Food & Rural Affairs. *2011 Rural–Urban Classification of Local Authorities and Other Geographies*, 1 May 2014. www.gov.uk/government/statistics/2011-rural-urban-classification-of-local-authority-and-other-higher-level-geographies-for-statistical-purposes

Department for Transport. Cycling and Walking Investment Strategy, 2017. www.gov.uk/government/uploads/system/uploads/attachment_data/file/603527/cycling-walking-investment-strategy.pdf

Department for Transport. *National Travel Survey:* 2019, 5 August 2020. www.gov.uk/government/statistics/national-travel-survey-2019

Dormann, Carsten F, Jane Elith, Sven Bacher, Carsten Buchmann, Gudrun Carl, Gabriel Carré, Jaime R García Marquéz et al. 'Collinearity: A review of methods to deal with it and a simulation study evaluating their performance'. *Ecography* 36, no. 1: 27–46, 2013.

Easton, Sue, and Ed Ferrari. 'Children's travel to school – The interaction of individual, neighbourhood and school factors'. *Transport Policy* 44: 9–18, 2015.

Ewing, Reid, William Schroeer, and William Greene. 'School location and student travel analysis of factors affecting mode choice'. *Transportation Research Record* 1895, no. 1: 55–63, 2004.

Fyhri, Aslak, Randi Hjorthol, Roger L Mackett, Trine Nordgaard Fotel, and Marketta Kyttä. 'Children's active travel and independent mobility in four countries: Development, social contributing trends and measures'. *Transport Policy* 18, no. 5: 703–710, 2011.

Giles-Corti, Billie, Gina Wood, Terri Pikora, Vincent Learnihan, Max Bulsara, Kimberly Van Niel, Anna Timperio, Gavin McCormack, and Karen Villanueva. 'School site and the potential to walk to school: The impact of street connectivity and traffic exposure in school neighborhoods'. *Health & Place* 17, no. 2: 545–550, 2011.

Ginja, Samuel, Bronia Arnott, Anil Namdeo, and Elaine McColl. 'Understanding active school travel through the Behavioural Ecological Model'. *Health Psychology Review* 12, no. 1: 58–74, 2018.

Giuliano, Genevieve, and Dhiraj Narayan. 'Another look at travel patterns and urban form: The US and Great Britain'. *Urban Studies* 40, no. 11: 2295–2312, 2003.

Helbich, Marco, Maarten J Zeylmans van Emmichoven, Martin J Dijst, Mei-Po Kwan, Frank H Pierik, and Sanne I de Vries. 'Natural and built environmental exposures on children's active school travel: A Dutch global positioning system-based cross-sectional study'. *Health & Place* 39: 101–109, 2016.

Hillier, Bill, and Iida, S. 'Network effects and psychological effects: A theory of urban movement'. In: Cohn, A.G., and Mark, D.M. (Eds.) *Spatial Information Theory. COSIT 2005*. Lecture Notes in Computer Science, vol 3693. Springer, Berlin, Heidelberg, 2005. https://doi.org/10.1007/11556114_30

Hillier, Bill, and Julienne Hanson. *The Social Logic of Space*. Cambridge University Press, 1984.

Hillier, Bill, Tao Yang, and Alasdair Turner. 'Normalising least angle choice in Depthmap and how it opens up new perspectives on the global and local analysis of city space'. *Journal of Space Syntax* 3, no. 2: 155–193, 2012.

Ikeda, Erika, Tom Stewart, Nicholas Garrett, Victoria Egli, Sandra Mandic, Jamie Hosking, Karen Witten et al. 'Built environment associates of active school travel in New Zealand children and youth: A systematic meta-analysis using individual participant data'. *Journal of Transport & Health* 9: 117–131, 2018.

Jarvis, Helen, and Seraphim Alvanides. 'School choice from a household resource perspective: Preliminary findings from a north of England case study'. *Community, Work & Family* 11, no. 4: 385–403, 2008.

Kerr, Jacqueline, Dori Rosenberg, James F Sallis, Brian E Saelens, Lawrence D Frank, and Terry L Conway. 'Active commuting to school: Associations with environment and parental concerns'. *Medicine & Science in Sports & Exercise* 38, no. 4: 787–793, 2006.

Kitamura, Ryuichi, Patricia L Mokhtarian, and Laura Laidet. 'A micro-analysis of land use and travel in five neighborhoods in the San Francisco Bay Area'. *Transportation* 24: 125–158, 1997.

Krizek, Kevin J. 'Residential relocation and changes in urban travel: Does neighborhood-scale urban form matter?' *Journal of the American Planning Association* 69, no. 3: 265–281, 2003.

Leung, Kevin YK, and Becky PY Loo. 'Determinants of children's active travel to school: A case study in Hong Kong'. *Travel Behaviour and Society* 21: 79–89, 2020.

Lin, Jen-Jia, and Hsiao-Te Chang. 'Built environment effects on children's school travel in Taipai: Independence and travel mode'. *Urban Studies* 47, no. 4: 867–889, 2010.

Living Streets. *Walk to School Rates Decline to the Lowest Level Ever*, 2016. www.livingstreets.org.uk/news-and-blog/press-media/number-of-primary-school-children-in-england-walking-to-school-falls

Living Streets. *Travel Tracker Data*, 2018. Unpublished raw data.

Living Streets. *WOW – The Year-Round Walk to School Challenge*, 2023. www.livingstreets.org.uk/products-and-services/projects/wow

Long, Robert. *The School Day and Year (England)*. House of Commons Library, (07148), 25 April 2023. https://researchbriefings.files.parliament.uk/documents/SN07148/SN07148.pdf

Lorenc, Theo, Ginny Brunton, Sandy Oliver, Kathryn Oliver, and Ann Oakley. 'Attitudes to walking and cycling among children, young people and parents: A systematic review'. *Journal of Epidemiology & Community Health* 62, no. 10: 852–857, 2008.

McMillan, Tracy E. 'Urban form and a child's trip to school: The current literature and a framework for future research'. *Journal of Planning Literature* 19, no. 4: 440–456, 2005.

Michail, Nafsika, Ayse Ozbil, Rosie Parnell, and Kyung Wook Seo. 'Children's route choice behaviour: Comparing the actual and metrically shortest routes for active journeys to school'. In *Proceedings 13th International Space Syntax Symposium*, Bergen, Norway, Western Norway University of Applied Sciences (HVL), pp. 1–23 June 2022.

Mitchell, Kit, and Terence Bendixson. *Planning for Walking.* Chartered Institution of Highways and Transportation, 2015. www.ciht.org.uk/media/4462/ciht_-_planning_for_walking_document-12pp_v2_singles.pdf

Mitra, Raktim. 'Independent mobility and mode choice for school transportation: A review and framework for future research'. *Transport Reviews* 33, no. 1: 21–43, 2013.

Murtagh, Elaine M, Martin Dempster, and Marie H Murphy. 'Determinants of uptake and maintenance of active commuting to school'. *Health & Place* 40: 9–14, 2016.

National Institute for Health and Care Excellence. *Physical Activity: Walking and Cycling Public Health Guideline [PH41]*, 28 November 2012. www.nice.org.uk/guidance/ph41

Ortegon-Sanchez, Adriana, Rosemary RC McEachan, Alexandra Albert, Chris Cartwright, Nicola Christie, Ashley Dhanani, Shahid Islam, Marcella Ucci, and Laura Vaughan. 'Measuring the built environment in studies of child health – A meta-narrative review of associations'. *International Journal of Environmental Research and Public Health* 18, no. 20: 10741, 2021.

Ozbil, Ayse, Demet Yesiltepe, Gorsev Argin, and Greg Rybarczyk. 'Children's active school travel: Examining the combined perceived and objective built-environment factors from space syntax'. *International Journal of Environmental Research and Public Health* 18, no. 1: 286, 2021.

Ozbil, Ayse, Gorsev Argin, and Demet Yesiltepe. 'Pedestrian route choice by elementary school students: The role of street network configuration and pedestrian quality attributes in walking to school'. *International Journal of Design Creativity and Innovation* 4, no. 2: 67–84, 2016.

Pont, Karina, Jenny Ziviani, David Wadley, Sally Bennett, and Rebecca Abbott. 'Environmental correlates of children's active transportation: A systematic literature review'. *Health & Place* 15, no. 3: 849–862, 2009.

Rojas Lopez, Maria Cecilia, and Yiik Diew Wong. '*Children's active trips to school: A review and analysis*'. *International Journal of Urban Sustainable Development* 9, no. 1:79–95, 2017.

Shaw, Ben, Ben Fagan-Watson, Bjorn Frauendienst, Andreas Redecker, Tim Jones, and Mayer Hillman. *Children's Independent Mobility: A Comparative Study in England and Germany (1971–2010)*. London Policy Studies Institute, 2013. https://westminsterresearch.westminster.ac.uk/item/8z178/children-s-independent-mobility-a-comparative-study-in-england-and-germany-1971-2010

Smith, Melody, Erika Ikeda, Greer Hawley, Suzanne Mavoa, Jamie Hosking, Victoria Egli, Jinfeng Zhao et al. 'An integrated conceptual model of environmental needs for New Zealand children's active travel to school'. *Journal of Transport & Health* 16:100814, 2020.

Sport England. *Active Lives Children and Young People Survey Academic Year 2017/18*, December 2018. www.sportengland.org/media/13698/active-lives-children-survey-academic-year-17-18.pdf

Su, Jason G, Michael Jerrett, Rob McConnell, Kiros Berhane, Genevieve Dunton, Ketan Shankardass, Kim Reynolds, Roger Chang, and Jennifer Wolch. 'Factors influencing whether children walk to school'. *Health & Place* 22: 153–161, 2013.

Timperio, Anna, Kylie Ball, Jo Salmon, Rebecca Roberts, Billie Giles-Corti, Dianne Simmons, Louise A Baur, and David Crawford. 'Personal, family, social, and environmental correlates of active commuting to school'. *American Journal of Preventive Medicine* 30, no. 1: 45–51, 2006.

Turner, Alasdair, and Friedrich, Eva. *Depthmap Software, Version 10.14.00b*. University College London, 2010. https://github.com/SpaceGroupUCL/depthmapX/

Zhao, Jia, Wei Su, Jiancheng Luo, and Jin Zuo. 'Evaluation and optimization of walkability of children's school travel road for accessibility and safety improvement'. *International Journal of Environmental Research and Public Health* 19, no. 1: 71, 2022.

11 The Effect of Building Façade Materials in Mitigating UHI Effect

An Urban City Centre Study

Shashwat Shashwat, Kishor Zingre, and Niraj Thurairajah

Introduction

Rapid urban development has caused a significant increase in air temperature (3–5 °C) as compared to adjacent rural areas, resulting in an urban heat island (UHI) effect (Arshad et al. 2021). Even in cities with high latitudes, such as London, global and urban warming has a negative influence on outdoor thermal comfort and causes buildings and infrastructure to overheat, resulting in many heat-related health conditions (Salvati et al., 2022). Additionally, the adverse impact also includes degraded air quality and additional cooling energy consumption, along with collateral social and economic costs (Liu et al., 2021). There have been multiple avenues of research undertaken to address these issues at different spatial levels, ranging from the building scale to the meteorological microscale and meteorological mesoscale (Toparlar et al., 2017). Mitigation strategies such as cool envelope materials, green roofs, insulation (Zingre et al., 2015), high reflective permeable pavement (Vujovic et al., 2021), shading by matured trees (Gupta & Anand, 2015), vertical greenery systems (Shuhaimi et al., 2022), and building form (Loh & Bhiwapurkar, 2022) can be effective measures to reduce the increasing outdoor urban temperature. Building façades in such circumstances become an important parameter that impacts the microclimate and urban comfort; in a mostly dense urban setting, pedestrian pathways are just adjacent to buildings (Shashwat et al., 2023).

Urban area comforts have also been investigated using various types of comfort indices (see Table 11.1) that consider different parameters, such as temperature and humidity, to provide a comprehensive evaluation of the built environment, which can assist in creating a more sustainable city. Some of these urban comfort indices act as a tool to measure the liveability of cities.

PET* is the improved version, sometimes referred as PET-reviewed, where changes are made to rectify assumptions and errors in the original PET equations (BIO-met, 2022; Walther & Goestchel, 2018).

Many computational tools are available to predict the thermal condition of a building façade, such as Energy-Plus, TRANSYS, and IES-VE (Gennaro et al., 2023). However, while most of the mentioned software can perform functions such as estimating heat transfer through the façade and thermal load of the building, they have limitations in conducting microclimate simulations (Forouzandeh, 2021).

DOI: 10.1201/9781003380559-11

Table 11.1 Overview of outdoor comfort indices and parameters

Indices	*Description*	*References*
Air temperature	Comparative analysis of the air temperature is used to analyse the mitigating effect.	(Cortes et al., 2022)
Universal Thermal Climate Index (UTCI)	Meteorological variables such as air temperature, relative humidity, solar radiation, and wind speed are used to provide a single output to represent the thermal comfort of a specific location. This index was proposed by EU COST Action 703.	(Jendritzky et al., 2012; Park et al., 2014)
Physiologically Equivalent Temperature (PET*)	Index uses air temperature, relative humidity, solar radiation, and wind speed to generate a value representing the air temperature perceived by the human body.	(Höppe, 1999; Manavvi & Rajasekar, 2022)
Standard Effective temperature (SET)	This is based on the ASHRAE 55-2013, which uses radiation, conduction, and evaporation for the calculation.	(Taleghani et al., 2015)

ENVI-Met is one of the few pieces of software available that can perform microclimate simulation with multiple features, as well as various urban environment attributes such as urban greenery profile, physical properties of urban surfaces (roads, pathways, building surfaces), and soils (Forouzandeh, 2021). It employs a user-defined three-dimensional grid box of the urban environment for simulating dynamic microscale climate calculations.

The size of the three-dimensional grid determines the granularity of the output, and the consumption of computational resources depends greatly on it. The computational model conducts simulations by considering the building structure (material, form), air temperature, air humidity, and other environmental parameters based on complex numerical calculations. The tool is based on three-dimensional equations, such as the non-hydrostatic Navier-Stokes, whose detailed description can be found in the simulation tool section of the research conducted on façade greening for mitigating the effect of heat waves (Alsaad et al., 2022). The energy balance of the outer wall, which forms the basis of the ENVI-met software, is described in Equation 1 and Figure 11.1, which shows how properties of the wall (such as solar emittance) play a role in the calculation.

$$Q^{abs}_{sw,net} + Q^{abs}_{sw,net} - \varepsilon\sigma T^{4}_{1,2} + h_{c,o}\left(T_{air} + T^{*}_{1}\right) + \frac{\lambda}{\Delta x}\left(T^{*}_{2} - T^{*}_{1}\right) = \frac{c_{wall}\,\rho\Delta x}{2\Delta t}\left(T^{*}_{1} - T_{1}\right) \quad [1]$$

(Huttner & Bruse, 2009)

Where,

$Q^{abs}_{sw,net}$ is incoming short-wave radiation [W]

$Q^{abs}_{sw,net}$ is incoming long-wave radiation [W]

ε is the emissivity [%]

σ is the Stefan-Boltzman constant

$h_{c,o}$ is the convective coefficient of the outside wall [W/(m²K)]

λ is the heat transfer coefficient [w/mk)]

Δx is the distance between two nodes [m]

c_{wall} is the heat capacity of the wall [J/(kgK)]

$T^{(*)}_{n}$ is the temperature at node n at present (T_n) of future (Tn^{*}) time step

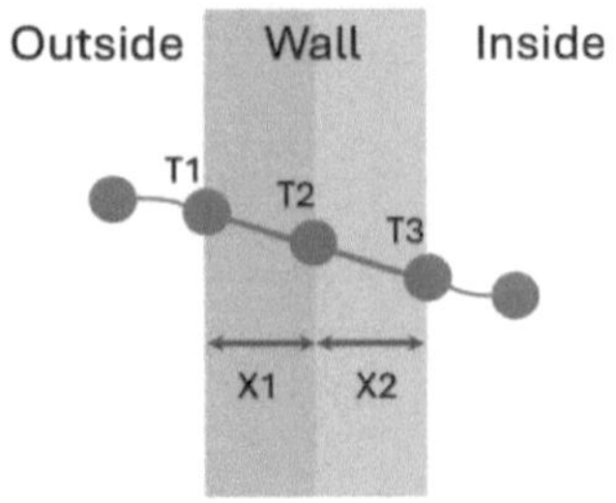

Figure 11.1 Schematic of three nodes.

There is a research gap in investigating the impact of building materials in enhancing the microclimate, especially for temperate climates such as the northeast of the United Kingdom (UK). This research investigates the impact of changes in façade material properties on the urban air temperature near the façade for a typical summer of UK weather. The analysis is also conducted using different urban comfort indices, as highlighted in Table 11.1.

Methodology

The methodology of the current research is highlighted in Figure 11.2. The research commences with the identification of the problem of the heat island effect in the urban city centre of Newcastle upon Tyne, located on the northeast side of the United Kingdom.

The city has a temperate climate with summer months ranging from May to September. After the identification of the problem, the site was selected, and various data were collected, such as typical annual meteorological data and measured data for the weather station as well as the urban centre collection point. The three-dimensional model was created using the open-source two-dimensional map extracted from the open street map (www.openstreetmap.org). The height of the building in the area was obtained through visual observation. Finally, the computational modelling method was utilised to perform the analysis.

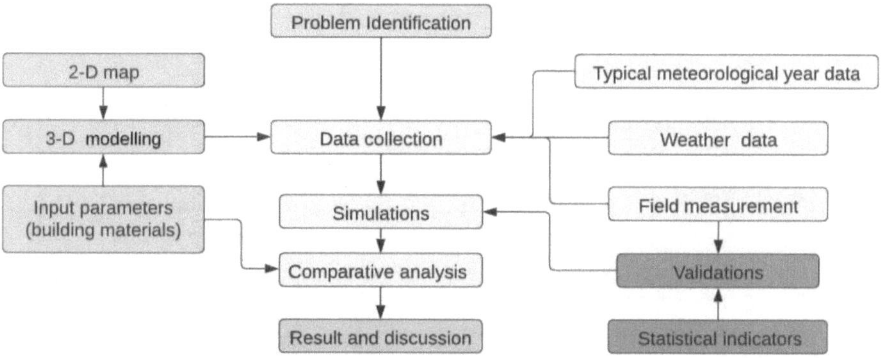

Figure 11.2 Various steps of the methodology adopted in this research.

Site selection

The site is located in the urban city centre of Newcastle upon Tyne, a large city in the northeast of the United Kingdom with high building density (see the satellite image from Google Earth in Figure 11.4). The geographical location coordinates in latitude and longitude are 54.9667 and 1.6000, respectively. Northumberland Road and College Street are two principal shopping streets in the urban city centre that are perpendicular to each other. Both streets have significant pedestrian movement throughout the daytime. The rectangular site area with the junction of two roads as the centre is selected for this research, with each site edge roughly 300 m in length.

The top right corner of the site (see Figure 11.3) is currently a large vacant area for the proposed development of a new commercial building. Most validated studies conducted for microclimate investigation have been done for park, residential, and academic buildings (Yang et al., 2021). College Street, which runs vertically, is an asphalt road with grey pathways on both sides; the left half of Northumberland Road is restricted to pedestrian movement, but on the right side, it constitutes an asphalt road, with walkways on both sides. Therefore, the selection of a city centre site for this research is relevant as these streets are significantly occupied by many pedestrians during the day.

Various storied buildings stand in the selected site, highlighted in different colours in Figure 11.3. The ground floor of most buildings at the selected site is a shop front with shopping or office space above it. Due to the high volume of pedestrian movement and the presence of high-density building structures, the site is ideal for conducting the current research on urban thermal comfort and urban heat mitigation strategies.

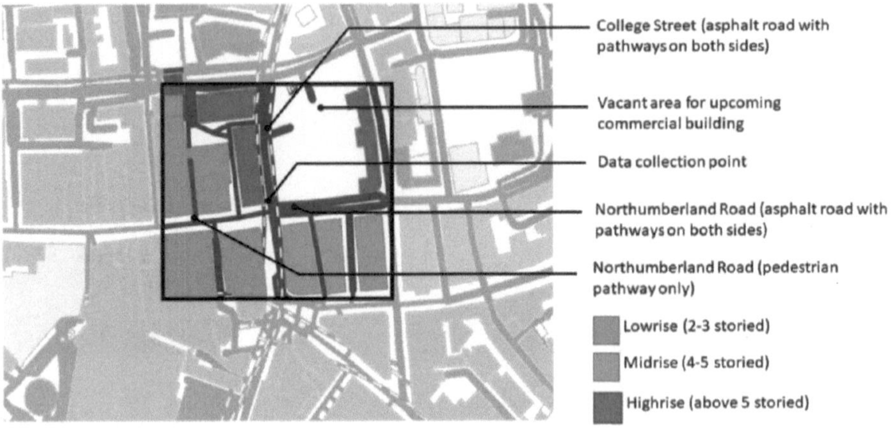

Figure 11.3 Two-dimensional image of the selected site with details.

Development of a computational model

The simulation tool ENVI-met (version 5.1.1) was used to conduct research to determine the intricate relationship between the wider urban settings. The model was created for commercial buildings located in the urban centre of Newcastle upon Tyne. Figure 11.4a shows the model created in the ENVI-met software, and Figure 11.4b shows a Google Maps screenshot of the buildings (GoogleEarth, 2023). The model is drawn with a grid size of 5m × 5m × 5m, and there are a total of 64 × 64 × 20 grids on x, y, and z axis, respectively, with an XY domain size of 320m × 320m. This size consists of an additional 20m area on each side of the site as per the requirement of software for the realistic application of boundary conditions.

Additionally, five vertical grids were generated for the lowest grid box in the model using the software features. However, all the models have the same boundary conditions. The choice of a grid size of 5m × 5m gives a detailed resolution with realistic results along with quick simulation time. A finer grid size of 2m × 2m was also tried, which drastically increased the simulation time of each run by AN estimated ~6 times for each simulation, while keeping all other parameters the same. The current model took around a 12-hour simulation period for each set of simulations.

The model consists of buildings, asphalt road networks, and concrete pathways that exist on the site. For this research, all buildings are assumed to have the same façade materials and properties for every case. The wall consists of three layers – outer brick (100 mm), mid insulation (120 mm), and inner hollow block (100 mm). The current model is assumed to be constituted of only opaque surfaces to analyse the maximum impact of the choice of building façade material. The overall U-value of the wall is 0.48 W/m^2K.

Figure 11.4 a) 3-D ENVI-met model overlapped on-site image. b) Google Earth image of the urban centre of Newcastle upon Tyne, United Kingdom.

Table 11.2 Details of building material properties

Scenario	*Details*	*Solar reflectance*	*Thermal emittance*	*Heat capacity (J/(Kg*K)*
Model validation		0.5	0.9	1500
Case 1	H. Ab	0.1	0.9	850
Case 2	H. Ref	0.9	0.9	850
Case 3	H. Em	0.1	0.1	850
Case 4	PCM	0.1	0.9	4000
Road	Asphalt	0.2	0.9	-
Pavement	Concrete	0.5	0.9	-

Table 11.2 shows the specification of exterior façade elements for the validation model. The material properties of Cases 1–4 comprise high absorptance, high reflectance, high emittance, and phase change materials (PCM), respectively. As the software does not have the capabilities for conducting the simulation for the dynamic behaviour of PCM material, the investigation was conducted with the assumption that the material has a high heat capacity (4000 J/Kg*K). The PCM material behaves as thermal energy storage if used as material in a building envelope that absorbs conductive heat (Lei et al., 2017). Each individual case was run for a representative summer day. The selection of representative summer days for simulation was made for 8 July 2021, which was considered after conducting multiple simulations during the validation process.

Validation of the computational model

The model is validated with the data recorded by the Urban Observatory group at Newcastle University. The Urban Observatory collects various types of data (weather, air quality, traffic, people, and water metrics) at different locations all around the city of Newcastle upon Tyne (Smith & Turner, 2019; UrbanObservatory, 2023). Weather data such as air temperature, air humidity, and wind direction and speed are four of the 50 data types that are collected and shared in real time for public access through their API services.

The data collection points located at the intersection of the two urban centre streets, highlighted in the plan shown in Figures 11.3 and 11.5, are used for validation of the model of the current studies. Air temperature and relative humidity are the two parameters used for validating the computational model for the study.

Weather data collected from the weather station collection point located around 6.4 km away from the city centre is used for full forcing as meteorological data in the simulation tool. These data were recorded at different temporal frequencies. For instance, air temperature and relative humidity are measured at one-minute intervals, but wind direction and speed are measured at 15-minute intervals. Data from the weather station was resampled and converted to 30-minute intervals for forcing in the simulation tool, which was done for each parameter using the Python script.

The Python script resamples the data by recalculating the mean of the data in each 30-minute interval against the timestamp at the beginning of each 30-minute interval (refer to Figure 11.6a). Similarly, the urban centre collected data was validated against the simulated output data (refer to Figure 11.6b). The simulated output data was produced for each hourly interval; therefore, the data sample was converted into an hourly sample using Python script. It is also observed from the comparison of the collected data that there is a difference of 1.3 °C between the

Figure 11.5 a) Image showing the sensor at the junction of the two streets of the selected site. b) Close-up image of the weather data collecting sensors.

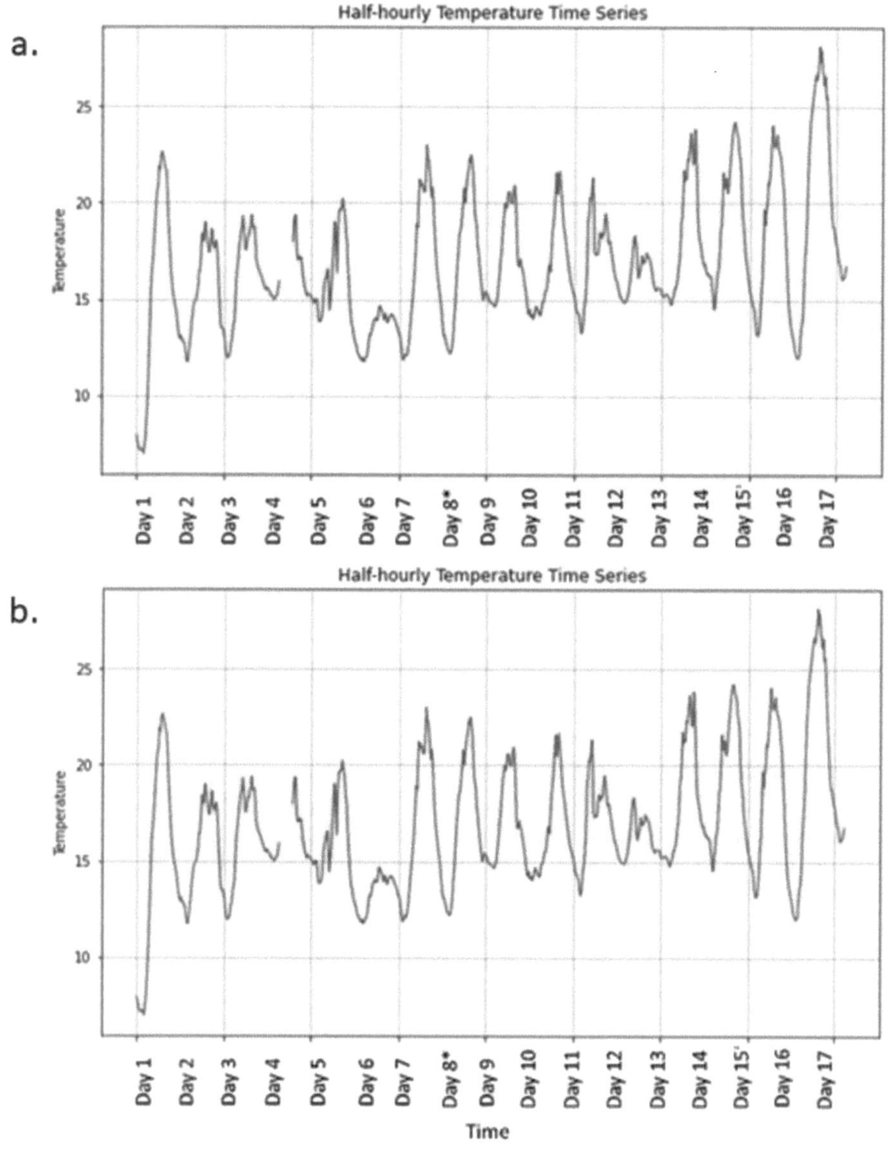

Figure 11.6 a) Half-hourly temperature time series data generated from a weather station for forcing in a simulation model. b) Hourly humidity data generated for statistical comparison from city centre collection points. The foremost simulation was conducted using all four data (air temperature, air humidity, wind direction, and wind speed) fully forced in the model, which resulted in impractical output. The primary reason for the unrealistic output is established to be the extremely low wind speed (the average wind speed recorded was 0.64 m/s on 8 July 2021) at the weather station point.

average air temperature of the weather station and city centre data collection point for a representative summer day (8 July 2021), showing the importance of mitigation of this UHI effect.

The low wind speed can be debated due to the 2m mounting height of the instrument set. This issue was resolved by replacing the wind speed and direction with the available free climate data from the Ladybug tool website (LadybugTool, 2023). These two parameters are generated from the historical data collection, so two representative summer dates (8 and 15 July 2021) were simulated to compare using the statistical indicators.

Computational model validation

Statistical indicators such as root mean square error (RMSE), mean absolute error (MAE), and R^2 correlation were used to analyse accuracy. Equation 2 shows the formula for calculating RMSE, whereas Equation 3 is for calculating MAE.

$$RMSE = \sqrt{\frac{\sum_{i=0}^{n} (Simulated - Measured)^2}{N}} \quad [2]$$

$$MAE = \frac{1}{n}\sum_{i=0}^{n} |Simulated - Measured| \quad [3]$$

'*Simulated*' is the hourly output of the potential air temperature near the building façade generated from the ENVI-met simulation, whereas '*Measured*' is the measured parameters from the city centre collection point (Figure 11.7) sampled into hourly series. *n* is the number of data samples that are compared. RMSE and MAE, both statistical indicators, have been used numerous times for validating ENVI-met modelling (Alsaad et al., 2022; Forouzandeh, 2018; Katsoulas et al., 2017; Yilmaz et al., 2021) for air temperature and relative humidity (Figure 11.8). With these statistical indicators, the lower the number, the higher the accuracy. However, the RMSE gives higher weightage to the greater difference.

Figures 11.7 and 11.8 present the comparison for validating the simulation model for two parameters, air temperature and relative humidity, respectively. The measured and simulated data follow the same trend, but during the daytime, the simulated data overestimates, while the measurement underestimates during the evening. The morning hours (first four hours) can be reasoned for the time to properly factor in all the boundary conditions of the simulation initiation. The value for RMSE in temperature is 2.08 °C, which is within a similar range as some of the previous literature following a similar methodology of validation (Alsaad et al., 2022). The MAE for the air temperature is 1.36 °C. Relative humidity simulated values underestimate the measured values during the day and early simulation hours.

The reason for the early hours may be similar to the air temperature stated above. However, the RMSE for relative humidity is slightly higher than 8.3%, and the MAE

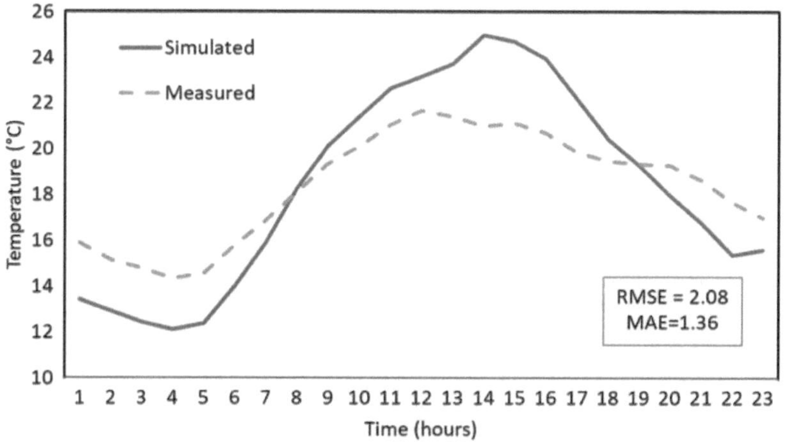

Figure 11.7 The measured air temperature compared with simulated air temperature for a representative summer day.

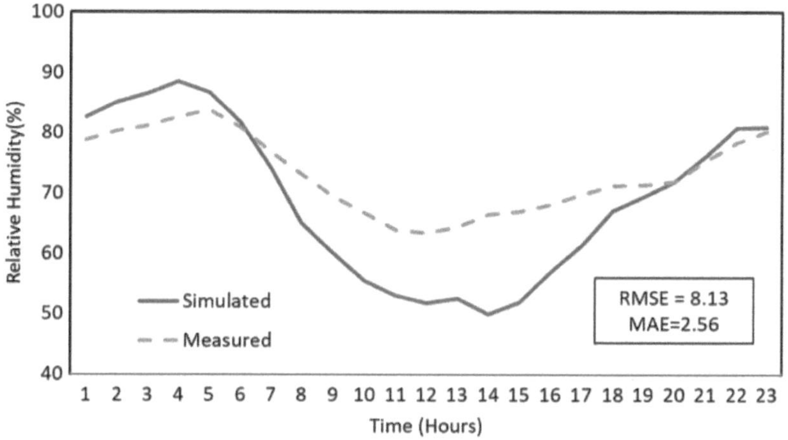

Figure 11.8 The measured relative humidity compared with simulated relative humidity for a representative summer day.

values are 2.56 for relative humidity. R^2 correlation was also investigated between the simulated and measured air temperature and relative humidity values. Figure 11.9 shows the result of the validation using the R^2 correlation for both parameters. It shows a strong correlation of 0.94 for air temperature and 0.95 for relative humidity.

Analysing the validated result concludes that the computational model shows a good agreement with the actual measurement. The validation established for this

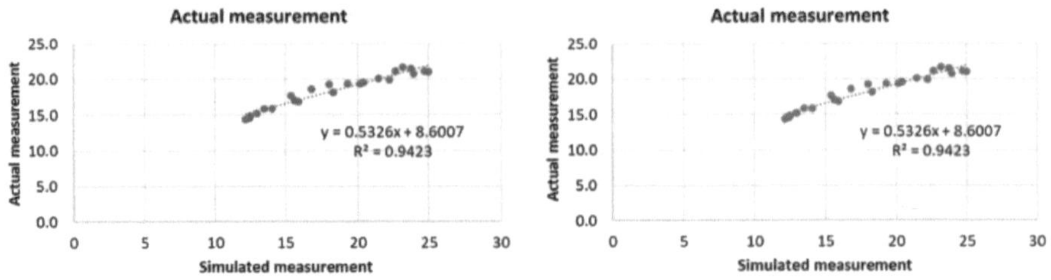

Figure 11.9 The scatterplot and correlation between the simulated and measured values of a) air temperature and b) relative humidity.

research is in the range of previously conducted similar research for both RMSE and MAE (Alsaad et al., 2022; Morakinyo et al., 2019). However, the existing difference can be attributed to the larger grid size, which was considered to reduce the computational time for the calculation.

Additionally, more detailed modelling with the incorporation of accurate massing of the building and fenestration may further enhance the calibration of the model. The higher detailing and finer grid size may result in frequent instability of the model. Measuring the wind speed, wind direction, cloud cover, and solar radiation will also help further calibrate the model.

Effect of building façade materials

To investigate the impact of the building façade material in the mitigation of the UHI effect, the results are analysed mainly for air temperature and relative humidity. Additionally, the impact of the building materials on various urban comfort indices (refer to Table 11.1), which has the maximum impact in minimising the air temperature on a representative summer temperature, is also conducted to check the effectiveness of the selected building façade material.

Air temperature and relative humidity

The simulations are conducted for 24 hours for a representative summer day (8 July 2021) to check the impact of the various building façade materials. Figure 11.10 presents the result of the air temperature next to the façade.

The selection of the building materials does have a considerable impact on the air temperature of the nearby area of the building façade. Figure 11.10 shows that case 2 (high solar reflectance) has the maximum impact in mitigating the outdoor air temperature. A peak reduction of 1.4 °C is achieved with the use of high solar reflectance properties on the building façade.

In Figure 11.10 it can also be seen that the building shows an effect of the thermal storage characteristic of the building façade. As compared to the measured values, which are conducted at the same location, it shows a peak air temperature lag of

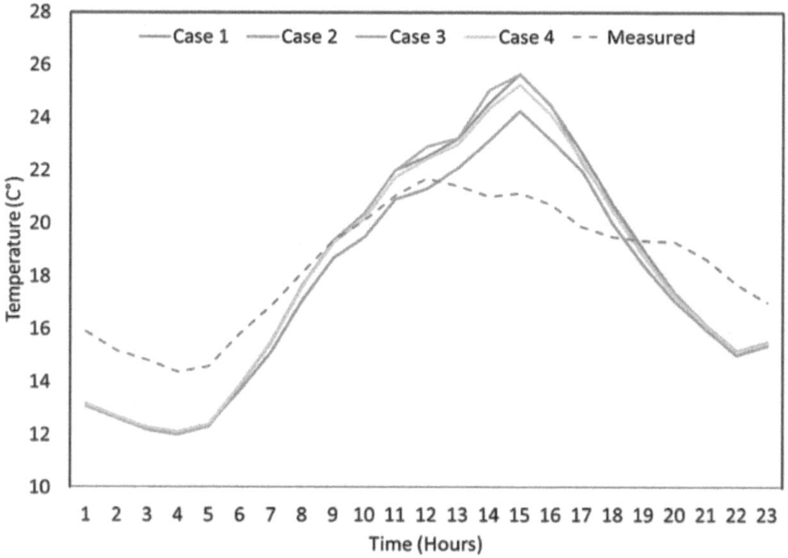

Figure 11.10 Air temperature of the grid next to the building façade for a representative summer day at 12 PM for a horizontal section at 1.5m above the ground surface.

around three hours. Figure 11.11 shows the different horizontal cross-sections of the comparison between case 1 and case 2. As is evident, there is a definitive difference. Near the façade, the temperature difference ranges from 1.05 °C to 1.45 °C.

The difference is due to the reduction in temperature because of the building façade material. Figure 11.11 shows that higher temperature differences in the two directions should not be considered due to the proximity to the physical boundary condition of the model. The rest of the cases (1, 3, and 4) follow a similar pattern, and the difference is minimal. The PCM (case 4) demonstrates as slightly better than the other two selections of façade materials.

In contrast to the air temperature, relative humidity shows the opposite trend. Relative humidity rises near the building façade due to its higher reflectance property (refer to Figure 11.12). The analysis shows that the higher reflective material effect reduces the outdoor air temperature, as shown in Figure 11.10.

This decrease can help in minimising the building heat gain, eventually helping in the reduction or abolition of the requirement of air conditioning in these cities. Nowadays, even in temperate climates located at high altitudes, people install air conditioning in buildings, which was not prevalent previously.

Outdoor thermal comfort indices

Outdoor thermal comfort indices play an essential role in analysing and understanding the impact of building material properties from the human perspective. This section

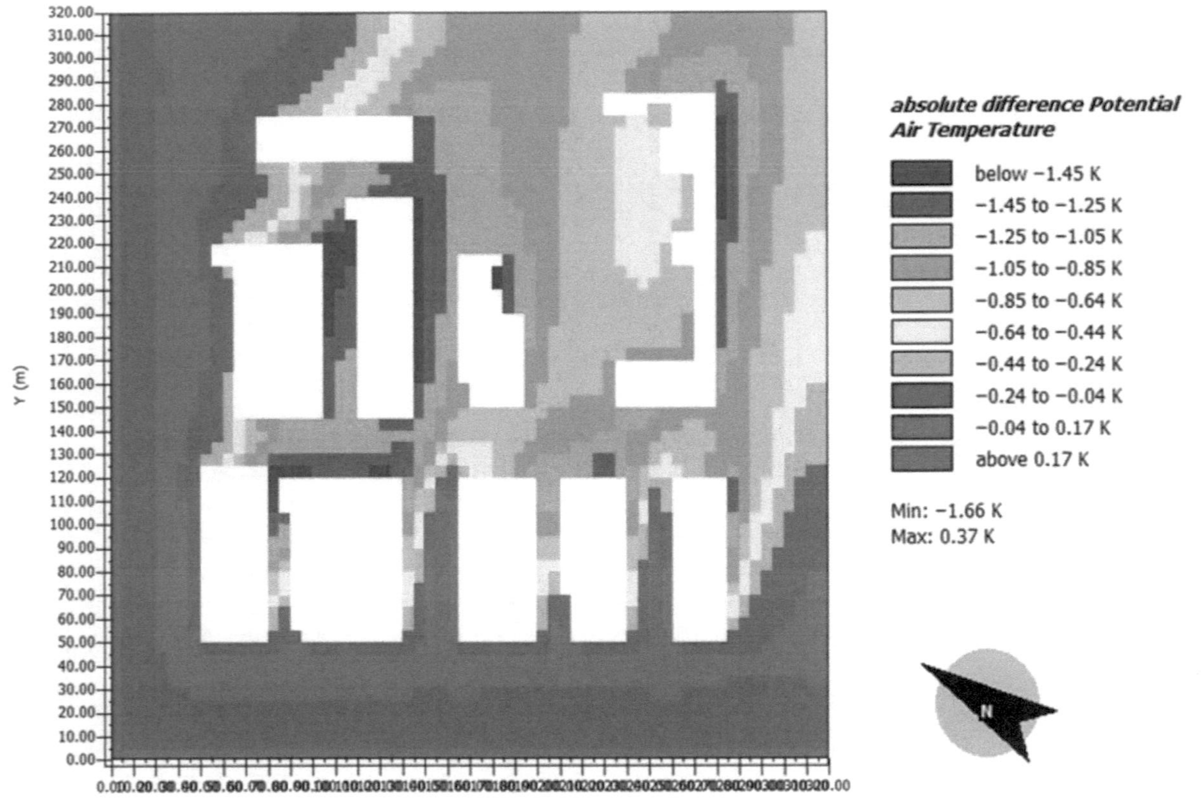

Figure 11.11 Air temperature difference of simulation case 2 (high reflectance) over case 1 (high absorptance) for a representative summer day at 12 PM at a horizontal section drawn at 1.5m above the ground surface.

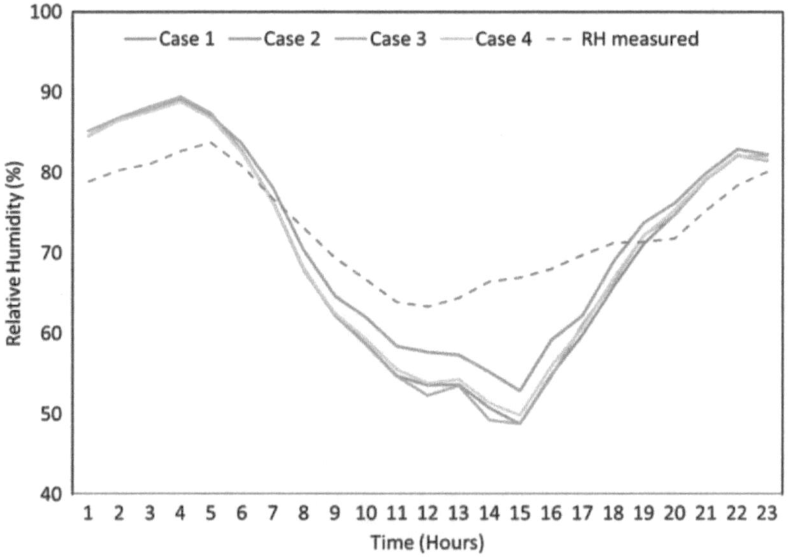

Figure 11.12 Relative humidity of the grid next to the building façade for a representative summer day at 12 PM for a horizontal section at 1.5m above the ground surface.

presents three types of outdoor thermal comfort indices, namely, UTCI, PET*, and SET, which are the most commonly used and are outlined in Table 11.1 (Binarti et al., 2020; Fang et al., 2019).

Figure 11.13 presents the absolute difference calculated for three thermal comfort indices extracted from the microclimate simulation software ENVI-met, providing two-dimensional visualisation (in Site Plan view) for each of them. It is the difference between two cases obtained for each thermal index at noon at 1.5 m above ground level, interpreted as a human comfort height. The visualisation represents the thermal comfort indices (UTCI in Figure 11.13a, PET* in Figure 11.13b, and SET in Figure 11.13c) under two different scenarios: Case 1 and Case 2 (refer to Table 11.2). The result for each respective thermal comfort index is made by examining the absolute difference for Case 2 (high reflectance) over Case 1 (high absorptance). Each of the three images is highlighted by a specific area under investigation, which corresponds to pedestrian walkways, indicated by an ellipse shape.

The impact temperature range is highlighted in the legend with an orange rectangle. As observed, the difference in thermal comfort indices is noticeable near the façade as well as beyond its immediate vicinity. For the pedestrian-occupied areas (marked by an ellipse shape in Figure 11.13), UTCI, PET*, and SET all show the difference (refer to Table 11.3).

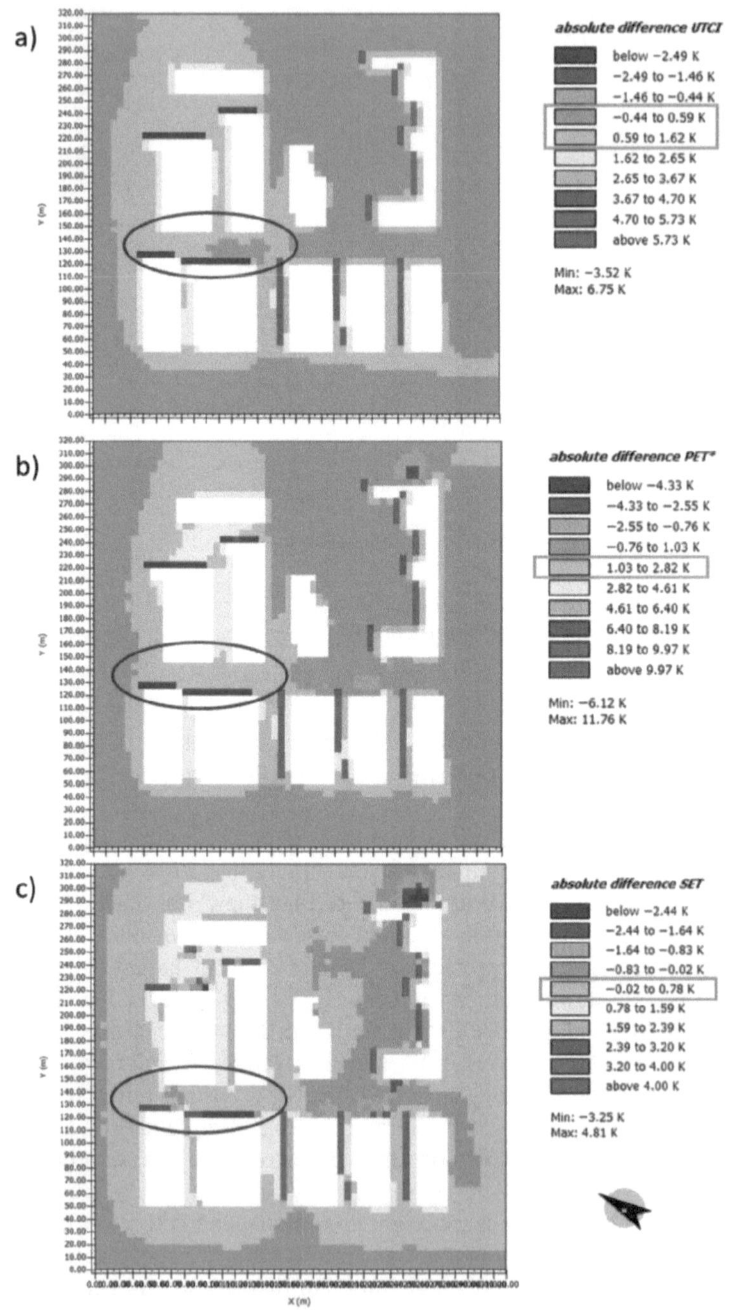

Figure 11.13 Analysis of the absolute difference of temperature between simulation case 2 (high reflectance) over case 1 (high absorptance) achieved by depicting various urban comfort indices (a. UTCI, b. PET*, and c. SET) for a representative summer day at 12 PM at a horizontal section drawn at 1.5m above the ground surface.

Table 11.3 Three outdoor thermal comfort indices (for definition refer to Table 11.1) absolute difference between case 2 over case 1

Thermal comfort indices	*Difference range (°C)*
UTCI	0.59 to 1.62
PET*	1.03 to 2.82
SET	-0.02 to 0.78

The PET* shows the highest difference with maximum impact ranging up to 2.8 °C, a considerable improvement. The UTCI and SET absolute difference ranges are up to 1.62 °C and 0.78 °C, respectively, which highlights the positive impact of the high reflectance surface to the nearby area from the façade.

Analysis of different types of urban outdoor thermal comfort indices demonstrates that the impact of the high reflectance is quite evident near the façade, and the usage of material with high reflectance of façade can assist in mitigating the UHI effect and increasing the city's liveability. The passive effect of the high reflectance of the building façade material can be easily achieved with minimal changes in the existing urban fabric.

Conclusions

This study investigated the effect of building façade materials on the mitigation of the UHI effect, which causes increased energy costs and summertime outdoor discomfort. A UHI of 1.3 °C is observed between the city centre and the weather station collection point (the average air temperature for a representative summer day). The evaluation of building façade material is conducted through the numerical simulation tool ENVI-met.

Firstly, validation is conducted for the measured and simulated values. The simulated air temperature values were found to be in good agreement with the measured values, with an RMSE of 2.08 °C and an MAE of 1.38 °C. It was slightly overestimated during the high temperature (daytime) and underestimated during the low temperature (nighttime). Additionally, the simulated relative humidity shows relatively less agreement with the measured values, which are underestimated during the daytime, with a higher RMSE of 8.13 °C and MAE of 2.56 °C. However, both air temperature and relative humidity show satisfactory R^2 correlation between simulated and measured with values of 0.94 and 0.95, respectively.

Outdoor thermal comfort indices, such as the urban thermal climate index, physiologically equivalent temperature, and standardised effective temperature were evaluated using the ENVI-met microclimate tool. Among the four simulation cases investigated post-validation process, the application of high reflectance material on the building façade impacts the microclimate by reducing the outdoor temperature the most. For a representative summer day, the peak reduction in the air temperature next to the façade was measured to be 1.4 °C for high reflectance façade material (Case 2) compared to high absorptance (Case 1).

The positive impact of the same can also be seen through the results of all three indices, which were conducted by analysing the difference in case 2 over case 1. For the pedestrian walkways next to the façade, the absolute difference in the urban thermal climate index between the two cases was in the range of 0.59 to 1.62 °C. Similarly, the physiologically equivalent temperature was in the range of 1.03 and 2.82 °C, and the standardised effective temperature was depicted in the range of −0.02 to 0.78 °C. The thermal comfort indices show that the impact is not only next to the façade but also extends to an area beyond the immediate vicinity of the façade. Adopting a high solar reflective façade has a positive impact on mitigating the effect of the UHI effect in the summer, which will help reduce heat gain through the façade. A time lag of around three hours is also observed, which can be attributed to the heat storage of the building façade. There is also an impact on relative humidity between the two cases, increasing by 3.9%.

Acknowledgements

This research is conducted as part of the PhD project funded jointly by the Research Development Fund (RDF) and Northumbria University. Appreciation is also extended to all the team members associated with the Urban Observatory project at Newcastle University for developing the open-source, publicly accessible data-sharing platform.

References

Alsaad, H., Hartmann, M., Hilbel, R., & Voelker, C. (2022). The potential of facade greening in mitigating the effects of heatwaves in Central European cities. *Building and Environment*, *216*, 109021. https://doi.org/10.1016/j.buildenv.2022.109021

Binarti, F., Koerniawan, M. D., Triyadi, S., Utami, S. S., & Matzarakis, A. (2020). A review of outdoor thermal comfort indices and neutral ranges for hot-humid regions. *Urban Climate*, *31*, 100531.

BIO-met. (2022, 2022/12/02). ENVI-met. https://envi-met.info/doku.php?id=apps:biomet_pet

Cortes, A., Rejuso, A. J., Santos, J. A., & Blanco, A. (2022). Evaluating mitigation strategies for urban heat island in Mandaue City using ENVI-met. *Journal of Urban Management*, *11*(1), 97–106. https://doi.org/10.1016/j.jum.2022.01.002

Fang, Z., Feng, X., Liu, J., Lin, Z., Mak, C. M., Niu, J., Tse, K.-T., & Xu, X. (2019). Investigation into the differences among several outdoor thermal comfort indices against field survey in subtropics. *Sustainable Cities and Society*, *44*, 676–690.

Forouzandeh, A. (2018). Numerical modeling validation for the microclimate thermal condition of semi-closed courtyard spaces between buildings. *Sustainable Cities and Society*, *36*, 327–345.

Forouzandeh, A. (2021). Prediction of surface temperature of building surrounding envelopes using holistic microclimate ENVI-met model. *Sustainable Cities and Society*, *70*, 102878.

Gennaro, G., Catto Lucchino, E., Goia, F., & Favoino, F. (2023). Modelling double skin façades (DSFs) in whole-building energy simulation tools: Validation and inter-software comparison of naturally ventilated single-story DSFs. *Building and Environment, 231*, 110002. https://doi.org/10.1016/j.buildenv.2023.110002

GoogleEarth. (2023). *Newcastle upon Tyne*. Retrieved 2023 from https://earth.google.com/web/

Gupta, S., & Anand, P. (2015). Improvement of outdoor thermal comfort for a residential development in Singapore. *International Journal of Energy & Environment, 6*(6), 567–586.

Höppe, P. (1999). The physiological equivalent temperature – A universal index for the biometeorological assessment of the thermal environment. *International Journal of Biometeorology, 43*, 71–75.

Huttner, S., & Bruse, M. (2009, June). Numerical modeling of the urban climate–a preview on ENVI-met 4.0. In *7th International Conference on Urban Climate ICUC-7*, Yokohama, Japan (Vol. 29).

Jendritzky, G., de Dear, R., & Havenith, G. (2012). UTCI – Why another thermal index? *International Journal of Biometeorology, 56*, 421–428.

Katsoulas, N., Antoniadis, D., Tsirogiannis, I., Labraki, E., Bartzanas, T., & Kittas, C. (2017). Microclimatic effects of planted hydroponic structures in urban environment: Measurements and simulations. *International Journal of Biometeorology, 61*, 943–956.

LadybugTool. (2023). EPW Map. Ladybug Tools. Retrieved February 2023 from www.ladybug.tools/epwmap/

Lei, J., Kumarasamy, K., Zingre, K. T., Yang, J., Wan, M. P., & Yang, E.-H. (2017). Cool colored coating and phase change materials as complementary cooling strategies for building cooling load reduction in tropics. *Applied Energy, 190*, 57–63. https://doi.org/10.1016/j.apenergy.2016.12.114

Liu, Z., Cheng, W., Jim, C. Y., Morakinyo, T. E., Shi, Y., & Ng, E. (2021). Heat mitigation benefits of urban green and blue infrastructures: A systematic review of modeling techniques, validation and scenario simulation in ENVI-met V4. *Building and Environment, 200*, 107939.

Loh, N., & Bhiwapurkar, P. (2022). Urban heat-mitigating building form and façade framework. *Architectural Science Review, 65*(1), 57–71.

Manavvi, S., & Rajasekar, E. (2022). Evaluating outdoor thermal comfort in urban open spaces in a humid subtropical climate: Chandigarh, India. *Building and Environment, 209*, 108659. https://doi.org/10.1016/j.buildenv.2021.108659

Morakinyo, T. E., Lai, A., Lau, K. K.-L., & Ng, E. (2019). Thermal benefits of vertical greening in a high-density city: Case study of Hong Kong. *Urban Forestry & Urban Greening, 37*, 42–55.

Park, S., Tuller, S. E., & Jo, M. (2014). Application of Universal Thermal Climate Index (UTCI) for microclimatic analysis in urban thermal environments. *Landscape and Urban Planning, 125*, 146–155. https://doi.org/10.1016/j.landurbplan.2014.02.014

Salvati, A., Kolokotroni, M., Kotopouleas, A., Watkins, R., Giridharan, R., & Nikolopoulou, M. (2022). Impact of reflective materials on urban canyon albedo, outdoor and indoor microclimates. *Building and Environment, 207*, 108459.

Shashwat, S., Zingre, K. T., Thurairajah, N., Kumar, D. K., Panicker, K., Anand, P., & Wan, M. P. (2023). A review on bioinspired strategies for an energy-efficient built environment. *Energy and Buildings, 296*, 113382.

Shuhaimi, N. D. A. M., Zaid, S. M., Esfandiari, M., Lou, E., & Mahyuddin, N. (2022). The impact of vertical greenery system on building thermal performance in tropical climates. *Journal of Building Engineering, 45*, 103429.

Smith, L., & Turner, M. (2019). Building the Urban Observatory: Engineering the largest set of publicly available real-time environmental urban data in the UK. In *EGU General Assembly Conference Abstracts*, 10456.

Taleghani, M., Kleerekoper, L., Tenpierik, M., & van den Dobbelsteen, A. (2015). Outdoor thermal comfort within five different urban forms in the Netherlands. *Building and Environment*, *83*, 65–78. https://doi.org/10.1016/j.buildenv.2014.03.014

Toparlar, Y., Blocken, B., Maiheu, B., & Van Heijst, G. (2017). A review on the CFD analysis of urban microclimate. *Renewable and Sustainable Energy Reviews*, *80*, 1613–1640.

UrbanObservatory. (2023). Newcastle University. Retrieved 19 April 2023 from https://newcastle.urbanobservatory.ac.uk/

Vujovic, S., Haddad, B., Karaky, H., Sebaibi, N., & Boutouil, M. (2021). Urban heat island: Causes, consequences, and mitigation measures with emphasis on reflective and permeable pavements. *CivilEng*, *2*(2), 459–484.

Walther, E., & Goestchel, Q. (2018). The PET comfort index: Questioning the model. *Building and Environment*, *137*, 1–10.

Yang, J., Hu, X., Feng, H., & Marvin, S. (2021). Verifying an ENVI-met simulation of the thermal environment of Yanzhong Square Park in Shanghai. *Urban Forestry & Urban Greening*, *66*, 127384. https://doi.org/10.1016/j.ufug.2021.127384

Yilmaz, S., Külekçi, E. A., Mutlu, B. E., & Sezen, I. (2021). Analysis of winter thermal comfort conditions: Street scenarios using ENVI-met model. *Environmental Science and Pollution Research*, *28*, 63837–63859.

Zingre, K. T., Yang, X., & Wan, M. P. (2015). Performance analysis of cool roof, green roof and thermal insulation on a concrete flat roof in tropical climate. *Evergreen*, *2*(2), 34–43.

12 Promoting Built Asset Flood Resilience for Sustainable Development through Data Advantage

Pavithra Rathnasiri, Onaopepo Adeniyi, and Niraj Thurairajah

Introduction

Sustainable communities and cities have particularly become important because of their ability to be conjoined with nature, people, and the built environment. Being interdependent and indivisible, sustainable communities and cities ensure balanced social, economic, and environmental goals offered by multifaceted and holistic sustainable developments (Pradhan et al., 2017). However, with the growing uncertainty of climate change and extreme weather events, increasing population and urbanisation, and expansion of urban agglomerations, there has been a dramatic increase in disaster risk around the globe, interrupting desired sustainable achievements (Sen, Dutta & Laskar, 2021; Yabe et al., 2022). A disaster is a sudden occurrence or natural catastrophe that may cause significant damage to infrastructure and the economy and may result in the loss of human lives (Bigueras et al., 2018). Floods stand out as the most frequent and devastating of different kinds of disasters, and the number of flood incidents has risen dramatically during the past three decades, with numerous environmental and socioeconomic consequences (Feloni et al., 2019). Catastrophes such as floods destroy properties, cause injuries and deaths, and disrupt utility services, affecting cities' normal operations and interrupting desired sustainable achievements (Hao & Wang, 2021; Yereseme et al., 2022). Floods have devastating effects on built assets, and the economic damages and losses associated with built assets are drastically growing worldwide (Fotovatikhah et al., 2018). Globally, the economic losses caused by flooding on built assets since 1980 are estimated at over $1 trillion (Edwards et al., 2017). The estimated economic losses as the result of flood damages to built assets were reported as US$49.8 billion in the United States in 2022, over £333 million in the United Kingdom, and US$25.4 million in Shanghai (Hao & Wang, 2021). These statistics indicate the challenge of flood damage mitigation for built assets, which hinders and prevents the ability to develop more sustainable cities and communities (Yereseme et al., 2022).

It is important to ensure the stability and proper performance of built assets to enhance the social, economic, and environmental sustainability of cities and communities (Arneson et al., 2017; Yin et al., 2022; Yereseme et al., 2022). However,

DOI: 10.1201/9781003380559-12

recurrent flood damages and losses to built assets have made achieving sustainable cities a challenge in terms of social, economic, and environmental developments (Yereseme et al., 2022). Existing resilience strategies for built assets are largely based on impact assessments. These conventional approaches are labour and time intensive and often delay response and recovery, leading to further losses for disaster-affected communities (Hao & Wang, 2021). Thus, current solutions to increase the resilience of built assets are based on ad hoc procedures (Dunn & Gonzalez-Otalora, 2021).

Recently, the use of data has been determined as an essential tool that can help enhance built asset flood resilience through context-specific policymaking, administration, research, and leadership (Sarker et al., 2020). Data is a fundamental part of resilience, and the use of data-driven approaches for decision-making gained prominence in built asset flood resilience (Towe et al., 2020). A data-driven approach is a method of harnessing and analysing a vast amount of data, with computational programmes and algorithms enabling new ways of empowering communities to address problems at the source (Carozza and Boudreault, 2021). It provides the ability for real-time reflexive awareness, advocating claims and evaluating risks and problems to manage potential damages and losses.

With such abilities, the prospect of data-driven knowledge production has been welcomed in informed decision-making of built asset flood resilience (Yabe et al., 2022). However, the existing data-driven flood resilience studies have merely focused on the use of different data-driven approaches and mainly on food-specific data for improving flood risk management through effective flood warnings and disaster relief systems (Fotovatikhah et al. 2018; Sarker et al., 2020).

Many authors have developed data-driven hydrological models for flood modelling and forecasting (Hapuarachchi et al., 2011; Ochoa-Rodriguez et al., 2015; Wang et al., 2015), urban catchment behaviours relevant to fast hydrologic responses, and hydraulic models for flood warnings (Skinner et al., 2015; Li et al., 2016). Even though there is a pool of data generated in the built environment, the use of data has been scattered into specific flood management aspects despite the huge data potential for enhancing built asset resilience towards sustainability. To address this, the existing context of data utilisation, challenges that impede data utilisation, and remedies to further encourage data utilisation should be identified with regard to new directions and pathways for strong sustainable communities.

Built Asset Flood Resilience

Resilience is the capacity of assets or the ability of a community to withstand and regain desirable performance after the occurrence of a natural disaster (Nofal & Lindt, 2020). Resilience can also be defined as: "the ability to recover from or resist being affected by some shock or disturbance", or the degree to which a system absorbs disturbances while continuing to function (Sarker et al., 2020). In the context of floods, resilience refers to being prepared, ready to respond, and able to cope and recover from flooding. The phrase "flood resilience" has recently gained

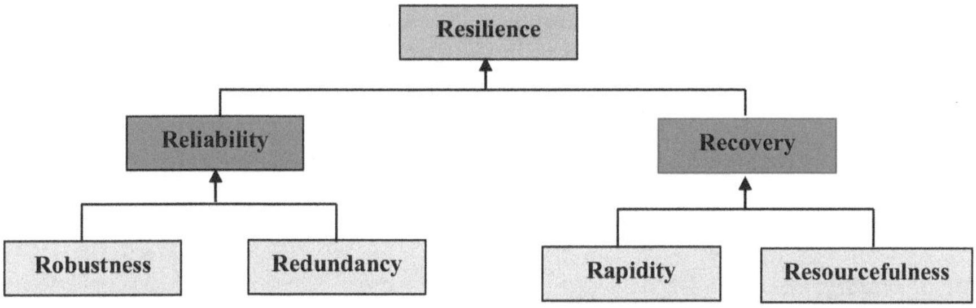

Figure 12.1 Key parameters of built asset flood resilience.
Source: Sen et al., 2021.

popularity and generally refers to the ability to withstand floods while minimising damages, which lowers the danger to people and built infrastructure and leaves enough space for flooding where appropriate.

Resilient buildings are designed and built with the capacity to avoid, prevent, and reduce the damage caused when flooding takes place. Contextually, the resilience of built assets can be characterised by its four goals: robustness, rapidity, resourcefulness, and redundancy (Figure 12.1) (Sarker et al., 2020). Robustness is the ability of a system to withstand a certain level of stress, while redundancy is a measure of the built-in sustainability of the system, i.e., the availability of replacements within the system. Resourcefulness refers to the availability of resources that aid in the detection, measurement, and survival of a hazard.

Rapidity is a measure of the time needed for the system to go back to its state before disruption. As further explained by Sen, Dutta, and Kabir (2021), the resilience of built assets comprises two key-dependent parameters, namely "reliability and recovery", made possible by various capabilities (Adeniyi et al., 2022).

Data-Driven Methodologies Deployed in Existing Flood Resilience Studies

Different methodologies are used in existing data-driven studies on flood resilience. These methods were mainly identified as machine learning techniques, artificial intelligence-based modelling, and hydrological flood simulations (Guidotti et al., 2019). Significantly, tremendous attention to and an increase in the use of machine learning techniques can be identified in recent studies on flood resilience.

Machine Learning

Machine learning can identify and learn patterns, discover trends and behaviours from large quantities of datasets, and generate and predict new data to improve its performance. Due to the ability of machine learning models to deal with large datasets, the complex interdependent attributes and hidden behaviours of data can be identified (Abdel-Mooty et al., 2021). With the proven capabilities and advantages of data handling and analysis of various machine learning techniques, a dramatic increase in the existing flood resilience studies can be identified as focused on flood prediction (Abdel-Mooty et al., 2021; Aguilar et al., 2013; Allahbakhshian-Farsani et al., 2020; Amininia & Saghebian, 2021; Araghinejad et al., 2018; Chen et al., 2019), flood monitoring (Ahmed et al., 2022), and flood damage assessments (Hao & Wang, 2021).

Artificial Intelligence

While machine learning techniques are undertaken in such studies as a part of artificial intelligence, some researchers have mainly focused on artificial intelligence-based modelling and algorithms such as neural networks, which are similarly applied with the benefit of modelling and analysing large datasets. Amininia and Saghebian (2021) and Araghinejad et al. (2018) conducted studies on river flow modelling using artificial intelligence algorithms to enhance flood predictions. A recent study was also undertaken by Dai and Cai (2021) using artificial intelligence for flood prediction. Thus, machine learning and artificial intelligence can be identified as a growing data-driven approach in the research field, especially for the prediction of floods.

Hydrological Simulations

Over the past years, credible attention has also been given to hydrological flood simulations, and studies have widely utilised flood simulations in terms of 2D and 3D flood inundation modelling. Flood simulation models were significantly utilised in flood monitoring (Adamovic et al., 2016), flood prediction (Aguilar et al., 2013; Chen, Chen, & Shen, 2019), flood risk assessments (Ahmadalipour & Moradkhani, 2019; Cian, Marconcini, & Ceccato, 2018), and flood damage assessments (Nofal & Lindt, 2020). Recently, advanced flood simulation software developments such as Citygml and OpenFloods can also be identified in existing studies (Adamovic et al., 2016; Aguilar et al., 2013; Ahmadalipour & Moradkhani, 2019; Nofal & Lindt, 2020). These studies mainly drew attention to flood risk and prediction at the urban and community level.

Building Information Modelling (BIM) and Geographic Information Systems (GIS)

Some of the existing data-driven studies focused on flood resilience in individual properties. Building information modelling (BIM), geographic information systems (GIS), and integrations of BIM and GIS with flood simulations in this regard were identified as popular data-driven techniques (Amirebrahimi et al., 2015; Nofal et al., 2020; Nofal & Lindt, 2020). The studies mainly aimed at facilitating comprehensive flood damage assessments for buildings; they are applicable for visualising vulnerability and property flood loss potentials. Few studies employed Bayesian network modelling (Sen, Dutta & Laskar, 2021) and the extreme learning machine (ELM) method (Chen et al., 2018) that aim at flood recovery processes and flood predictions, respectively.

Sources of Data

Different data sources have been identified in recent literature that are widely used in data-driven flood resilience studies. The sources are mainly categorised as internal/exposure data, open-source data, satellite Earth observation (EO) data, and big data sources.

Insurance Internal Source

Internal data include data typically derived from clients, for example, insurance claims and building exposure data. These data are mainly handled and maintained by property or flood insurers. In addition, these data include location, type of building, mode of construction, occupancy information, and year built. Alongside exposure and vulnerability information, policy insights are also important datasets; they provide information on aspects including premiums, cancellations, and coverage gaps (Rumson & Hallet, 2019).

Open-Source

Open-source data provides opportunities for data reuse; open data are collected for one purpose but are freely available for another use. The UK is considered to be making progress in this area, particularly regarding datasets provided by the Environment Agency (EA), which, for instance, has a thorough mapping programme in place involving the gathering of airborne floodplain and coastal LiDAR data that is openly distributed. This can be used to acquire floods and building footprints for flood-related uses. Flood modellers have identified the Risk of Flooding from Rivers and Seas (RFRS) dataset from the EA as an important input (Rumson & Hallet, 2019).

Earth Observation (EO)

Many opportunities for satellite Earth observation data have arisen, creating various data products through web-mapping interfaces and API feeds, through which it is possible to leverage vast volumes of EO data. The capabilities of EO data as a source are growing over the last ten years. Terrain mapping is now derived from higher resolution EO data such as stereo imagery techniques (SITs; DeWitt et al., 2017). SITs are another emerging approach and data source. Google Earth Engine combines a vast database of EO data from several sources based on need. These sources are being applied to flood risk assessment, exposure management, disaster response, and flood risk insurance (Mashaly & Ghoneim, 2018).

Big Data Sources

There are systems set up to pull a variety of data in terms of types and forms. Increased computing power is enabling various flood analysis with such data at different spatial scales. Further potentials are being imagined. This source category is a pool of data from satellite imagery, aerial imagery and videos, wireless sensor web network, light detection and ranging (LiDAR), simulation data, spatial data, crowdsourcing, social media, mobile-based GPS, and record of call data (Rumson & Hallet, 2019).

Data Types Used in Existing Data-Driven Flood Resilience Studies

Data types are usually distinguished by flood hazard characteristics, attributes of the location, and physical infrastructure attributes. Damage and resilience can be influenced by several variables, some of which are event-related (over-floor water depth, flow velocity, rate of water rise, duration of inundation, and timing). Some are specific to built assets (type of structure and construction, materials used and their drying characteristics, the age and condition of the building) (Wang et al., 2015). The data types from the earlier described sources and related to the variables influencing damage and resilience can be classified as flood-specific data, built assets-specific data and geographical data. Data relating to building characteristics and construction can be identified as built asset-specific data; data directly relating to flood characteristics corresponding to flooded locations can be identified as flood-specific data; and data referring to geographic attributes of locations can be identified as geographical data (Towe et al., 2020).

Built Asset-Specific Data

The data variables found in this category in the data-driven studies include building geometry, topology, type of buildings, materials, building use, exterior and interior components, and characteristics, building value, number of stories, type of foundations, elevations (Nofal & Lindt, 2020) damages and cost data, building size, type, quality, insurance information, and economic value (Amadio et al., 2019).

Locational/Geographical Data

Topographic data includes the type of area (construction site, educational building, etc.), size of the area, land use, land cover (Bucar & Hayeri, 2022), geology, topographic data (Adamovic et al., 2016; Bucar & Hayeri, 2022), soil type, soil moisture data, distance from roads, distance from rivers (Khosravi et al., 2020), water accumulation capacity, soil temperature (Cesarini, et al., 2021), and ground cover, as identified from existing studies.

Locational Flood Data

Finally, locational flood data includes flood type, flood duration and frequency, flood depth, flood volume, flood velocity, rate of rising, rainfall intensity, flood duration and patterns, water level and flow pathways, water channel structure, watershed morphometry data, frequency inundations, flood probabilities, water level and flow direction, water channel structure, rivers' length and slope (Chen et al., 2018; Bucar & Hayeri, 2022), and catchment delineation (Amininia & Saghebian, 2021).

Overview of the Existing Data-Driven Studies on Flood Resilience

The existing data-driven studies are mainly focuses on flood prediction, flood monitoring, flood risk assessments, and flood damage assessments, using the emerging data modelling, handling, and analysing techniques of machine learning, artificial intelligence, flood simulations, building information modelling, and geographic information systems. The growing use of machine learning techniques and models was identified for flood prediction, flood monitoring, flood risk assessments, flood damage assessments, and flood insurance. Deep learning techniques such as artificial neural networks have similarly become significant in flood-related studies for flood predictions and flood risk assessments. Hydrologic flood simulations have also gained prominence in existing studies, especially in flood hazard mapping and flood risk assessments (Nofal et al., 2020). As an integrated approach, BIM and GIS are used in existing data-driven studies, especially for flood damage assessments. Bayesian network modelling is also gaining attention in current disaster-related studies; however, little research is available specifically regarding flood resilience.

Generally, the use of these techniques are largely limited to flood predictions, flood monitoring, flood risk assessments, and damage assessments, while there is still the ability for further use in structural and non-structural measures planning, preparation, asset performance rating and mitigations, investment allocation, flood response, comprehensive insurance, potential recovery procedure and processes (Henry & Ramirez-Marquez, 2012; Tran et al., 2017), and issues on building back better, as well as overall built asset safety while thinking resilience. The summary of existing studies with key findings is shown in Figure 12.2.

Challenges of Data Utilisation for Flood Resilience

The desired outcomes of built asset flood resilience are greatly ensured using pertinent data and information. However, there are still various challenges to using data and analytics that need to be resolved. This section of the chapter identifies various data challenges prevailing in the literature with potential strategies that can be used to overcome them.

As often mentioned in the existing literature, the lack of data mainly hinders flood resilience strategies, and as a result, research in data-limited scenarios are mainly restricted to either building exposure assessment or the identification of susceptibility indicators without further linkage to potential damage (Malgwi et.al., 2021). The collection of data relating to the flood resilience of built assets should include various data sources as required data are not available in a single source in public or private domains, which is challenging. Hence, significant efforts need

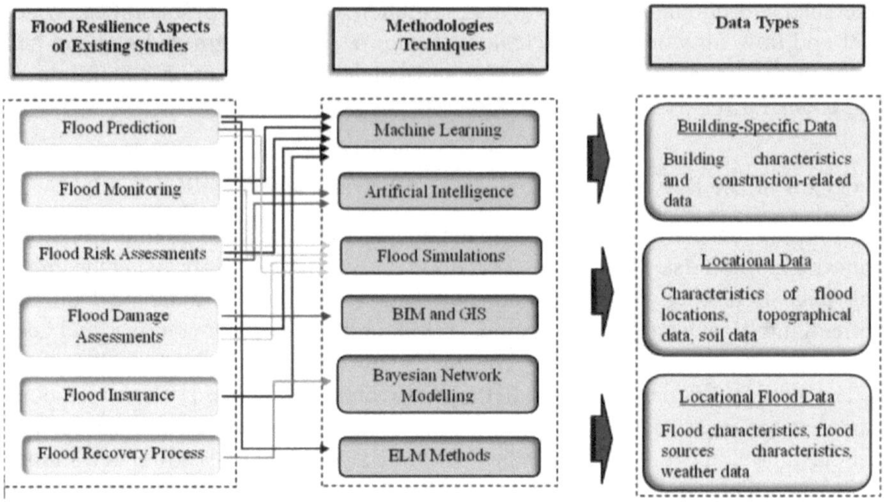

Figure 12.2 Summary of existing data-driven studies.

to be taken to facilitate the collection of data at the built assets level. The need for extensive effort, the involvement of experts, and the need for both qualitative and quantitative data have also been identified as challenges. Also, it is important to identify the relationships of existing data for flood resilience; however, non-linear relationships between the dependent parameters/data and a lack of mathematically proven equations or studies to represent these relationships have discouraged the use of data for flood resilience of built assets (Nofal & Lindt, 2020; Sen, Dutta, Kabir, 2021). The different challenges mentioned in the literature were mainly identified relating to data acquisition, classification, management, and utilisation, as shown in Table 12.1.

Strategies to Overcome the Challenges

The feasible strategies for the existing data challenges were identified from different sources of literature and are discussed in this section. Challenges of structuring and standardising data capture, data cleaning, and archiving can be addressed using software tools, including Hadoop, MapReduce, Apache Spark, Nifi, HBase, Hive, MongoDB (Li et al., 2017; De Bruijn et al., 2018; Gritta et al., 2018; Roth et al., 2018; Wang et al., 2018; Rumson & Hallett, 2019). Several authors mentioned the challenge of analysing data records in the form of narratives. To address this challenge, natural language processing (NLP) was proposed (Marz & Warren, 2015) to simplify and facilitate the data processing and conversation of qualitative data for quantitative data.

Outsourcing data processing tasks are suggested to address the challenges of the inherent complexity of open data outputs (De Bruijn et al., 2018; Gritta et al., 2018;

Table 12.1 Challenges of data for flood resilience

Challenges		*Source*
Data acquisition	Scarcity of data	Wang et al. (2018); Nofal and Lindt (2020); Sen, Dutta, Kabir (2021)
	Data are not available in a single source in public or private domains	Nofal and Lindt (2020); Sen, Dutta, Kabir (2021)
	Need extensive effort and the involvement of experts	Nofal and Lindt (2020); Sen, Dutta, Kabir (2021)
	Problems of openly available data can vary significantly depending on levels of respective government funding	De Bruijn et al. (2008); Gritta et al. (2018); Roth et al. (2018); Wang et al. (2018)
	Higher cost of openly available data	De Bruijn et al. (2008); Gritta et al. (2018); Roth et al. (2018); Wang et al. (2018)
	Limited availability of free data	De Bruijn et al. (2008); Gritta et al. (2018); Roth et al. (2018); Wang et al. (2018)
	The complexity of open data outputs	De Bruijn et al. (2008); Gritta et al. (2018); Roth et al. (2018); Wang et al. (2018)
	Obtaining data with the appropriate level of detail	Rumson and Hallett (2019); Wang et al. (2018)
	Restrictions made by governments to open data due to problems with complexity	Rumson and Hallett (2019); Wang et al. (2018)
	Lack of willingness of builders to share information with flood modelling companies	Rumson and Hallett (2019); Wang et al. (2018)
	Extracting data from open data sources like social media	Wang et al. (2018); Rumson and Hallett (2019)
	Challenges related to the ethical and political use of data	Sarker et al. (2020)
	A lack of coordination of data sharing exists among the organisations	Sarker et al. (2020)
Data classification	Data records in the form of narratives are difficult to handle when many data need to be analysed	de Bruijn et al. (2008); Gritta et al. (2018); Roth et al. (2018); Wang et al. (2018)
	Range of uncertainties and spurious and incomplete data	Rumson and Hallett (2019); Wang et al. (2018)
	Invalid data associated with social networking and crowdsourcing sites	Sarker et al. (2020)

(*Continued*)

Table 12.1 (Continued)

Challenges		*Source*
Data management	Inaccuracies, incompleteness, and presented in an unusable form	Li et al. (2017); Sarker et al. (2020); Rumson and Hallett (2019)
	Issues of data granularity and compatibility	Rumson and Hallett (2019); Libarikian et al. (2017)
	Issues of data quality	De Bruijn et al. (2008); Gritta et al. (2018); Roth et al. (2018); Wang et al. (2018)
	Data pre-processing, structuring and standardising data capture, data cleaning and archiving	Gritta et al. (2018); Roth et al. (2018); Wang et al. (2018)
	Issues of data privacy and validation of data	Li et al. (2017); Sarker et al. (2020)
	Reliability of data sources	De Bruijn et al. (2008); Gritta et al. (2018); Roth et al. (2018); Wang et al. (2018)
	Difficulties in processing the data language, which needs diverse translation requirements and additional time and resources for data inputs	Rumson and Hallett (2019); Libarikian et al. (2017)
	Developing patterns between data	Chang et al. (2013); Joseph and Kakade (2014)
Data utilisation	Need for using both qualitative and quantitative data	Nofal and Lindt (2020); Sen, Dutta, Kabir (2021)
	To identify the relationships between existing data	Nofal and Lindt (2020); Sen, Dutta, Kabir (2021)
	Non-linear relationships between the dependent parameters/data	Nofal and Lindt (2020); Sen, Dutta, Kabir (2021)
	Lack of mathematically proven equations or studies to represent these relationships	Nofal and Lindt (2020); Sen, Dutta, Kabir (2021)
	Data sources available with different standards	Li et al. (2017); Rumson and Hallett (2019)
	Adoption of the most appropriate data techniques	Rumson and Hallett (2019); Libarikian et al. (2017)
	Gaps exist between human and technological capacity that create challenges in handling information	Sarker et al. (2020)

Roth et al., 2018; Wang et al., 2018). To overcome the challenges in developing patterns between data, Li et al. (2017) and Rumson and Hallett (2019) suggested artificial neural networks for spotting patterns and understanding relationships between data variables. To address the challenges of accessing data sources (e.g., social media), NLP, geocoding, and computer vision were further proposed (Wang

Table 12.2 Existing strategies to overcome the challenges

Strategies		*Source*
Data acquisition	NLP, geocoding, and computer vision to access the data	Wang et al. (2018); Rumson and Hallett (2019)
	Data protection, privacy, and safeguarding data rights	
	Making free access to the public and ethical frameworks for data usage	Nofal and Lindt (2020); Sen, Dutta, Kabir (2021)
	Set up clear cross-government accountability, governance, and funding for data to support the delivery of the data	Davies (2022)
	Use data strategies to identify and address the barriers to better use of data	Davies (2022)
Data classification	Development of rules, standards, and common ways to collect, store, record, and manage data	Davies (2022)
	Making standardised ways of recording data	Han et al. (2011)
	Develop a consistent approach to balancing competing demands between standardisation and local data requirements	Chen and Zhang (2014)
Data management	Natural language processing (NLP) to simplify and facilitate the data processing and converting qualitative data for quantitative data	Chen and Zhang (2014), Han et al. (2011)
	Outsourcing data processing tasks	De Bruijn et al. (2008); Gritta et al. (2018); Roth et al. (2018); Wang et al. (2018)
	Development of an open-source framework that allows distributed storage, processing, and analysis of large datasets	Han et al. (2011)
	Use of software tools including Hadoop, MapReduce, Apache Spark, Nifi, HBase, Hive, and MongoDB to overcome the challenges of data structuring and standardising data capture, data cleaning and archiving	Li et al. (2017); De Bruijn et al. (2008); Gritta et al. (2018); Roth et al. (2018); Wang et al. (2018); Rumson and Hallett (2019)
	Artificial neural networks for spotting patterns and understanding relationships between data variables	Li et al. (2017); Rumson and Hallett (2019)
Data utilisation	Implement guidance for front-line staff for handling data	Chen and Zhang (2014); Han et al. (2011)
	Designing a system important in data sourcing and incorporating more datasets and variables	Chen and Zhang (2014); Han et al. (2011)
	Making a valid single source of information	Nofal and Lindt (2020); Sen, Dutta, Kabir (2021)
	Outsourcing technically expert people	Sarker et al. (2020)

et al., 2018; Rumson & Hallett (2019). Table 12.2 shows the strategies suggested in the existing literature to overcome the identified data challenges.

When data sources are available with different sources, causing issues for the users, new standards were suggested to harmonise different data standards (Li et al., 2017; Rumson & Hallett, 2019). Table 12.2 shows the summary of strategies available in the existing literature to overcome the data challenges. The strategies are separately mentioned under each category of data challenges.

Conclusion

Flooding has made a considerable demand on built asset resilience for improving the sustainability of communities. Due to the intrinsic technological ability to comprehend the dynamics on built assets and communities of flood-induced consequences, data-driven approaches are rapidly growing in today's context of flood resilience. Data is becoming a powerful tool that helps to understand how community factors and built asset features affect resilience and recovery for improving the societal and environmental sustainability of communities.

Built environments generate large amounts of data that have a huge potential in improving built asset resilience through improved and informed resilience-related decisions. Despite the wide spectrum of data and its potential to improve flood resilience, current studies often focus on limited data usage for built asset resilience. This chapter, therefore, synthesised existing literature on data-driven built asset flood resilience to discover the prospects of data utilisation, challenges, and available strategies to enhance the deployment of data potentials for built asset flood resilience towards sustainability.

In summary, the existing context of and use of data for flood resilience aspects, emerging data-driven technologies, data sources, and various data types are discussed in this chapter together with data challenges and potential existing strategies. These findings are significant for decision-makers to create data-inspired plans for resilient built assets to enhance sustainable outcomes in communities.

References

Abdel-Mooty, M. N., Yosri, A., El-Dakhakhni, W., & Coulibaly, P. (2021). Community flood resilience categorization framework. *International Journal of Disaster Risk Reduction, 61*, 102349. https://doi.org/10.1016/j.ijdrr.2021.102349

Adamovic, M., Branger, F., Braud, I., & Kralisch, S. (2016). Development of a data-driven semi-distributed hydrological model for regional scale catchments prone to Mediterranean flash floods. *Journal of Hydrology, 541*, 173–189. https://doi.org/10.1016/j.jhydrol.2016.03.032

Adeniyi, O., Rathnasiri, P., Oladinrin, O. T., & Rana, M. Q. (2022). Strategic capabilities for enhancing the flood resilience of business premises: An expert review and lessons from case studies. *Sustainability, 14*(16), 9814.

Aguilar, J. V., Langarita, P., Linares, L., Gómez, M., & Rodellar, J. (2013). An adaptive predictive approach for river level forecasting. *Journal of Hydroinformatics, 15*(2), 232–245. https://doi.org/10.2166/hydro.2012.172

Ahmadalipour, A., & Moradkhani, H. (2019). A data-driven analysis of flash flood hazard, fatalities, and damages over the CONUS during 1996–2017. *Journal of Hydrology, 578*, 124106. https://doi.org/10.1016/j.jhydrol.2019.12410

Ahmed, A. N., Yafouz, A., Birima, A. H., Kisi, O., Huang, Y. F., Sherif, M., Sefelnasr, A., & El-Shafie, A. (2022). Water level prediction using various machine learning algorithms: A case study of durian Tunggal River, Malaysia. *Engineering Applications of Computational Fluid Mechanics, 16*(1), 422–440. https://doi.org/10.1080/19942060.2021.2019128

Allahbakhshian-Farsani, P., Vafakhah, M., Khosravi-Farsani, H., & Hertig, E. (2020). Regional flood frequency analysis through some machine learning models in semi-arid regions. *Water Resources Management, 34*(9), 2887–2909. https://doi.org/10.1007/s11269-020-02589-2

Amadio, M., Scorzini, A. R., Carisi, F., Essenfelder, A. H., Domeneghetti, A., Mysiak, J., & Castellarin, A. (2019). Testing empirical and synthetic flood damage models: The case of Italy. *Natural Hazards and Earth System Sciences, 19*(3), 661–678. https://doi.org/10.5194/nhess-19-661-2019

Amininia, K., & Saghebian, S. M. (2021). Uncertainty analysis of monthly river flow modeling in consecutive hydrometric stations using integrated data-driven models. *Journal of Hydroinformatics, 23*(4), 897–913. https://doi.org/10.2166/hydro.2021.142

Amirebrahimi, S., Rajabifard, A., Mendis, P., & Ngo, T. (2015). A framework for a microscale flood damage assessment and visualization for a building using BIM–GIS integration. *International Journal of Digital Earth*, 9(4), 363–386. https://doi.org/10.1080/17538947.2015.1034201

Araghinejad, S., Fayaz, N., & Hosseini-Moghari, S. (2018). Development of a hybrid data driven model for hydrological estimation. *Water Resources Management, 32*(11), 3737–3750. https://doi.org/10.1007/s11269-018-2016-3

Arneson, E., Deniz, D., Javernick-Will, A., Liel, A., & Dashti, S. (2017). Information deficits and community disaster resilience. Natural Hazards Review, *18*(4). https://doi.org/10.1061/(asce)nh.1527-6996.0000251

Bigueras, R. T., Torio, J. O., & Palaoag, T. D. (2018). A data-driven architectural framework for LGUs in disaster preparedness and management system. *International Journal of Machine Learning and Computing*, *8*(3), 256–261. https://doi.org/10.18178/ijmlc.2018.8.3.696

Bucar, R. C., & Hayeri, Y. M. (2022). Quantitative flood risk evaluation to improve drivers' route choice decisions during disruptive precipitation. *Reliability Engineering & System Safety, 219*, 108202. https://doi.org/10.1016/j.ress.2021.108202

Carozza, D. A., & Boudreault, M. (2021). A global flood risk modeling framework built with climate models and machine learning. *Journal of Advances in Modeling Earth Systems, 13*(4). https://doi.org/10.1029/2020ms002221

Cesarini, L., Figueiredo, R., Monteleone, B., & Martina, M. L. (2021). The potential of machine learning for weather index insurance. *Natural Hazards and Earth System Sciences, 21*(8), 2379–2405. https://doi.org/10.5194/nhess-21-2379-2021

Chen, L., Sun, N., Zhou, C., Zhou, J., Zhou, Y., Zhang, J., & Zhou, Q. (2018). Flood forecasting based on an improved extreme learning machine model combined with the backtracking search optimization algorithm. *Water, 10*(10), 1362. https://doi.org/10.3390/w10101362

Chen, Z., Luo, J., Chen, N., Xu, R., & Shen, G. (2019). RFim: A real-time inundation extent model for large floodplains based on remote sensing big data and water level observations. *Remote Sensing, 11*(13), 1585. https://doi.org/10.3390/rs11131585

Chen, & Zhang, C. Y. (2014). Data-intensive applications, challenges, techniques and technologies: A survey on Big Data. *Information Sciences*, *275*, 314–347.

Cian, F., Marconcini, M., & Ceccato, P. (2018). Normalized difference flood index for rapid flood mapping: Taking advantage of EO big data. *Remote Sensing of Environment*, *209*, 712–730. https://doi.org/10.1016/j.rse.2018.03.006

Dai, W., & Cai, Z. (2021). Predicting coastal urban floods using artificial neural network: The case study of Macau, China. *Applied Water Science*, *11*(10). https://doi.org/10.1007/s13201-021-01448-8

Davies, G. (2022, August 26). *Challenges in using data across government*. National Audit Office. www.nao.org.uk/insights/challenges-in-using-data-across-government/

De Bruijn, K., Mens, M., & Klijn, F. (2008). A method for developing long-term strategies for flood risk management. *Flood Risk Management: Research and Practice*, 793–801. https://doi.org/10.1201/9780203883020.ch90

DeWitt, P. D., Schuler, M. S., Visscher, D. R., & Thiel, R. P. (2017). Nutritional state reveals complex consequences of risk in a wild predator–prey c*ommunity. Proceedings of the Royal Society B: Biological Sciences, 284*(1858), 20170757. https://doi.org/10.1098/rspb.2017.0757

Dunn, S., & Gonzalez-Otalora, S. (2021). Development of an adaptive solution to increase infrastructure system resilience based upon a location-allocation methodology. *Journal of Infrastructure Systems*, *27*(1). https://doi.org/10.1061/(asce)is.1943-555x.0000585

Edwards, L., Mullagh, L., Towe, R., Nundloll, V., Dean, C., Dean, G., & Simm, W. (2017). Data-driven decisions for flood risk management. Retrieved from www.ensembleprojects.org/wp-content/uploads/2017/11/Conference-paper-Data-driven-decisions-for-flood-risk-management-1.pd

Feloni, E., Mousadis, I., & Baltas, E. (2019). Flood vulnerability assessment using a GIS-based multi-criteria approach—The case of Attica region. *Journal of Flood Risk Management*, *13*(S1). https://doi.org/10.1111/jfr3.12563

Fotovatikhah, F., Herrera, M., Shamshirband, S., Chau, K., Faizollahzadeh Ardabili, S., & Piran, M. J. (2018). Survey of computational intelligence as basis to big flood management: Challenges, research directions and future work. *Engineering Applications of Computational Fluid Mechanics*, *12*(1), 411–437. https://doi.org/10.1080/19942060.2018.1448896

Gritta, M., Pilehvar, M. T., & Collier, N. (2018). Which Melbourne? Augmenting geocoding with maps. Proceedings of the 56th Annual Meeting of the Association for Computational Linguistics (Volume 1: Long Papers). https://doi.org/10.18653/v1/p18-1119

Guidotti, R., Gardoni, P., & Rosenheim, N. (2019). Integration of physical infrastructure and social systems in communities' reliability and resilience analysis. *Reliability Engineering & System Safety*, *185*, 476–492. https://doi.org/10.1016/j.ress.2019.01.008

Han, E. Haihong, Le, G., & Du, J. (2011). *Survey on NoSQL database*. Proceedings of IEEE 6th International Conference on Pervasive computing and Applications (ICPCA), pp. 363–366.

Hao, H., & Wang, Y. (2021). Assessing disaster impact in real time: Data-driven system integrating humans, hazards, and the built environment. *Journal of Computing in Civil Engineering*, *35*(5). https://doi.org/10.1061/(asce)cp.1943-5487.0000970

Hapuarachchi, H. A., Wang, Q. J., & Pagano, T. C. (2011). A review of advances in flash flood forecasting. *Hydrological Processes*, *25*(18), 2771–2784. https://doi.org/10.1002/hyp.8040

Henry, D., & Emmanuel Ramirez-Marquez, J. (2012). Generic metrics and quantitative approaches for system resilience as a function of time. *Reliability Engineering & System Safety*, *99*, 114–122. https://doi.org/10.1016/j.ress.2011.09.002

Khosravi, K., Panahi, M., Golkarian, A., Keesstra, S. D., Saco, P. M., Bui, D. T., & Lee, S. (2020). Convolutional neural network approach for spatial prediction of flood hazard at national scale of Iran. *Journal of Hydrology*, *591*, 125552. https://doi.org/10.1016/j.jhydrol.2020.125552

Li, W., Xu, B., & Wen, J. (2016). Scenario-based community flood risk assessment: A case study of Taining County town, Fujian province, China. *Natural Hazards*, *82*(1), 193–208. https://doi.org/10.1007/s11069-016-2187-2

Li, T., Xie, N., Zeng, C., Zhou, W., Zheng, L., Jiang, Y., Yang, Y., Ha, H., Xue, W., Huang, Y., Chen, S., Navlakha, J., & Iyengar, S. S. (2017). Data-driven techniques in disaster information management. *ACM Computing Surveys, 50*(1), 1–45. https://doi.org/10.1145/3017678

Marz, J., & Warren, N. (2015). *Big data: Principles and best practices of scalable realtime data systems*. Simon & Schuster.

Mashaly, J., & Ghoneim, E. (2018). Flash flood hazard using optical, radar, and stereo-pair derived DEM: Eastern desert, Egypt. *Remote Sensing*, *10*(8), 1204. https://doi.org/10.3390/rs10081204

Malgwi, M. B., Schlögl, M., & Keiler, M. (2021). Expert-based versus data-driven flood damage models: A comparative evaluation for data-scarce regions. *International Journal of Disaster Risk Reduction*, *57*, 102148. https://doi.org/10.1016/j.ijdrr.2021.102148

Nofal, O. M., & Van de Lindt, J. W. (2020). Understanding flood risk in the context of community resilience modeling for the built environment: Research needs and trends. *Sustainable and Resilient Infrastructure*, 1–17. doi:10.1080/23789689.2020.1722546

Nofal, O. M., Van de Lindt, J. W., Do, T. Q. (2020). Multi-variate and single-variable flood fragility and loss approaches for buildings. *Reliability Engineering and System Safety*, *202*, 106971.

Ochoa-Rodriguez, S., Wang, L., Gires, A., Pina, R. D., Reinoso-Rondinel, R., Bruni, G., Ichiba, A., Gaitan, S., Cristiano, E., Van Assel, J., Kroll, S., Murlà-Tuyls, D., Tisserand, B., Schertzer, D., Tchiguirinskaia, I., Onof, C., Willems, P., & Ten Veldhuis, M. (2015). Impact of spatial and temporal resolution of rainfall inputs on urban hydrodynamic modelling outputs: A multi-catchment investigation. *Journal of Hydrology*, *531*, 389–407. https://doi.org/10.1016/j.jhydrol.2015.05.035

Pradhan, P., Costa, L., Rybski, D., Lucht, W., & Kropp, J. P. (2017). A systematic study of sustainable development goal (SDG) interactions. *Earth's Future*, *5*(11), 1169–1179. https://doi.org/10.1002/2017ef000632

Rumson, A. G., & Hallett, S. H. (2019). Innovations in the use of data facilitating insurance as a resilience mechanism for coastal flood risk. *Science of the Total Environment*, *661*, 598–612. https://doi.org/10.1016/j.scitotenv.2019.01.114

Roth, D., Vink, M., Warner, J., & Winnubst, M. (2018). Watered-down politics? Inclusive water governance in The Netherlands. *Ocean & Coastal Management, 150*, 51–61. https://doi.org/10.1016/j.ocecoaman.2017.02.020

Sen, M. K., Dutta, S., & Kabir, G. (2021). Flood resilience of housing infrastructure modeling and quantification using a Bayesian belief network. *Sustainability*, *13*(3), 1026. https://doi.org/10.3390/su13031026

Sen, M. K., Dutta, S., & Laskar, J. I. (2021). A hierarchical Bayesian network model for flood resilience quantification of housing infrastructure systems. *ASCE-ASME Journal of*

Risk and Uncertainty in Engineering Systems, Part A: Civil Engineering, *7*(1). https://doi.org/10.1061/ajrua6.0001108

Sarker, M. N., Peng, Y., Yiran, C., & Shouse, R. C. (2020). Disaster resilience through big data: Way to environmental sustainability. *International Journal of Disaster Risk Reduction*, *51*, 101769. https://doi.org/10.1016/j.ijdrr.2020.101769

Skinner, C. J., Bellerby, T. J., Greatrex, H., & Grimes, D. I. (2015). Hydrological modelling using ensemble satellite rainfall estimates in a sparsely gauged river basin: The need for whole-ensemble calibration. *Journal of Hydrology*, *522*, 110–122. https://doi.org/10.1016/j.jhydrol.2014.12.052

Tran, V., Porcher, R., Tran, V., & Ravaud, P. (2017). Predicting data saturation in qualitative surveys with mathematical models from ecological research. *Journal of Clinical Epidemiology*, *82*, 71–78.e2. https://doi.org/10.1016/j.jclinepi.2016.10.001

Towe, R., Dean, G., Edwards, L., Nundloll, V., Blair, G., Lamb, R., Hankin, B., & Manson, S. (2020). Rethinking data-driven decision support in flood risk management for a big data age. *Journal of Flood Risk Management*, *13*(4). https://doi.org/10.1111/jfr3.12652

Wang, L., Ochoa-Rodríguez, S., Van Assel, J., Pina, R. D., Pessemier, M., Kroll, S., Willems, P., & Onof, C. (2015). Enhancement of radar rainfall estimates for urban hydrology through optical flow temporal interpolation and Bayesian Gauge-based adjustment. *Journal of Hydrology*, *531*, 408–426. https://doi.org/10.1016/j.jhydrol.2015.05.049

Wang, R., Mao, H., Wang, Y., Rae, C., & Shaw, W. (2018). Hyper-resolution monitoring of urban flooding with social media and crowdsourcing data. *Computers & Geosciences*, *111*, 139–147. https://doi.org/10.1016/j.cageo.2017.11.008

Yabe, T., Rao, P. S., Ukkusuri, S. V., & Cutter, S. L. (2022). Toward data-driven, dynamical complex systems approaches to disaster resilience. *Proceedings of the National Academy of Sciences*, *119*(8). https://doi.org/10.1073/pnas.2111997119

Yereseme, A. K., Surendra, H. J., & Kuntoji, G. (2022). Sustainable integrated urban flood management strategies for planning of smart cities: A review. *Sustainable Water Resources Management*, *8*(3). https://doi.org/10.1007/s40899-022-00666-5

Yin, J., Ren, X., Liu, R., Tang, T., & Su, S. (2022). Quantitative analysis for resilience-based urban rail systems: A hybrid knowledge-based and data-driven approach. *Reliability Engineering & System Safety*, *219*, 108183. https://doi.org/10.1016/j.ress.2021.108183

13 Post-Occupancy Evaluation for Sustainable Performance of Green Buildings: A Case Study of a Selected Project

Oluwadamilare Olamide Ilesanmi, Ayodeji Emmanuel Oke, John Aliu, and Oluwapelumi Ruth Oluwaleye

Introduction

The global construction industry increasingly recognizes the importance of sustainable practices in mitigating environmental degradation and promoting resource efficiency (Mehta et al., 2023). Sustainable building design, especially green buildings, represents a significant advancement, aiming to balance structural, environmental, and community needs (Gil-Ozoudeh et al., 2023). However, buildings that fail to meet performance expectations post-occupancy can impose unintended social and economic repercussions, such as reduced occupant health and productivity (Boissonneault and Peters, 2023). For green buildings, where substantial resources are invested to achieve optimal indoor environmental quality (IEQ), resource efficiency, and sustainability, evaluating whether these buildings perform as intended after construction is essential (Yahia et al., 2024). Thus, post-occupancy evaluation has emerged as a crucial process for assessing the real-world performance of green buildings (Elsayed et al., 2023).

Despite the widespread adoption of post-occupancy evaluation frameworks globally, research reveals that green buildings may not always meet anticipated performance standards, especially regarding IEQ and energy efficiency (Agyekum et al., 2023; Elforgani et al., 2023). Studies have found that green buildings may underperform in some cases compared to non-green counterparts (Laiche et al., 2021; Yahia et al., 2024). While post-occupancy evaluation research has expanded internationally, there remains a significant gap in its application within developing countries, particularly in Nigeria, where limited studies examine whether green buildings meet sustainable and user-centric performance goals post-occupancy. This lack of empirical evaluation impedes the construction industry's understanding of green building effectiveness in the Nigerian context, where environmental, social, and economic challenges intersect with sustainable development goals.

DOI: 10.1201/9781003380559-13

In Nigeria, the need for rigorous post-occupancy evaluation is underscored by the country's unique socio-environmental context, which influences how sustainable buildings perform under local conditions (Etudaiye et al., 2024; Mba et al., 2023). Although green building projects have gained traction in Nigerian urban centers, post-delivery evaluations of these buildings are infrequent, limiting opportunities to correct design shortcomings and assess the satisfaction levels of occupants (Adedeji et al., 2020). Unlike in developed countries, where comprehensive post-occupancy evaluation frameworks like Leadership in Energy and Environmental Design (LEED) for Building Operations and Maintenance support the evaluation of green buildings, Nigeria's construction industry lacks systematic post-occupancy evaluation practices (Adegoke et al., 2021; Zhao et al., 2024). This gap not only restricts the growth of sustainable practices in Nigeria but also hinders researchers and industry professionals from understanding how Nigerian green buildings align with global sustainability standards.

This study seeks to address these gaps by conducting a post-occupancy evaluation of Heritage Place, a landmark green building in Lagos, Nigeria. Specifically, this study investigates three research questions:

1 How does Heritage Place perform in terms of indoor environmental quality, water efficiency, and materials and resources?
2 What is the level of user satisfaction with Heritage Place's indoor environmental quality, building facilities, water efficiency, and materials and resources?
3 What is the relationship between the building's performance across the three criteria and user satisfaction?

By exploring these questions, this research contributes to a limited body of knowledge on green building performance in Nigeria, providing valuable insights into the operational effectiveness of such buildings. The findings from this study will inform Nigerian construction professionals, policymakers, and sustainability advocates about the long-term impacts of green buildings on user satisfaction and resource efficiency. In addition, this study offers a foundation for future post-occupancy evaluation studies in Nigeria, where sustainable building practices are essential to meeting sustainable development goals and promoting resilience in urban infrastructure.

Literature Review

Post-Occupancy Evaluation and Green Buildings

The concept of post-occupancy evaluation plays a vital role in the sustainable assessment of buildings, especially in evaluating the extent to which these buildings meet their design goals and user satisfaction (Boissonneault and Peters, 2023). Initially, building performance assessments were informal, and the results were applied sporadically in subsequent constructions (Mba et al., 2023). This lack of formal evaluation led to repetitive design flaws and performance issues. The emergence of post-occupancy evaluation represents a structured approach to assessing a building's operational effectiveness and aligning it with the expectations of users and clients (Zhao et al., 2024). For instance, Elsayed et al. (2023) regard post-occupancy evaluation as an invaluable tool that enables systematic assessment, focusing on both building performance and user satisfaction (Yahia et al., 2024). It is particularly relevant in the context of green buildings, where sustainability goals necessitate continuous performance monitoring (Laiche et al., 2021; Adedeji et al., 2020). In developed and developing countries, post-occupancy evaluations have become integral to green building projects (Zhao et al., 2024; Etudaiye et al., 2024). Developed nations such as the United States employ advanced techniques to assess green building performance, focusing on energy efficiency, IEQ, and resource use (Wang et al., 2024). Conversely, in developing nations, the approach often focuses on basic aspects such as indoor air quality and thermal comfort, with limited resources for in-depth assessments (Fakhabi et al., 2024). This contrast highlights the need for tailored evaluation practices across different contexts, which becomes evident when examining green building initiatives in Nigeria (Adedeji et al., 2020). For example, in the Nigerian context, post-occupancy evaluation can address issues of resource efficiency and sustainability in a rapidly urbanizing environment, making it an essential tool for advancing the green building agenda (Adegoke et al., 2021).

The Role of Post-Occupancy Evaluation in Green Buildings

Green buildings are designed with principles of sustainability, aiming to reduce environmental impact, conserve energy, and enhance user comfort (Agyekum et al., 2023). Post-occupancy evaluations are essential for measuring the performance of these buildings in terms of energy efficiency, water use, and IEQ (Elforgani et al., 2023). By assessing these factors, stakeholders can identify discrepancies between the intended and actual performance of green buildings, ensuring ongoing improvements (Laiche et al., 2021). For example, Gil-Ozoudeh et al. (2023) point out that even buildings awarded high green ratings may fail in specific areas, such as IEQ, if proper evaluations and maintenance are not conducted. In this way, post-occupancy evaluations are not just a check on current performance but are vital for informing future green building standards (Yahia et al., 2024). Globally, post-occupancy evaluations have highlighted the adaptability of green building principles to different climates and cultural contexts (Elsayed et al., 2023). However, developing countries may face challenges such as limited resources and expertise, making post-occupancy evaluation even more critical. In Nigeria, where green building practices are relatively new, post-occupancy evaluations are crucial to adapting green building principles effectively, considering local conditions and resource constraints (Etudaiye et al., 2024; Mba et al., 2023).

Heritage Place: A Model of Sustainable and Resilient Green Building Design in Ikoyi, Lagos, Nigeria

Heritage Place, located in Lagos, Nigeria, is an exemplary green building that integrates advanced sustainable features to minimize environmental impact while enhancing occupant comfort. This environmentally certified commercial building holds LEED certification, reflecting its commitment to energy efficiency and sustainable design. Heritage Place's power infrastructure includes three synchronized 1100 KVA generators, providing full redundancy, with diesel tanks capable of supporting operations for seven days. A water storage system ensures a four-day supply of potable water, complemented by a dedicated sprinkler water reserve for fire safety. The building's design optimizes natural light and ventilation, reducing energy demands for cooling and heating through strategic orientation and the use of high-efficiency glazing. Fresh air is supplied at 10 liters per second per person, facilitated by a central air system that incorporates carbon dioxide removal risers. A sophisticated building management system oversees airflow, cooling, and lighting, with VAV boxes and thermostats enabling personalized climate control. Water conservation is emphasized through rainwater harvesting, condensate recovery, and water recycling for landscaping and flushing. Motion sensor faucets reduce potable water usage, while stormwater attenuation helps manage runoff for up to a 20-year storm event. Sustainable construction materials, including fly ash to reduce embodied energy, further reflect Heritage Place's commitment to environmentally responsible practices. By combining these technical, design, and green features, Heritage Place stands as a model for sustainable commercial buildings in Nigeria.

Building Performance Criteria in Post-Occupancy

Evaluation studies on IEQ in green buildings often focus on key factors like air quality, lighting, thermal comfort, and acoustic performance (Laiche et al., 2021; Wang et al., 2024). These factors are critical to ensuring occupants' well-being and comfort (Etudaiye et al., 2024). Air quality, for instance, influences health outcomes, while thermal comfort affects energy efficiency (Arowoiya et al., 2024). A study by Amoah and Smith (2024)) reveals that while green buildings generally perform better than traditional buildings in terms of IEQ, some areas still require improvement. Gaps exist in consistently measuring IEQ across different building types and climates (Ilesanmi et al., 2024). More attention should be given to the long-term effects of ventilation systems and the integration of natural lighting, as well as acoustic treatments, which are often overlooked in current assessments (Arowoiya et al., 2024). Water efficiency is a vital aspect of green buildings, with studies highlighting innovative practices such as rainwater harvesting, greywater reuse, and low-flow fixtures (Iwuanyanwu et al., 2024). These measures not only conserve water but also enhance building performance by reducing water bills and energy consumption (Gil-Ozoudeh et al., 2023). Despite these benefits, water efficiency is a relatively understudied area, particularly in regions like Nigeria (Umoh et al., 2024). While some research has focused on water conservation in green buildings, more studies are needed to assess how such practices impact user satisfaction and building performance in different climates (Wang et al., 2024). There is a clear opportunity for further exploration into water efficiency solutions tailored to local conditions in developing countries (Adedeji et al., 2020). The use of sustainable materials plays a critical role in green building performance (Yahia et al., 2024). Omopariola et al. (2024) argued that materials with low environmental impact, such as recycled content or locally sourced products, contribute to a building's overall sustainability. However, studies often overlook the long-term durability and environmental consequences of these materials (Umoh et al., 2024). In developing countries, including Nigeria, there is a noticeable gap in research on the durability of green building materials (Adedeji et al., 2020). This gap could be addressed by exploring the performance of sustainable materials over time and their impact on resource conservation, building longevity, and user satisfaction.

User Satisfaction in Post-Occupancy Evaluation

User satisfaction with IEQ is an essential aspect of green buildings, with research emphasizing factors like air quality, thermal comfort, and lighting (Zhao et al., 2024). Hassanain et al. (2024) mentioned that users in green-certified buildings tend to report higher satisfaction levels with these aspects compared to those in traditional buildings. However, differences in user perceptions often arise, especially regarding air quality, as some users may still experience discomfort due to ventilation system inefficiencies (Arowoiya et al., 2024). While comfort and lighting in green buildings generally meet expectations, further research could address how well these systems adapt to varying climatic conditions and user needs, ensuring a more personalized experience (Wang et al., 2024). Satisfaction with building facilities is another significant factor in green buildings. Halicioglu and Gurel (2024) highlight that users are generally satisfied with amenities like energy-efficient appliances and advanced building systems, but a gap often exists between the intended quality of these facilities and users' actual experiences. Issues such as maintenance problems, usability, or insufficient performance of certain features may undermine user satisfaction (Rasheed et al., 2024). Post-occupancy evaluations frequently show that while green buildings aim to offer superior facilities, these amenities sometimes fall short in practice, affecting overall user experience (Yahia et al., 2024). User perceptions of water efficiency and sustainable materials are critical in green buildings. Although studies reveal that users value water-saving technologies and eco-friendly materials, there is often a gap between their awareness of these features and their actual effectiveness (Iwuanyanwu et al., 2024). In some cases, users may not fully recognize the benefits of these systems unless they see direct results in their daily usage (Onubi et al., 2024). In regions like Nigeria, there is a lack of user satisfaction studies in green buildings. Economic, social, and environmental factors unique to developing countries may influence satisfaction levels (Etudaiye et al., 2024; Mba et al., 2023). These regional gaps highlight the importance of examining user experiences in such contexts to better understand local needs and improve green building practices.

Relationship between Building Performance and User Satisfaction

Studies analyzing the relationship between building performance and user satisfaction in green buildings often utilize methodologies like surveys, performance monitoring, and statistical analysis to establish connections (Halicioglu and Gurel, 2024). Surveys are frequently used to gather user feedback on factors like comfort, lighting, air quality, and amenities, while performance monitoring tracks energy efficiency, IEQ, and water usage (Wang et al., 2024). Statistical analysis helps in understanding how these factors correlate with user satisfaction (Rasheeed et al., 2024). According to Yahia et al. (2024), there is a positive relationship between high building performance and user satisfaction in green buildings, with improved energy efficiency, air quality, and thermal comfort contributing to higher satisfaction levels. However, some studies indicate occasional disconnects, where users may not fully appreciate the benefits of high-performance features unless they directly impact their daily lives or comfort (Onubi et al., 2024; Mba et al., 2023). Additionally, cultural and environmental factors unique to developing countries can alter how users perceive and value building performance, highlighting the importance of understanding these dynamics in the context of Nigeria to enrich existing literature (Alohan et al., 2023).

Research Gaps and Contribution of This Study

There is limited research on post-occupancy evaluations of green buildings in Nigeria, especially regarding how building performance affects user satisfaction. Many studies focus on developed countries, leaving a gap in empirical data for developing nations. Additionally, there is insufficient understanding of the relationship between building performance and user satisfaction in the Nigerian context. This study addresses these gaps by providing real-world data on the post-occupancy performance of a green building in Nigeria and examining its impact on user satisfaction. The findings will offer new insights into sustainability practices, inform policymaking, and contribute to the green building certification process in Nigeria, helping to improve both the design and management of green buildings in the country.

Research Methodology

This study aims to assess the sustainable performance of Heritage Place, a green building in Lagos, Nigeria, through a post-occupancy evaluation examining both technical aspects and user-centered experiences. Heritage Place in Ikoyi, Lagos, was selected due to its recognition as one of Nigeria's most prominent green buildings.

As a leading economic hub in West Africa, Lagos significantly influences the construction, financial, technology, and finance sectors (Oke et al., 2023). Primary data was collected using a well-structured questionnaire that effectively reached a broad audience of building users, offering a reliable data source (Ilesanmi et al., 2024). A quantitative approach was chosen for its ability to quantify findings and support generalizations through statistical analysis (Sürücü and Maslakci, 2020).

The study used convenience sampling to select participants based on accessibility, availability, and proximity, aligning with Stratton (2021). Heritage Place building occupants responded to a questionnaire created from an extensive literature review. Out of 55 distributed questionnaires, 47 were accurately completed and used for data analysis. Before distribution, a pilot study involving 15 experienced project and construction managers with expertise in green buildings and smart technologies was conducted to ensure the survey's clarity and effectiveness (Quiles et al., 2019). Their feedback refined the final questionnaire, which comprised four sections: (1) respondent background, (2) sustainable performance assessment of Heritage Place using LEED criteria (IEQ, water efficiency, and materials and resources), (3) user satisfaction evaluation with a focus on IEQ, building facilities, water efficiency, and materials and resources, and (4) the relationship between building performance and user satisfaction, examining how technical features influence occupant satisfaction.

For the first objective, the sustainable performance of Heritage Place was assessed based on three LEED criteria: IEQ, water efficiency, and materials and resources. To evaluate the level of performance, a 5-point Likert scale was used, with scores ranging from 5 for "Excellent" to 1 for "Very Poor." This scale provided a clear measure of the building's sustainability performance based on the responses from participants. For the second objective, which evaluates user satisfaction, the study used a satisfaction-focused 5-point Likert scale. The respondents rated their satisfaction with aspects like IEQ, building facilities, water efficiency, and materials and resources, with a scale ranging from 5, representing "Very Satisfied," to 1, representing "Very Dissatisfied." This scale allowed for an in-depth understanding of the user experience and satisfaction with various features of the building. The third objective, which analyzes the relationship between building performance and user satisfaction, employed the same 5-point Likert scale to ensure compatibility in evaluating both performance and satisfaction. For building performance, the scale ranged from 5, indicating "Excellent," to 1, indicating "Very Poor," while for satisfaction, the scale ranged from 5 for "Very Satisfied" to 1 for "Very Dissatisfied." By using the same scale for both aspects, the study ensured a consistent and comprehensive analysis of how performance correlates with user satisfaction.

Statistical Package for the Social Sciences software was utilized to analyze the data. The first objective, measuring the sustainable performance of Heritage Place, was evaluated using mean ranking analysis, where mean scores for each LEED criterion provided insights into performance levels. For the second objective, RSI, ASSac, and MSS metrics offered detailed insights into user satisfaction across different building features. Finally, the third objective, analyzing the relationship between performance and satisfaction, was assessed using Kendall's tau correlation statistics to evaluate the strength and direction of relationships between building performance variables and user satisfaction levels.

Results

Level of Building Performance by LEED Criteria

Table 13.1 presents the performance level of the green building based on LEED criteria: IEQ, water efficiency, and materials and resources. The highest-performing items are natural light during the daytime (4.7234), cleanliness of the building (4.7234), and ventilation (4.7021) under IEQ, indicating high satisfaction. For water efficiency, water storage (4.8298) and reduction of indoor water use (4.5106) received the highest rankings, showing good performance in water management. However, in materials and resources, waste management (4.3617) and the storage of recyclables (4.0638) scored lower, suggesting that improvements are needed

Table 13.1 Level of Performance of the Green Building by LEED Criteria

LEED Criteria	*Items*	*Mean*	*Ranking Within Group*	*Level of Performance on LEED Criteria*
Indoor Environmental Quality	Natural light during daytime	4.7234	1	4.5957
	Cleanliness of the building	4.7234	2	
	The ventilation in the building	4.7021	3	
	Air quality inside the building	4.6809	4	
	Paints used for the building	4.6809	5	
	The temperature of office space in the cold season	4.6170	6	
	Artificial light during daytime	4.5532	7	
	Automatic presence detectors	4.5106	8	
	Sound insulation quality	4.5106	9	
	Vegetation maintenance	4.4894	10	
	The temperature of office space in the dry season	4.3617	11	
Water Efficiency	water storage	4.8298	1	4.3526
	Reduction of indoor water use	4.5106	2	
	Motion sensor-controlled hand-wash basins	4.4043	3	
	Water harvesting	4.3617	4	
	Rainwater management	4.2979	5	
	Motion sensor-controlled urinals	4.1702	6	
	Water re-use	3.8936	7	
Materials and Resources	Waste management	4.3617	1	3.9681
	Storage of recyclables	4.0638	2	
	Collection of recyclables	4.0426	3	
	Re-use of some materials	3.4043	4	
	Pooled			4.3055

in resource management. Overall, the building performs best in IEQ and water efficiency but requires enhancements in materials and resource management to improve sustainability.

Level of User Satisfaction

To determine the level of user satisfaction, data was obtained from the users and the results are shown below.

Level of Satisfaction with Indoor Environmental Quality

The user satisfaction with IEQ in the green building in Table 13.2 reveals high ratings across most items. The cleanliness of the building, the type of paints used, and sound insulation quality received some of the highest satisfaction scores, with RSI values above 0.90 and MSS ratings above 4.8, indicating a strong level of user approval. Other factors, such as the temperature in office spaces during both wet and dry seasons, as well as natural light during daytime, also scored highly, further demonstrating user satisfaction. The air quality inside the building and the ventilation system were also well rated, with mean satisfaction scores ranging from 4.57 to 4.66. Overall, the analysis shows that building users are generally very satisfied with the IEQ, with the most prominent areas of satisfaction being cleanliness and the quality of materials used in the building's design.

Table 13.2 Level of Satisfaction of Users on Indoor Environmental Quality

S/N	*Items*	*RSI*	*ASSac*	*MSS*
1	Cleanliness of the building	0.97	227	4.809
2	The kind of paints used for the building	0.94	220	4.830
3	Sound insulation quality	0.93	219	4.809
4	The temperature of your office space during the wet season	0.91	213	4.681
5	Natural light during daytime	0.91	213	4.660
6	Air quality inside the building	0.91	213	4.574
7	Artificial light during daytime	0.91	215	4.574
8	Air quality inside the office	0.90	211	4.723
9	Ventilation in the building	0.89	210	4.660
10	The temperature of your office space during the dry season	0.89	208	4.574
11	Distraction from noise coming from outside the office	0.89	208	4.574
12	Vegetation in the building	0.87	205	4.383
13	Design of office	0.86	202	4.596
14	Distraction from noise coming from outside the building	0.85	199	4.574
15	Automatic presence detectors	0.85	199	4.404
16	Size of your office	0.84	197	4.468
	Pooled	0.89	210	4.618

Level of Satisfaction with Building Facilities

The level of user satisfaction with building facilities in Table 13.3 indicates generally high ratings across all items, with some aspects receiving particularly favorable responses. Building design, security access control systems, and high-efficiency lighting materials achieved the highest satisfaction scores, with mean scores exceeding 4.7, reflecting positive user perceptions. Essential facilities such as the passenger lift, restroom, and parking spaces also garnered high satisfaction levels, with scores around 4.6 to 4.7. Power redundancy systems, air-conditioning, and water storage tanks were also well received, indicating that users are content with the functional aspects of the building. However, satisfaction with office space, furniture, and motion sensor-controlled hand-wash urinals showed slightly lower scores, indicating potential areas for improvement. Overall, users expressed strong satisfaction with the building's facilities, with the pooled mean satisfaction score of 4.61 showing general approval of the building's functionality.

Level of Satisfaction with Water Efficiency

The level of user satisfaction with water efficiency features in Table 13.4 shows moderate satisfaction with most aspects. Items related to rainwater management, harvesting, and storage received satisfactory ratings, with mean scores ranging from 4.13 to 4.22, indicating positive user perceptions. The highest satisfaction

Table 13.3 Level of Satisfaction with Building Facilities

S/N	*Item*	*RSI*	*ASSac*	*MSS*
1	Building design	0.82	193	4.851
2	Security access control system	0.81	191	4.809
3	High-efficiency lighting materials	0.81	190	4.787
4	Passenger lift	0.80	188	4.787
5	Power redundancy standby generators	0.80	187	4.766
6	Building safety	0.80	187	4.723
7	Automatic presence detectors	0.80	187	4.723
8	Generator diesel tanks	0.79	185	4.681
9	Sheltered car drop-off	0.78	184	4.660
10	Restroom	0.77	181	4.596
11	Parking space	0.77	180	4.574
12	Internet services	0.77	180	4.574
13	Air conditioning	0.77	180	4.574
14	Water storage tank	0.77	180	4.574
15	Motion sensor-controlled hand-wash basins	0.75	176	4.489
16	Office space	0.74	175	4.468
17	Furniture	0.70	164	4.234
18	Motion sensor-controlled hand-wash urinals	0.68	160	4.149
	Pooled	0.77	182	4.612

Table 13.4 Level of Satisfaction of Users on Water Efficiency

S/N	*Items*	*RSI*	*ASSac*	*MSS*
1	Re-use of harvested water	0.68	159	4.128
2	Management of rainwater	0.69	162	4.191
3	Rainwater harvesting	0.69	163	4.213
4	Water storage	0.69	163	4.255
5	Reduction of indoor water use	0.72	169	4.340
	Pooled	0.69	163	4.226

Table 13.5 Level of Satisfaction with Materials and Resources

S/N	*Items*	*RSI*	*ASSac*	*MSS*
1	Waste management	0.61	143	3.787
2	Re-use of recyclables	0.51	121	3.234
3	Collection of recyclables	0.52	123	3.277
4	Storage of recyclables	0.56	131	3.489
	Pooled	0.55	130	3.447

score was associated with the reduction of indoor water use, which scored 4.25, suggesting strong approval from users in this area. The pooled mean score of 4.23 reflects general contentment with water efficiency practices in the building. Overall, users show a favorable response to the building's water efficiency measures, with some room for improvement in areas like re-use of harvested water and rainwater management.

The level of user satisfaction with materials and resources is relatively low as indicated in Table 13.5, with all items scoring below 4. The highest satisfaction score was for waste management, with a mean of 3.79, indicating moderate approval. The reuse and collection of recyclables received lower satisfaction scores, with means ranging from 3.23 to 3.49. The pooled mean score of 3.45 reflects a general tendency toward dissatisfaction with materials and resources management, pointing to areas where improvements may be necessary to enhance user satisfaction in this category.

Relationship between Building Performance and User Satisfaction

The correlation analysis between building performance and occupant satisfaction in Table 13.6 reveals very strong positive relationships across all three key performance areas. IEQ shows a very strong positive correlation with both water efficiency (0.973) and materials and resources (0.959), indicating that improvements in environmental quality significantly enhance satisfaction with water efficiency and resource management. Similarly, water efficiency has a strong positive correlation with materials and resources (0.969), suggesting that users satisfied with water management practices

Table 13.6 Correlation Analysis Between Building Performance and Occupant Satisfaction

Variables for Building Performance	*Indoor Environmental Quality*	*Water Efficiency*	*Materials and Resources*
Correlation Coefficient	.973**	.959**	.969**
Variables for User Satisfaction	Indoor Environmental Quality	Water Efficiency	Materials and Resources

are also likely to be satisfied with the building's resource management features. Overall, the analysis shows that the building's performance in these sustainability areas is closely linked to occupant satisfaction, with strong interrelations suggesting that improving one aspect is likely to enhance satisfaction across the others.

Discussions and Implication of Findings

The background information of the research population has the green building which is the Heritage Place users as the respondents for this study and since the population is finite, it involved every one of them that was available.

Determining the Level of Building Performance of the Heritage Place

The building generally performs well in IEQ and water efficiency, with strong performance in areas such as natural light, air quality, and cleanliness. However, materials and resource management, particularly waste handling and the storage of recyclables, require improvement to enhance sustainability. This finding aligns with previous studies, such as those by Iwuanyanwu et al. (2024), who noted the significant role of IEQ in occupant satisfaction, as seen in this building's high ratings. Similarly, Umoh et al. (2024) highlighted the importance of water efficiency in sustainable buildings, a strength also reflected in the current study. Furthermore, Liache et al. (2021) pointed to the challenges in managing materials and resources in green buildings, which corresponds with the need for improvements in waste and resource management in the buildings under study. Thus, for design teams, including architects and facility managers, this finding emphasizes the need for a more integrated approach to materials and resource management during the design and operational phases of green buildings. Enhancing waste management systems and incorporating sustainable materials will be critical in optimizing the building's overall sustainability. The construction industry can use these insights to refine green building standards, focusing on improving materials management alongside IEQ and water efficiency. Policymakers should also consider these findings when updating regulations and incentives for green building projects, ensuring they address all aspects of sustainability, including resource management.

Determining the Level of Users' Satisfaction

The second objective assesses user satisfaction across four critical areas: IEQ, building facilities, water efficiency, and materials and resources. The results show that users are most satisfied with IEQ, particularly in aspects like cleanliness, air quality, and natural light. Satisfaction with building facilities, such as security systems, lifts, and restrooms, is also high, while water efficiency and materials management received relatively lower satisfaction scores, especially in areas like waste management and recyclables reuse. This aligns with the findings of Li et al. (2024), who highlighted that air quality and natural light are key drivers of occupant satisfaction in green buildings. Kineber et al. (2023) also emphasized that satisfaction with building amenities, including security and lifts, plays a significant role in overall user contentment. On the other hand, Harimi et al. (2023) noted that water efficiency and material reuse often lag behind other green building features, a trend observed in this study as well. These findings present a valuable opportunity for construction teams to prioritize the aspects that matter most to users. The focus should be on maintaining and enhancing IEQ, ensuring that natural light, air quality, and cleanliness are top-notch. In parallel, improving building facilities and addressing users' concerns about water efficiency and materials management can significantly elevate overall satisfaction. Policymakers have a unique opportunity to shape regulations that could drive improvements in water efficiency and material management, ensuring that these areas are given equal attention in the green building certification process, creating a more holistic approach to sustainable building practices.

Relationship between the Level of Building Performance and Users' Satisfaction

The third objective reveals a correlation between building performance and occupant satisfaction, particularly in the areas of IEQ, water efficiency, and materials and resources. These results suggest that improvements in these areas directly contribute to higher satisfaction levels among users, highlighting the importance of prioritizing them in future building design and operations. Previous studies corroborate this finding. For example, Li et al. (2024) found that enhanced IEQ, such as better air quality and natural lighting, directly influenced occupants' comfort and overall satisfaction. Similarly, a study by Halicioglu and Gurel (2024) demonstrated that efficient water use and waste management practices not only reduce operational costs but also improve occupant perceptions of sustainability. Furthermore, Yahia et al. (20224) emphasized the importance of materials and resources management, noting that sustainable material use significantly impacts both environmental outcomes and occupant contentment. These findings are instrumental for design teams, and building developers, who should focus on optimizing building performance in IEQ, water efficiency, and resource management to boost user satisfaction. By addressing these elements, they can ensure that green buildings not only meet sustainability targets but also deliver the high-quality environments that occupants value. In the construction industry, emphasis should be on designing buildings that prioritize the user experience while maintaining sustainability, ensuring that occupant satisfaction and building performance go hand in hand for future developments.

Conclusions and Recommendations

This study contributes to building sustainable communities through digital transformation by evaluating the post-occupancy performance of green buildings, an essential step to ensure these structures meet sustainability and user satisfaction goals. Sustainable building practices encompass efficient use of resources, adoption of renewable energy, reduction of pollution and waste, and encouragement of reuse and recycling. Post-occupancy evaluation offers valuable insights into these areas, assessing the performance of buildings after they are occupied and identifying areas for improvement in newly introduced environmentally friendly structures.

In this study, Heritage Place, a green building in Ikoyi, Lagos, was assessed to determine the overall building performance, user satisfaction levels, and the relationship between these factors. The evaluation focused on three LEED criteria: IEQ, water efficiency, and materials and resources, using data collected from building users via questionnaires. The results align with findings from other developing countries, confirming that post-occupancy evaluations help address client satisfaction, meet user needs, refine design practices, and inform regulatory processes to support sustainable goals. Findings indicate that Heritage Place demonstrates a strong overall performance, fulfilling its role as a sustainable environment and contributing to the construction sector's global sustainable practices. Users expressed high satisfaction with IEQ and water efficiency but lower satisfaction with materials and resources, suggesting that, while the building largely achieves its sustainability goals, further improvements in materials and resource management would enhance overall performance and user satisfaction.

Based on the findings of this study, to address the identified gap in materials and resource management, future projects should emphasize sustainable material selection, waste management, and recycling practices to align with environmental goals and boost occupant satisfaction. Additionally, developing streamlined waste collection systems within green buildings can enhance user satisfaction and set a new standard for sustainability in the industry. Future studies should explore more comprehensive approaches to materials management in green buildings and investigate the impacts of user-driven feedback on building performance adjustments, providing further insights to inform sustainable construction practices.

References

Adedeji, J.A., Fadamiro, J.A. and Odeyale, T.O. (2020). Design toolkits for campus open spaces from post-occupancy evaluations of federal universities in South-west Nigeria. *Built Environment Project and Asset Management*, *10*(2), pp. 296–311.

Adegoke, A.S., Ajayi, C.A., Oladokun, T.T. and Ayodele, T.O. (2021). A post-occupancy evaluation of students' halls of residence in Obafemi Awolowo University, Ile-Ife, Nigeria. *Property Management*, *39*(2), pp. 163–179.

Agyekum, K., Akli-Nartey, E.E.K., Kukah, A.S. and Agyekum, A.K. (2023). Importance-performance analysis (IPA) of the indoor environmental quality (IEQ) of an EDGE-certified building in Ghana. *International Journal of Building Pathology and Adaptation*, *41*(1), pp. 73–95.

Alohan, E.O., Oyetunji, A.K., Amaechi, C.V., Dike, E.C. and Chima, P. (2023). An agreement analysis on the perception of property stakeholders for the acceptability of smart buildings in the Nigerian built environment. *Buildings*, *13*(7), p. 1620.

Amoah, C. and Smith, J. (2024). Barriers to the green retrofitting of existing residential buildings. *Journal of Facilities Management*, *22*(2), pp. 194–209.

Arowoiya, V.A., Onososen, A.O., Moehler, R.C. and Fang, Y. (2024). Influence of thermal comfort on energy consumption for building occupants: The current state of the art. *Buildings*, *14*(5), p. 1310.

Boissonneault, A. and Peters, T. (2023). Concepts of performance in post-occupancy evaluation post-probe: A literature review. *Building Research & Information*, *51*(4), pp. 369–391.

Elforgani, M.S.A., Alabsi, A.A. and Alwarafi, A. (2024). Strategic approaches to design teams for construction quality management and green building performance. *Buildings*, *14*(7), p. 2020.

Elsayed, M., Romagnoni, P., Pelsmakers, S., Castaño-Rosa, R. and Klammsteiner, U. (2023). The actual performance of retrofitted residential apartments: Post-occupancy evaluation study in Italy. *Building Research & Information*, *51*(4), pp. 411–429.

Etudaiye, A.E., Ahmad, W. and Sada, B.M. (2024). Post-occupancy evaluation of design quality in higher education: A case study of TETFUND buildings in the Federal Polytechnic, Damaturu. *Fane-Fane International Multi-Disciplinary Journal*, *8*(1), pp. 112–124.

Fakhabi, M.M., Hamidian, S.M. and Aliehyaei, M. (2024). Exploring the role of the Internet of Things in green buildings. *Energy Science & Engineering*, *12*(9), pp. 3779–3822.

Gil-Ozoudeh, I., Iwuanyanwu, O., Okwandu, A.C. and Ike, C.S. (2023). Sustainable urban design: The role of green buildings in shaping resilient cities. *International Journal of Applied Research in Social Sciences*, *5*(10), pp. 674–692.

Halicioglu, F.H. and Gurel, K. (2024). The perceived quality of green buildings: From a review of occupant satisfaction towards a conceptual framework. *International Journal of Quality & Reliability Management*, *41*(3), pp. 757–777.

Hassanain, M.A., Sanni-Anibire, M.O. and Mahmoud, A.S. (2024). An assessment of users' satisfaction with a smart building on university campus through post-occupancy evaluation. *Journal of Engineering, Design and Technology*, *22*(4), pp. 1119–1135.

Ilesanmi, O.O., Oke, A.E., Moyanga, D.T. and Aliu, J. (2024). Unveiling the imperatives: Drivers for incorporating smart building technologies in the construction sector in Nigeria. In: Chakir, A., Bansal, R. and Azzouazi, M. (eds.), *Industry 5.0 and Emerging Technologies*. Studies in Systems, Decision and Control, vol. 565. Springer, Cham. doi:10.1007/978-3-031-70996-8_10

Iwuanyanwu, O., Gil-Ozoudeh, I., Okwandu, A.C. and Ike, C.S. (2024). Cultural and social dimensions of green architecture: Designing for sustainability and community well-being. *International Journal of Applied Research in Social Sciences*, *6*(8), pp. 1951–1968.

Karimi, H., Adibhesami, M.A., Bazazzadeh, H. and Movafagh, S. (2023). Green buildings: Human-centered and energy efficiency optimization strategies. *Energies, 16*(9), p. 3681.

Kineber, A.F., Oke, A., Aliu, J., Hamed, M.M. and Oputu, E. (2023). Exploring the adoption of cyber (digital) technology for sustainable construction: A structural equation modeling of critical success factors. *Sustainability*, *15*(6), p. 5043.

Laiche, A.B., Dahlan, N.D., Shari, Z. and Jaafar, M.F.Z. (2021). A review of the global evidence: Comparison of occupant satisfaction between green and non-green certified buildings. *Journal of Sustainable Tropical Design Research and Practice*, *14*(2), pp. 1–11.

Li, M., Xu, K. and Huang, S. (2024). Evaluation of green and sustainable building project based on extension matter-element theory in smart city application. *Computational Intelligence*, *40*(1), p. 12286.

Mba, E.J., Okeke, F.O., Ezema, E.C., Oforji, P.I. and Ozigbo, C.A. (2023). Post occupancy evaluation of ventilation coefficient desired for thermal comfort in educational facilities. *Journal of Human, Earth, and Future, 4*(1), pp. 88–102.

Mehta, A., Shah, T. and Thomas, R.J. (2023). Strategies to address environmental degradation in developing nations: A multifaceted approach. *Journal of Intelligent Connectivity and Emerging Technologies*, *8*(2), pp. 17–34.

Oke, A.E., Aliu, J., Oni, S.B. and Ilesanmi, O.O. (2023). The complexities of embracing mechatronics in the Nigerian construction industry. *Construction Innovation*. doi:10.1108/CI-07-2023-0154.

Omopariola, E.D., Olanrewaju, O.I., Albert, I., Oke, A.E. and Ibiyemi, S.B. (2024). Sustainable construction in the Nigerian construction industry: Unsustainable practices, barriers and strategies. *Journal of Engineering, Design and Technology*, *22*(4), pp. 1158–1184.

Onubi, H.O., Carpio, M. and Hassan, A.S. (2024). Job satisfaction in green construction projects: Antecedent roles of green work climate, pro-environmental construction practice and green human capital. *Engineering, Construction and Architectural Management*, *31*(5), pp. 1857–1878.

Quiles, N.N., McCullough, A.K. and Piao, L. (2019). Validity and reliability of the exercise vital sign questionnaire in an ethnically diverse group: A pilot study. *Journal of Primary Care and Community Health,* 10, 2150132719844062. doi: 10.1177/2150132719844062.

Rasheed, E.O., Khoshbakht, M. and Baird, G. (2024). "Do office workers' comments corroborate the ratings they give their buildings?" A qualitative analysis of comments on operational factors in the workplace. *Facilities*, *42*(3/4), pp. 274–320.

Stratton, S.J. (2021). Population research: Convenience sampling strategies. *Prehospital and Disaster Medicine*, *36*(4), pp. 373–374.

Sürücü, L. and Maslakci, A. (2020). Validity and reliability in quantitative research. *Business & Management Studies: An International Journal*, *8*(3), pp. 2694–2726.

Umoh, A.A., Nwasike, C.N., Tula, O.A., Adekoya, O.O. and Gidiagba, J.O. (2024). A review of smart green building technologies: Investigating the integration and impact of AI and IOT in sustainable building designs. *Computer Science & IT Research Journal*, *5*(1), pp. 141–165.

Wang, X., Zhu, L., Tang, Y., Deng, H. and Wang, H. (2024). Multiple paths to green building popularization under the TOE framework—A qualitative comparative analysis of fuzzy sets based on 26 Chinese cities. *Sustainability*, *16*(21), p. 9360.

Yahia, A.K.M., Rahman, D.M.M., Shahjalal, M. and Morshed, A.S.M. (2024). Sustainable materials selection in building design and construction. *International Journal of Science and Engineering*, *1*(04), pp. 106–119.

Zhao, J., Abdul Aziz, F., Deng, Y., Ujang, N. and Xiao, Y. (2024). A review of comprehensive post-occupancy evaluation feedback on occupant-centric thermal comfort and building energy efficiency. *Buildings*, *14*(9), p. 2892.

14 Evaluation of Historical Buildings With Heritage Building Information Modelling

Merve Anaç, Yusuf Arayici, and Bülent Haznedar

Introduction

Historical buildings are unique and constitute cultural heritage because they reflect the construction system, material properties, aesthetic understanding, and socio-cultural characteristics of the period. The most basic feature required for the transfer and protection of this heritage to future generations is to carry out correct restoration works, to document and archive them with the right methods (Anaç & Arun, 2024). The correct determination of the current status of the historical building works has an impact on the intervention to be taken on and the maintenance and repair decisions to be made about the structures. Various technologies have been developed to determine the current situation and, in the study of historical buildings, are utilised in order to determine the current status of the buildings accurately and in a short time. These include automated close-range photogrammetry (SfM) and more recently terrestrial laser scanning (TLS) and especially personal laser scanning (PLS) (Pavelka, 2022). However, these systems create point cloud data that can be used for the current condition assessment of historic buildings and are data for project design processes. However, much more needs to be done to understand, evaluate, decide on interventions, and archive historic buildings.

Historical buildings have a heterogeneous information structure, including semantic and geometric information due to interventions, time, and various technical features. It is important that this information is contained in a single file format to prevent loss and storage. Also, in order to manage so much information, many domains (e.g. restorers report the state of conservation of materials and elements, historians contribute to the understanding of different historical phases, etc.) (Previtali et al., 2020) must be able to work together. For this reason, a system developed for the storage, management, and evaluation of information on historical buildings has been sought.

Building information management (BIM) is an innovative methodology for managing all information about a building to optimise the building life cycle for new buildings (Arayici, 2008; Aygün, 2011). In other words, BIM describes a digital representation of the physical and functional characteristics of a built asset (Martinelli et al., 2022). The fact that BIM systems can control information in an

DOI: 10.1201/9781003380559-14

organised and systematic way raises the question of whether it can be used for historical buildings. In the literature, the evaluation of historical buildings with BIM systems is called historical building information management (HBIM). However, since BIM is a system developed for new buildings, there are difficulties in using it for historical architectural heritage. For this reason, its use is not yet widespread. Although the HBIM methodology is not yet fully applicable to historic building restoration, it has revolutionised the entire panorama of cultural heritage documentation since 2009 (Diara, 2022). The term HBIM refers to the recording, modelling, and management of monuments and buildings of cultural significance and has so far received more academic interest than practical applications. HBIM studies for different purposes, such as restoration of historical buildings, energy saving, or structural monitoring, are defined as a possible strategy since they are not yet widely used (Gargaro et al., 2018).

It was discovered that historical building survey, restitution, and restoration projects drawn in 2D with traditional methods can be drawn in 3D through the Autodesk Revit program. It is obvious that the preparation of 3D models of historical buildings will facilitate easier understanding of the structures. However, this is not enough for historical building studies. The recording of historical building heritage requires the recording of all semantic information and the observation of its changes over time. In this respect, the documentation of the buildings in question requires multifaceted descriptions with time layers (Li et al., 2019). It was discovered that the time dimension of historical buildings can be integrated into 3D models through HBIM (Arayici et al., 2017). This means that with HBIM, historic buildings can be not only modelled in 3D, but also that the models can be combined with time layers to characterise them in 4D (Brusaporci et al., 2023). HBIM systems allow these time layers to be displayed in a single format. In this way, this is an innovative method that will allow us to not only understand the current state of the buildings but also to evaluate the processes they have gone through as a whole.

Within the scope of this study, a field study will be carried out to explain HBIM systems with a holistic approach by combining them with time layers in the historical structure. The building chosen for the field study is preferably a building that has changed many times. For the field study, a traditional Gaziantep house was selected and the preparation of the 4D HBIM model and the advantages of the HBIM model were explained through the model.

In this chapter, studies in the field of HBIM are summarised and the uniqueness of this study is revealed. Then, the characteristics of Gaziantep traditional houses are presented. HBIM models, including the general characteristics of Gaziantep traditional houses, and the building element libraries of these houses are prepared. These libraries are used in HBIM models. Finally, the presentation features and conveniences of HBIM models are listed.

Background

The transfer of historical buildings to future generations has intensified in the context of the digital archiving process with developing technologies (Pauwels et al.,

2008). An exemplary organisation conducting active and extensive research into possible techniques for this purpose is the World Heritage Centre (WHC) of the United Nations Educational, Scientific, and Cultural Organization (UNESCO) (Quintero et al., 2002). These digitisation processes were first carried out in 2D. The format differences between the geometric representation and semantic information of 2D drawings prepared by traditional methods cause various difficulties and information disconnections in data analyses. In order to understand historical buildings in the most accurate way and to transfer them to future generations, 3D digitisation is needed (Anaç & Arun, 2024). The widespread use of 3D models of historical buildings and the increase in the accuracy of these models can be associated with the widespread use of laser scanning devices. The 3D geometric features of historical buildings can be revealed with the point clouds produced as a result of scanning the historical building and its surroundings with various laser scanning devices (Özdemir & Remondino, 2019). These data provide detailed geometric information for historic buildings but are raw data lacking semantic information. However, when the literature studies are examined, it is seen that studies involving parametric objects using laser scanning data have been carried out (Delis et al., 2017; Lanzara et al., 2022).

It is possible to have both geometric information and semantic information together in 3D modelling of historical buildings with BIM-based programs. In this context, HBIM elements produced in BIM-based programs consist of parametric objects. The parametric object involves the application of a predetermined set of interrelated rules (algorithms) that set out the formal and dimensional constraints of an element and the data (parameters) that contain various semantic information and allow the controlled exchange of this information (Caetano et al., 2020). This parametric technique tries to integrate variable restrictions and dimensions to improve the geometric representation of building elements when modelling historical architecture (Osello et al., 2018; Quattrini et al., 2015), as well as to extend the bounds of such models by incorporating alphanumeric information. This means that HBIM models are accessible, implementable, assessable, manageable, and controllable.

In the literature, there are various studies regarding the conversion of laser scanning data used in 3D data generation into HBIM models. One of the first studies on this subject was carried out by Baik et al. in 2013. In the study, laser scanning data were re-modelled by means of BIM-based programs and a model consisting of parametric objects was obtained with a case study. The study concluded that 3D models can enable remote inspection of indoor and outdoor spaces with a better understanding than that shown in 2D flat drawings (Baik et al., 2013). BIM-based programs are much faster than traditional methods for the design of projects for new buildings. One of the primary reasons for this is the building element libraries within the BIM-based programs.

This library consists of building elements currently used in the market. However, since historical buildings are unique structures, the elements in BIM-based programs cannot be used in HBIM models. Special and unique elements must be produced for each structure. For this reason, libraries consisting of parametric objects were produced in many studies in the literature. Various programs were used in the libraries produced. Brusaporci et al. used laser scanning data as a base and converted structural elements into parametric objects in their study in 2018. Dynamo application was used to convert models into parametric objects in the production of the library (Brusaporci et al., 2018). A similar study was carried out by Klapa and Gawronek. In the study, BIM of the historic building was used to obtain not only high quality three-dimensional data but also an object documentation consisting of parametric objects (Klapa & Gawronek, 2023). As seen from the studies, laser scans alone are not sufficient to produce HBIM models. HBIM models also need semantic information.

In order for the geometric and semantic information in HBIM systems to be used in a common way by various stakeholders, it is necessary to store the information in a systematic way (Lee et al., 2016). However, it is not possible to access this data due to the commercial concerns of the producers of BIM-based programs. Therefore, open access software was needed (Diara, 2022). There are also efforts to manage the information of historical buildings (Celli & Ottoni, 2023). However, when all the studies in the literature are examined, it is seen that the time parameter is excluded. However, for historical buildings, the historical development period of the structures is a very important parameter for the interventions to be made.

This study will use a field study example to explain how the time layer can be used in HBIM models. In this way, it is an original and significant study for the use of this temporal change, which is important for historical buildings.

Case Study

In the study, traditional civil residences in the Gaziantep region were selected. The area designated as the Gaziantep conservation zone consists of historical buildings that are largely authentic in terms of characteristic structure. Gaziantep is a region where many cultures live together, and these cultural differences are reflected in architectural works. The intensive migration experienced in recent years led to the use of idle historical buildings and resulted in unconscious intervention to these historical buildings. In time, due to changes in lifestyle and user needs, new buildings were built and historical buildings have changed their functions. However, these changes have distorted the city centre both texturally and visually, making it difficult to understand the cultural characteristics of the city. For these reasons, conservation, renovation, restoration, intervention, and archiving systems have become much more important to transfer historical buildings to future generations.

The fact that many historical buildings were damaged and even destroyed in the Kahramanmaraş earthquake, which is called as the disaster of the century, on 6 February 2023, proved once again that these buildings should be protected and archived. When the historical buildings damaged after the earthquake are analysed, it is seen that those that underwent restoration were much more damaged than those that had not undergone intervention. This situation shows that the decisions or decision methods taken in the restoration of historical buildings to this point should be reconsidered. Structures that have not yet been drawn or archived but have been completely demolished herald the destruction of cultural heritage.

Features of Traditional Gaziantep Traditional Residences

The relationship of Gaziantep houses with the street is in the form of windowless walls on the lower floors and walls with windows on the upper floors, as in traditional Anatolian Turkish houses. One of the most important reasons for designing the buildings in this way is the privacy perception of Turkish society. The buildings are not on a straight line due to the narrow streets and unevenness of the terrain, depending on the climate (Kuban, 2001). To make good use of the plot location and topographical features of the residential buildings, it is very difficult to distinguish the designed houses according to their plan forms.

When the general characteristics of Gaziantep traditional residential buildings are analysed, it is seen that they have high walls and courtyards. These houses are shaped according to history, beliefs, culture, tradition, geographical conditions, and economy (Kanalıcı, 2012). According to various studies, one of the important factors in the formation of the urban texture is the region's climatic characteristics. Due to the hot climate, the streets are narrow and the houses are designed inwards to create shade in courtyards and ensure the comfort of daily life (Yıldırım & Çağdaş, 2017).

Typological Evaluation

Gaziantep houses are built in three different architectural styles as single-storey, two-storey, or three-storey houses, generally mansion-type houses, houses with open iwan, and houses with elevation differences between floors. Gaziantep houses generally consist of two buildings in a large garden. One of these buildings is orientated to the north and the other to the south. The building orientated to the north is referred to as such, while the building directed to the south is called the qibla house.

In Gaziantep traditional houses, the north house in the north direction is used in the summer and spring months and the qibla house in the south direction is used in the winter and autumn months, allowing the building to be used effectively according to the climate. The storey height of the qibla house is higher than that of the hub house, which increases the amount of shading in the courtyard and makes the use of the courtyard favourable in the summer months (Atalar, 2004). In Günkut Akın's (1984) doctoral thesis, in which the typologies and meanings of historical houses in Eastern and Southeastern Anatolia are analysed (Akın, 1984), it is written that Gaziantep houses fit the iwan house type, as seen in Figure 14.1.

Depending on the karstic nature of the region's geological structure, there may be various cavities under the ground. In traditional residential architecture, these caves are called "hazna" and were generally utilised as cellars (Ünal, 1998) (Figure 14.2).

In Gaziantep traditional houses, toilets are generally located in the courtyard and close to the entrance door. However, with the change of user needs in buildings that undergone restoration over time, the toilets were moved inside the building. When the families living in Gaziantep in the past were interviewed, it was found that the house were generally created by adding buildings due to the increase in

Figure 14.1 Plan typology of traditional Gaziantep residences.

the number of family members and needs. In this case, it was determined that there were elevation differences between the new and old buildings. Due to the difference between the floors, the spaces were connected to each other with a few steps (Figure 14.3). The level difference between floors is important data for the Gaziantep traditional building typology.

Structural Assessment

WALL

Keymık stone, havara stone, and karataş were used in the walls of traditional Gaziantep houses. Keymık stone has a white colour and is a very hard stone type. It has been used in courtyard walls and floors due to its high strength. Havara stone is a less durable and softer stone type compared to keymık stone; it is easy to cut and process. It is used in interior walls due to its low resistance to external influences (Gül, 2005). The external walls of the building are in the form of a crate wall formed by filling the area between two rows of cut stones with a material obtained from a mixture of lime, rubble, ash, white keymık powder, and hessian fibres (Kanalıcı, 2012) (Figure 14.4). The stone used on the external part of the wall is always keymık stone, while the stone used on the internal part is sometimes keymık and sometimes havara stone, depending on the economic status of the owner of the house.

The keymık and havara stones used in the building are not plastered. It is possible to come across examples where only the exterior surface of the garden wall

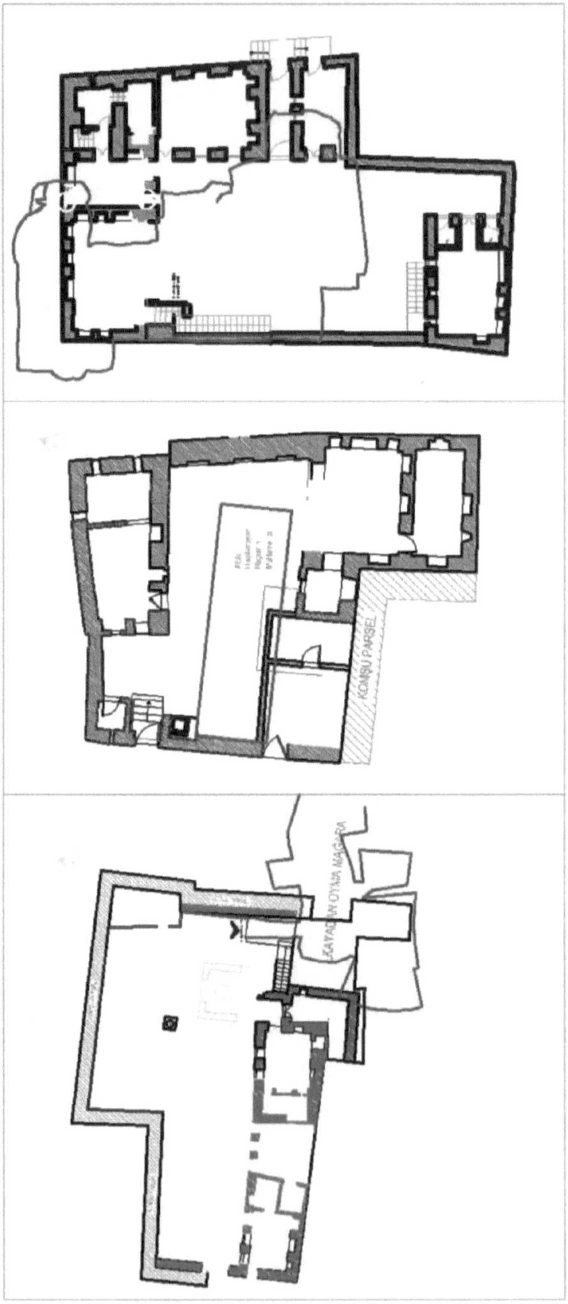

Figure 14.2 Examples of buildings with underground caves.

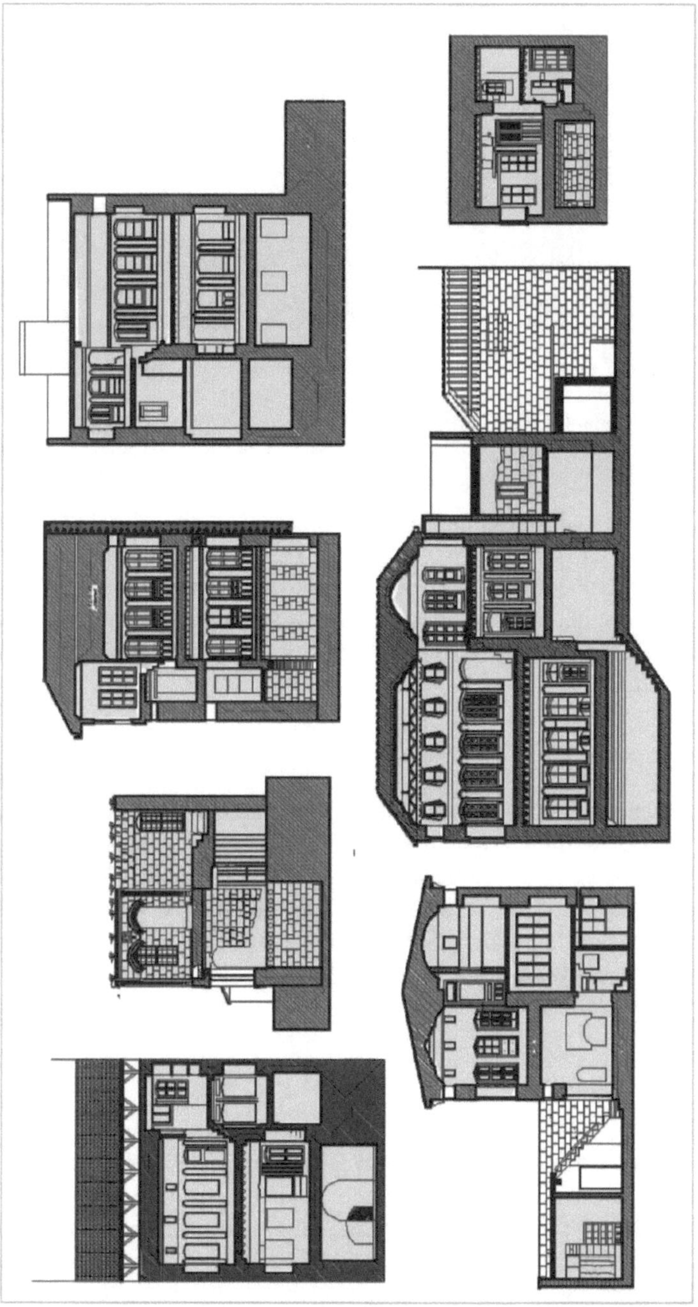

Figure 14.3 Buildings with levelling difference between floors.

Figure 14.4 Traditional Gaziantep residence masonry technique.

facing the street was plastered in order to protect from icing or rain in winter. The walls are covered with woodwork, not plaster. These coverings may consist of special carvings or ornaments and form a continuous whole along with the cabinet and window doors throughout the entire structure. In addition to creating aesthetic and original features, the coatings also affect the spaces in terms of thermal insulation and thermal comfort.

STAIRS

In the Gaziantep traditional construction system, stairs can be found in the courtyard or inside the building. High-strength keymık stone was generally used for the stairs. The dimensions of the stones used in the stairs are generally 30 cm in width, 21 cm in height, and 130 cm in length. It is possible to see various motifs on the lower surfaces of the stair stones. Stair positions and shapes vary from building to building. As seen in Figure 14.5, stairs can be designed parallel, perpendicular, or both according to the courtyard facade.

FLOORING

In Gaziantep traditional building types, the ground floors are made of stone material, but the upper floors are made of stone or wood coating on wooden beams placed on stone walls. When the flooring details are examined, a covering board can be seen on the wooden beams; there is rubble filling on the covering board and a flat floor is obtained with black plaster, which is today's screed on the rubble. Finally, the flooring is completed with floor covering elements. In this way, the thickness of the floor sometimes reaches up to 40 cm (Figure 14.6).

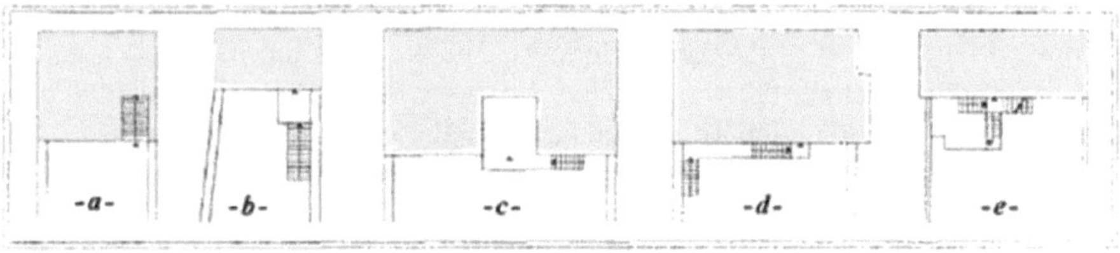

Figure 14.5 Relationship between courtyard, staircase, and building.

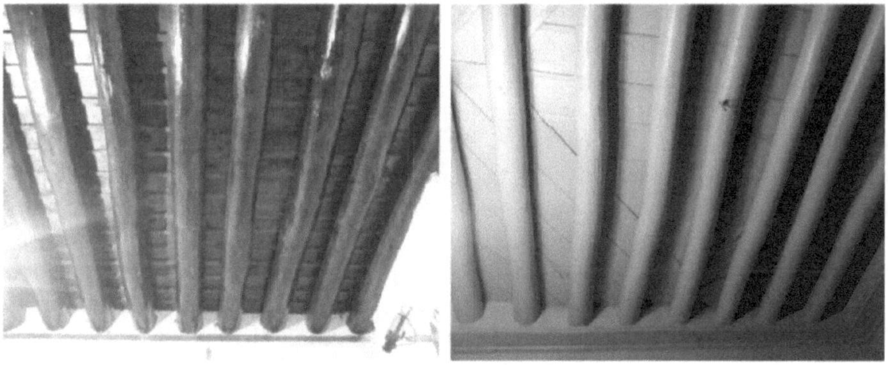

Figure 14.6 Flooring examples in traditional Gaziantep residence.

DOORS

In traditional Gaziantep housing types, it is possible to distinguish doors as room doors and garden doors as these differ from each other in terms of material, form, and ornamentation. It is possible to see arched and flat weft doors in both door types. When Gaziantep traditional housing types are analysed, one striking feature is the special embroideries of the doors opening to the courtyard. Exterior doors are usually wooden, but in order to prevent the doors from being affected by weather conditions, they are covered with sheet metal, whose traditional name is tudya, to protect the wood from external influences. The tudya is attached to the wooden doors with headed nails, and the order, number, and direction of these nails form the ornamentation on the door.

WINDOW

In Gaziantep traditional housing types, windows can be classified as basement, sofa, room, roof, and bird window according to their location. The number of windows facing the courtyard is higher than the number of windows facing the street. There are wrought iron windows outside the surfaces of the room windows facing both the street and the courtyard. There are also examples where wooden doors are used on the inward facing sides of the room windows. These covers form a whole together with the claddings along the wall surface. Covers and claddings increase the aesthetic and privacy perception as well as the thermal comfort of the spaces.

Almost every building has bird windows at the upper level of the windows, which are designed without frames, with frames, or with wooden doors. These are smaller compared to other windows, and the local people call them bird windows because they create safe areas for birds. When evaluated in terms of function, bird windows are designed to provide air circulation between the windows based on the principle of rising heated air. It is also seen that the bird windows in the region are sometimes shaped according to the religious beliefs of the house owner.

According to the study, the general characteristics, original architectural features, plan typologies, facades, and element features of Gaziantep traditional civil buildings have been assessed. As a result of the evaluation, it was determined that all the original qualities of civil buildings emerged from the climatic and cultural data of the region. As a result of the assessment, it was concluded that these elements should be preserved and archived due to their original and aesthetic qualities. In existing archiving systems, architectural drawings of these elements and semantic information are stored in separate formats. This situation often causes errors due to people during the transfer of data, such as data loss. The next section explains how to store both semantic and geometric information of buildings through HBIM models and how to use the time layer in HBIM models.

HBIM Model of Aynur Khan's Mansion

The building selected to be produced with the HBIM model is located on 1080 block 174 parcels. The building, which was first used as a mansion, is currently used as a boutique hotel registered to Aynur Zorkirişçi. There are two buildings connected by a courtyard within the parcel. There is no written record or inscription indicating the buildings' year of construction. However, when compared to the other buildings in the urban conservation area, it is thought that building number 1 in Figure 14.7 was built in the 19th century and building number two was built at a later date, approximately in the 1940s.

A cave was found under the house during the application. However, within the scope of this study, only the above ground structures were included in the model.

According to the survey, restitution, and restoration studies approved by the Regional Board for the Protection of Cultural Assets, the building's first restitution period occurred between 1890 and 1940, the second between 1940 and 1970, and the third between 1970 and 2005. Finally, the building survey and restoration work were carried out in 2015. These detected years are defined in the Autodesk Revit program, as shown in Figure 14.8. The time layer is defined through the phase tool under the manage tab in the Autodesk Revit program.

The completed time layers are seen under the phasing tab, which is the lower layer of the properties section, as shown in Figure 14.8. When the drawn-up stage is selected in this section, the building survey drawings taken in 2015

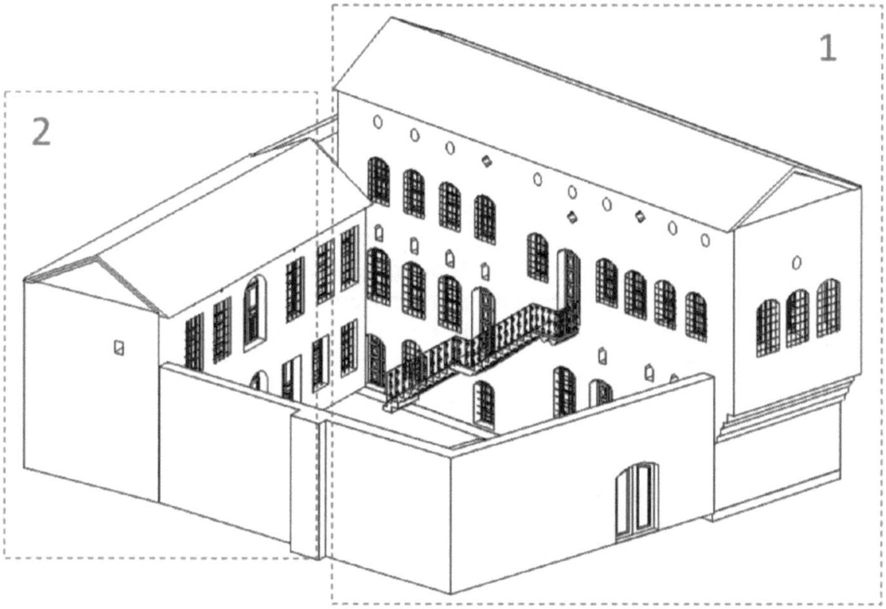

Figure 14.7 Current status model of Aynur Hanım Mansion.

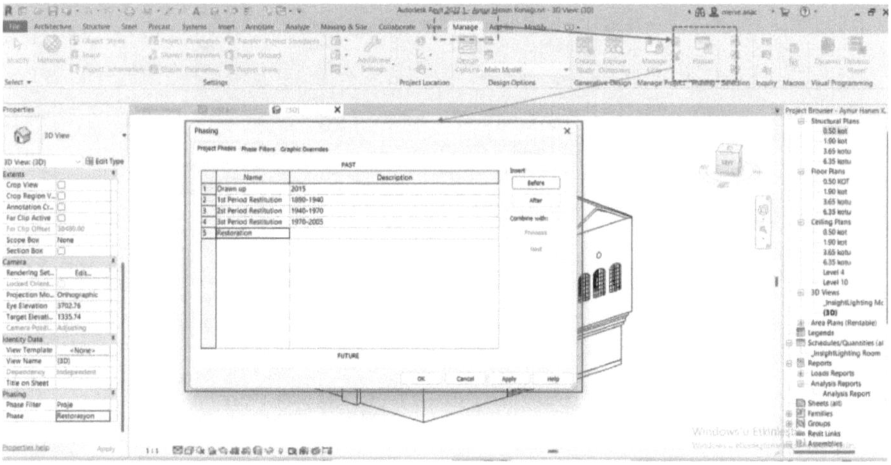

Figure 14.8 Identification of time layers.

are displayed on the screen. This building, which was constructed in different numbers of storeys using cut stone, consists of two building masses. There is a cave under the building opposite the main entrance (building number 2). In the construction of the house, keymık stone was used on the exterior walls and keymık and havara stone were used together on the interior walls. These stones are a traditional building material extracted from the region. Although the same wall materials were used in both buildings, differences in the system and material types used in the flooring prove that the buildings were built in different periods. The floor of building 2 is reinforced concrete and the floor of building number 1 is wooden. The stair systems of both buildings also differ. Building 1 has a stone staircase with access to the upper floors from the shower, as in traditional Gaziantep houses, while building 2 has a wooden staircase with access to the upper floors from inside.

In traditional Gaziantep houses, the stones used generally are extracted from the parcel, shaped, and then used in the building. Extraction of material from the ground results in the space under the ground forming a cave. During the use of the building, these caves serve as a cellar used for storing supplies. As can be seen in Figure 14.9, according to the draw-up of the building in 2015, the door and window joinery of the building completely disappeared. The wrought iron in front of the

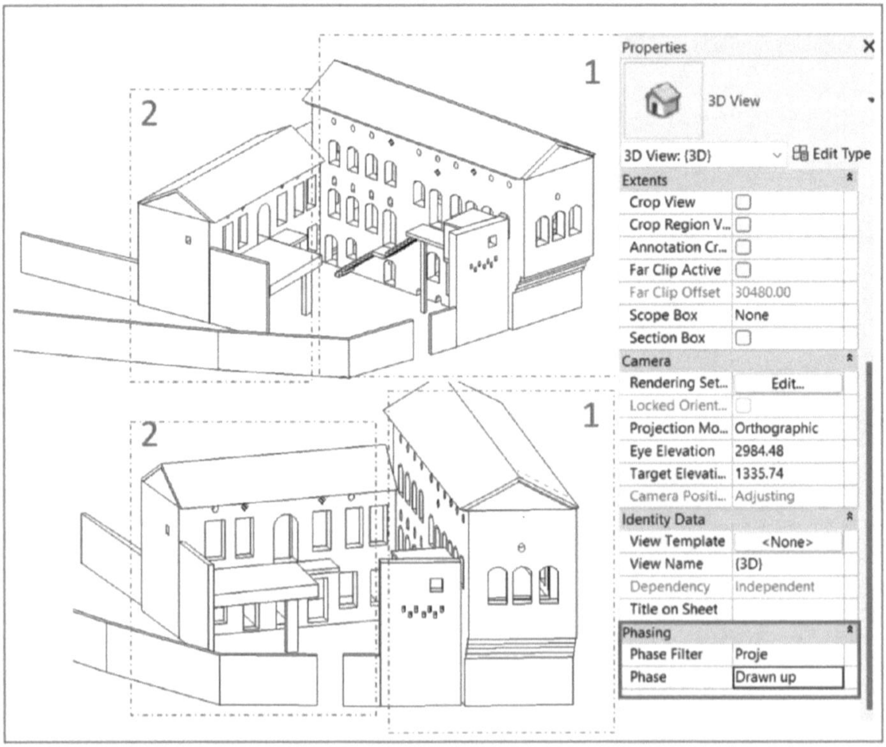

Figure 14.9 Aynur Hanım Mansion HBIM building draw-up work.

windows were removed. The workmanship (wooden coverings) inside the building disappeared. There are ventilation windows in different geometries on the facades of the building.

FIRST PERIOD RESTITUTION (PERIOD BETWEEN 1890 AND 1940)

As can be seen in Figure 14.10, when the "phase" section of the Revit program screen is marked as "1st Period restitution," a building group in which three different buildings are located around a single courtyard appears on the screen. The outer walls of building 1 are made of minced stone, which is the traditional construction material of Gaziantep, and the inner walls are built of havara stone masonry. Access to the spaces is provided directly through the courtyard door on the ground floor and by a stone staircase from the courtyard on the upper floors.

The stone staircase has a special cast iron railing. The building has wooden beam floors and the level differences between the floors were solved with several stairs.

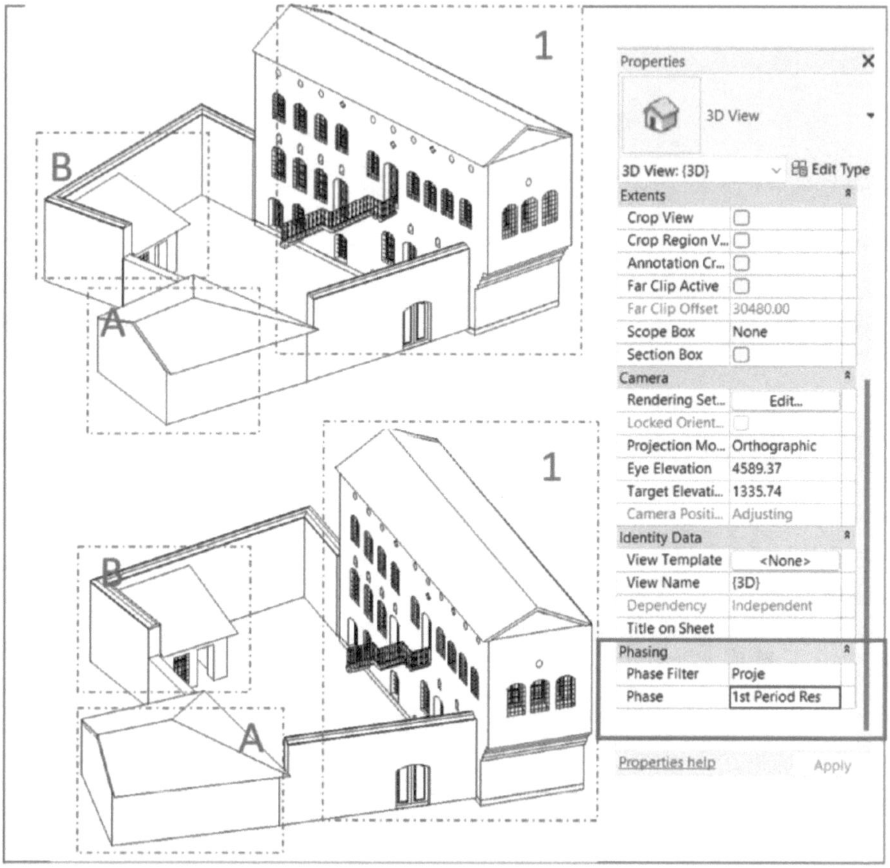

Figure 14.10 Aynur Hanım Mansion HBIM first period restitution.

There is a projection on the street facing the side of the building. This projection is gradually supported by three rows of stones parallel to the wall. Building 1 has survived to the present despite damage during the interim periods, but buildings "A and B" were destroyed. According to the restitution reports, it is understood that the buildings identified as "A" and "B" were built as a woodshed and toilet in the first period.

SECOND PERIOD RESTITUTION (PERIOD BETWEEN 1940 AND 1970)

According to the second period restitution studies of the building, it appears that the one-room building B was removed and building 2, which is located directly opposite the entrance, was built instead. Building 2 has keymık walls, reinforced concrete floors, and a reinforced concrete balcony on the facade of the building. The different flooring types indicate that building 2 was built later. The roof of

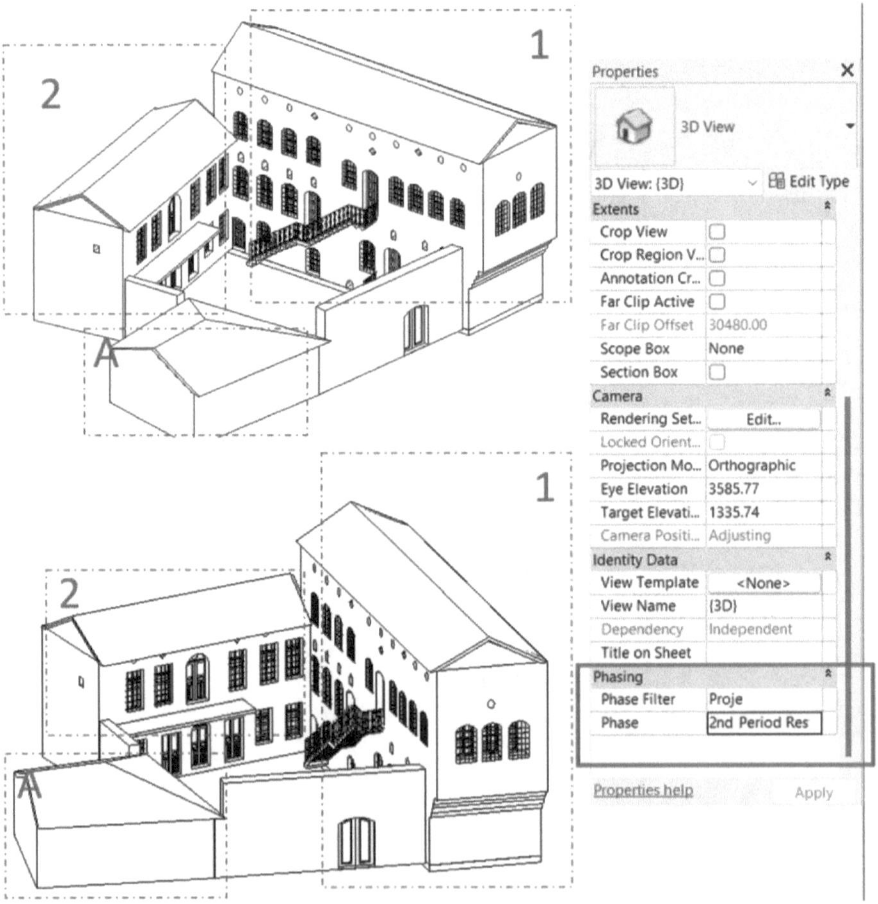

Figure 14.11 Aynur Hanım Mansion HBIM second period restitution.

both buildings is a gable roof (Figure 14.11). The building has two floors with two rooms on the lower floor and two rooms on the upper floor. The rooms on the lower floor are accessed directly from the courtyard and the rooms on the upper floor are accessed from inside via a reinforced concrete staircase. The most distinctive feature between the buildings is that the windows of building 2 are flat while the windows of building 1 are arched. According to the restitution reports, the interior walls of the building have ragged workmanship.

THIRD PERIOD RESTITUTION (PERIOD BETWEEN 1970 AND 2005)

From the third period onwards, the building underwent a serious transformation process and unqualified additions were made. Structure A, shown in Figure 14.11, was demolished, as were the walls delimiting the courtyard, and a wall was built with rubble stones (Figure 14.12). On the upper floor of the building the staircase landing was extended in reinforced concrete and supported on the rubble wall, and

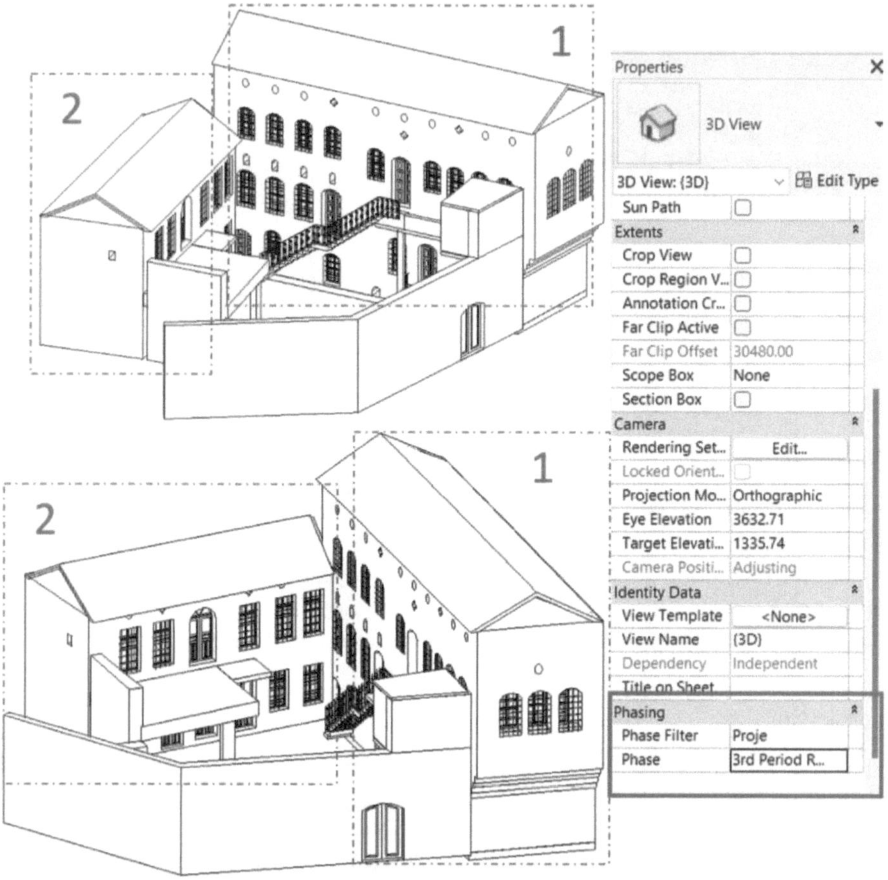

Figure 14.12 Aynur Hanım Mansion HBIM third period restitution.

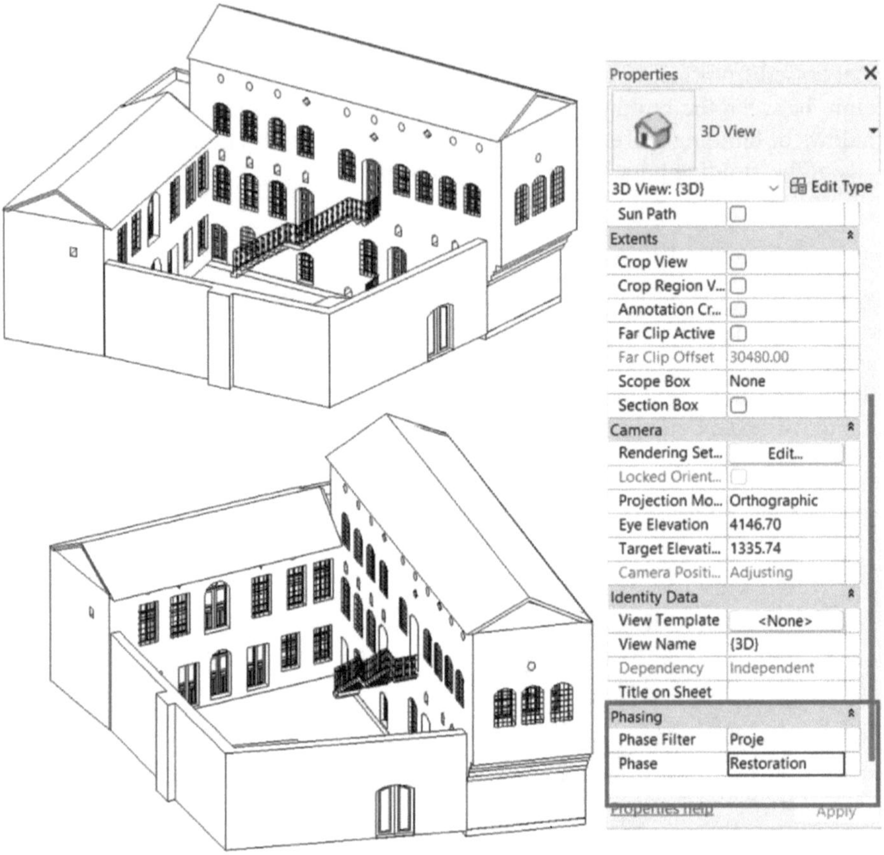

Figure 14.13 Aynur Hanım Mansion HBIM restoration work.

a toilet was built on it with masonry technique made of briquette material. This space negatively affects the architectural aesthetics, authenticity, and material unity of the building. The reinforced concrete balcony built in front of building 2 during the second restitution period was expanded in the same material, supported by a wall on one side and a column on the other. During this period, various makeshift additions were made to meet the needs of those living in the house.

RESTORATION

When the "phase" section in the file is marked as "restoration," the restoration project prepared in 2015 and reflecting the geometric features of the building in the current situation is displayed (Figure 14.13). In the restoration project, the aim has been to preserve the original values of the building in terms of its design, architectural features, construction system, and material use, and the study was carried out in light of the third period restitution data. The aim was to preserve the original

Table 14.1 Floor plans, elevations, and sections according to time layers for Aynur Hanım Mansion

First Period Restitution (1890–1940)	*Second Period Restitution (1940–1970)*

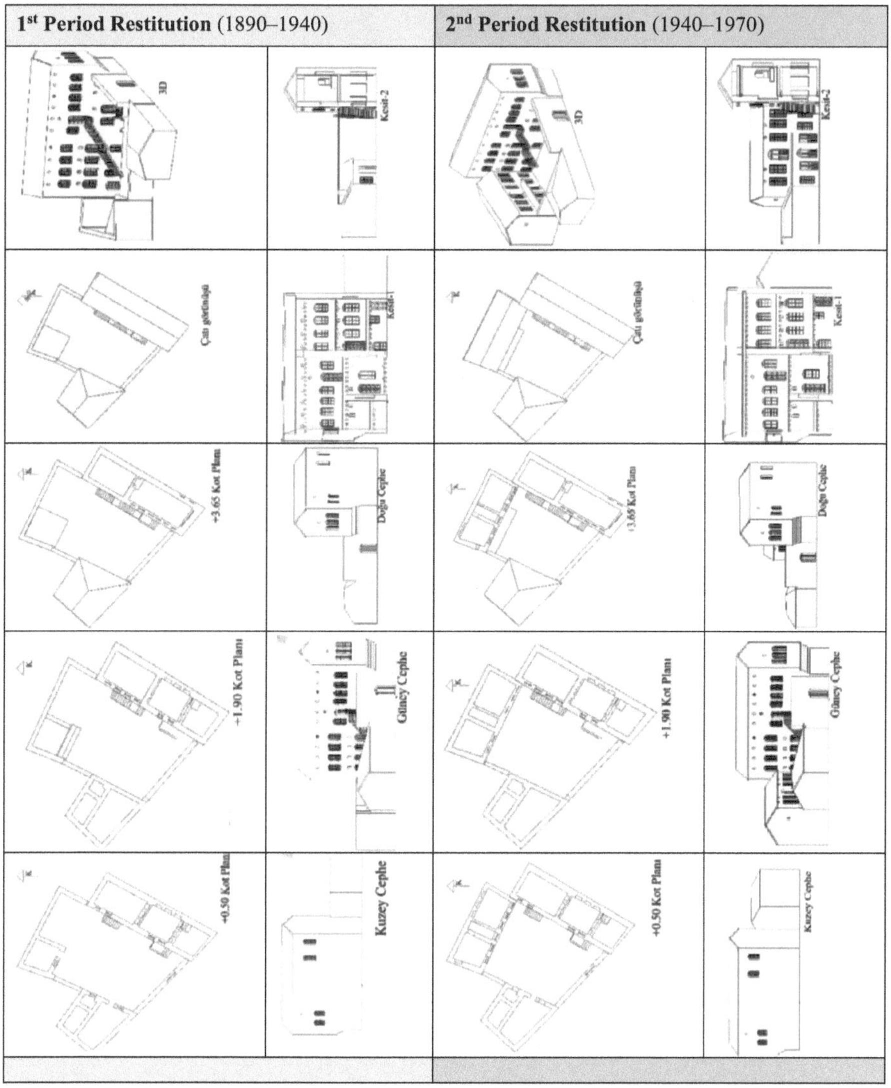

(*Continued*)

Table 14.1 (Continued)

Third Period Restitution (1970–2005)	*Draw up*

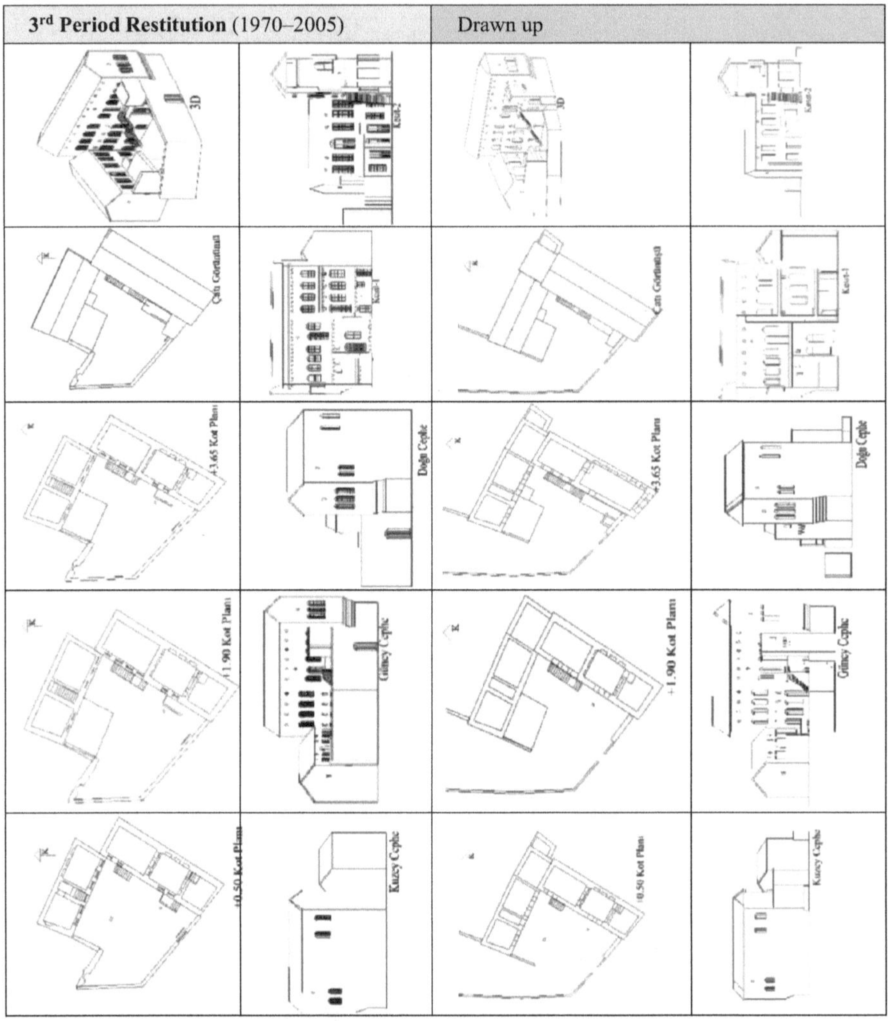

Table 14.1 (Continued)

Restoration

Restoration

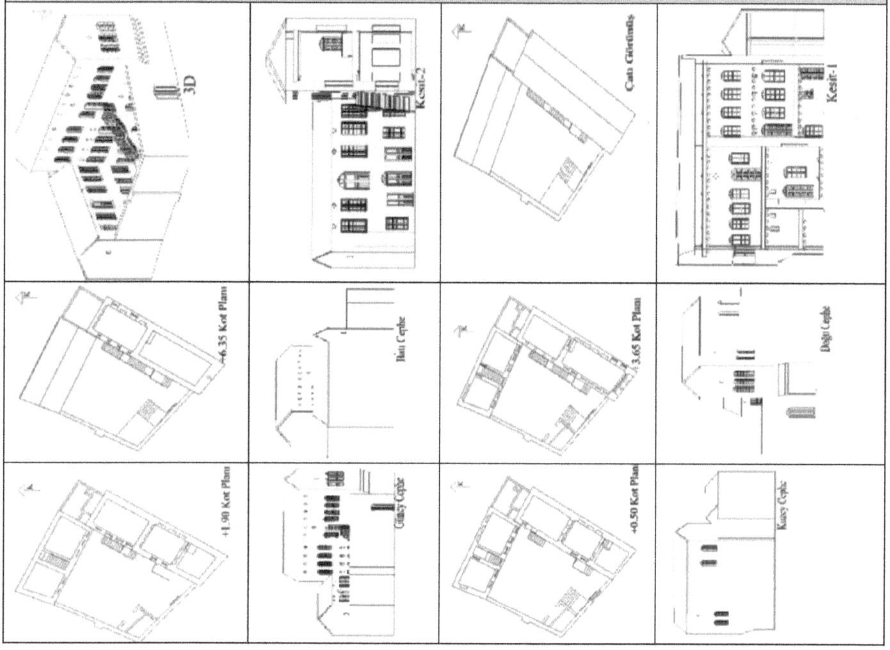

material in situ, and as a principle of restoration, the goal of not intervening more than necessary was adopted. However, in cases of necessity, the deteriorated original material was replaced with new material.

During the restoration project, the unqualified additions that were added to the building in the historical process and that adversely affected the building visually and structurally were cleaned. Architectural elements such as doors, windows, workmanship, and wrought iron were made in accordance with the original methods.

HBIM models not only provide 3D images but also make it possible to access plan sections and elevations from a single platform. In Table 14.1, the plans, sections, and elevations in the model for Aynur Hanım Mansion are given in a single table. It is possible to access all the specified visuals from a single platform and generate the data automatically. Since any update in the building is automatically updated in all planes, it is advantageous in terms of time to process any information while documenting.

All the plan planes given in the table are derived automatically from BIM-based programs and provide convenience in presentation. Any adjustment made in any drawing plane is automatically changed in other planes.

HBIM Element Library

In order to prepare HBIM models of historical buildings, the most important step is to prepare a HBIM element library consisting of original building elements. In BIM systems, which are generally used in the production of new buildings, there is no need to prepare a new element library while modelling, because the library within the BIM-based programs are suitable elements for new buildings. If different elements are preferred, it is possible to access the BIM element models prepared by the related companies. These elements, together with the semantic and geometric information they contain, form digital twins of real building elements.

This library is specific for HBIM models prepared for historical buildings and can be used for different buildings in the same area because cultural, technological, and aesthetic parameters are effective in the creation of historical buildings. However, the creation of HBIM models in a different region is desired, a new element library is needed to reflect the characteristics of that region.

HBIM Wall Model

This study started with the definition of wall elements. The wall tool cannot be used directly because the wall thickness and mesh form of historical buildings are different from new buildings. Instead, the Component tool is selected in the Revit program and then the "Wall" parameter is selected to reflect the parametric

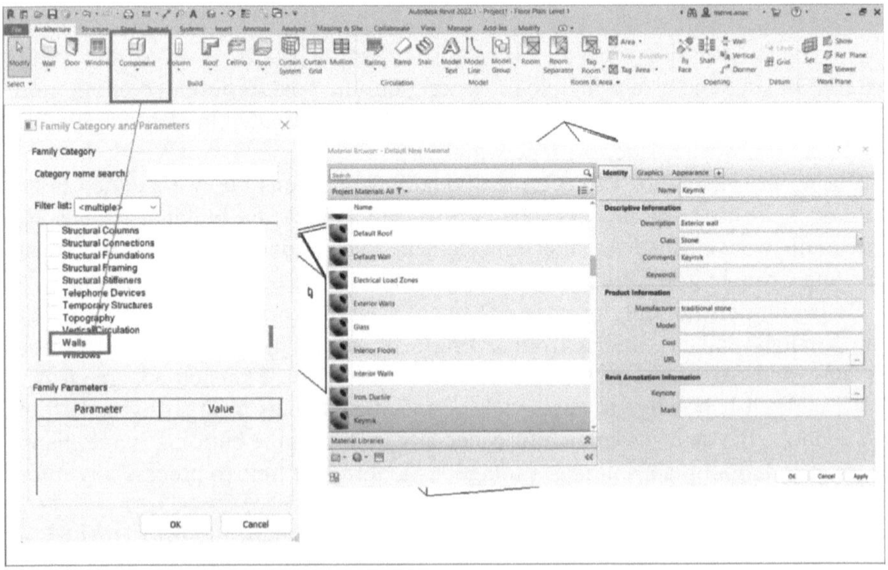

Figure 14.14 HBIM semantic and geometric information management platform for wall elements.

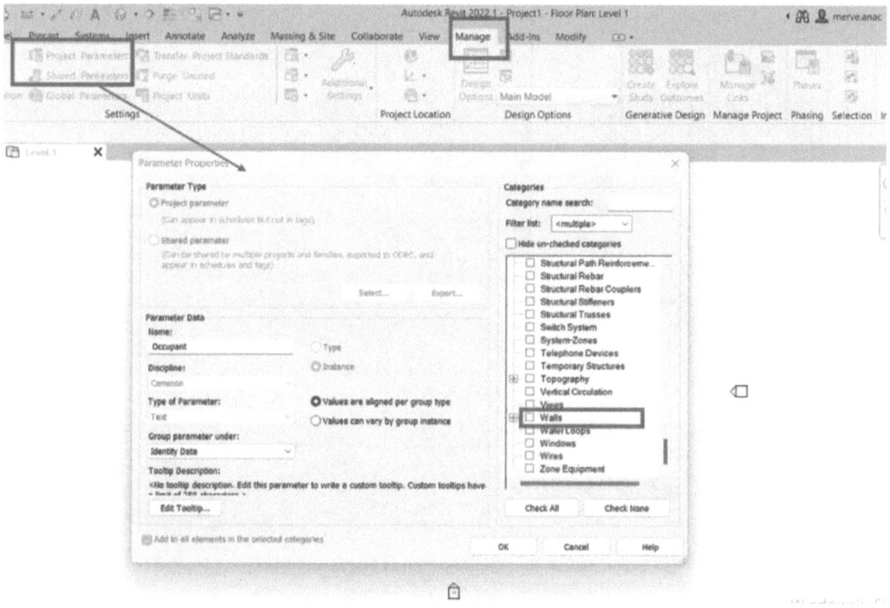

Figure 14.15 The parameter generated for the identification of damages in case of damages.

properties of the wall elements. This process is shown in Figure 14.14. Then the wall elements of the structure are drawn. After the drawing of the wall element is completed, all semantic information and material information are defined. In addition, the time layer for the construction of the wall must also be defined.

Further, it is also possible to define special semantic information about the historical building. In the Revit program, under the "manage" tab, new parameters of the historical building can be generated with "share parameters." Parameters can be defined in "text file," "pdf" format, or "jpeg" format. Figure 14.15 shows the generation of a "text" parameter to indicate the damage. These parameters can be customised for not only the wall element but also each object. These systems will provide superiority over the projects prepared with traditional methods since they contain the information of the whole historical building. Since it contains both geometric information and semantic information, the unique features of the building can be seen by the stakeholders and these data can be taken into consideration in the decisions made about the building.

HBIM Window and Door Models

BIM-based systems are used not only in the design phase of buildings but also for information management during the utilisation process. BIM systems contain semantic information that will allow various analyses, including energy consumption of the building or comfort requirements of the users. In case of entering data

Figure 14.16 Example of a window drawn in Revit.

to be used in various analyses for all produced building elements, it will be possible to evaluate the energy gains and losses of the historical building before and after restoration. These analyses and data management of the historic building are beyond the scope of this study. However, semantic data entries are shown in Figure 14.16.

The elements in the HBIM building element library prepared for use in HBIM models can be shown in plan, section, view, and 3D presentation method. These visuals are not drawn separately; the plan section and views of the elements drawn in 3D are created automatically. Changes made in any plane in the system are also updated in other planes. In Table 14.2, a few examples of the plan, section, view, and 3D visuals of the elements prepared for the HBIM library are given.

In Gaziantep traditional houses, the walls are covered with wood. This covering is a type of woodwork called "nacar" in traditional language. These nacars, which continue along the wall, gain a new function by acting as a cover in windows and niches. The nacar affects not only the aesthetic characteristics of the spaces but also the thermal comfort of and the amount of energy consumed in the spaces. Since these elements are special and unique elements for traditional Gaziantep houses, these elements were also prepared and integrated into the buildings within the HBIM element library models. These elements are important in intervention or thermal comfort studies to be carried out in buildings. Nacar coatings are also given in Table 14.2.

Table 14.2 Examples of Gaziantep HBIM building elements

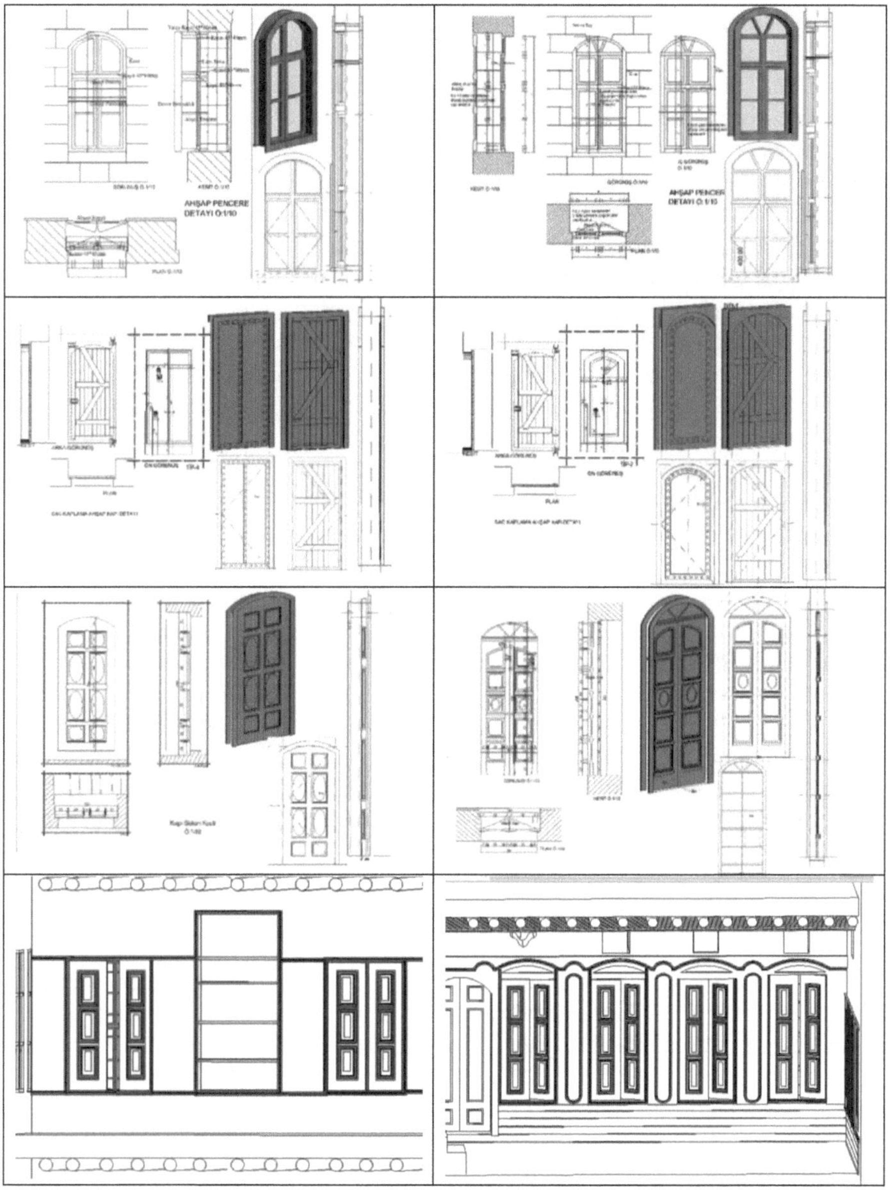

Results

Historical buildings consist of a heterogeneous information network formed by the combination of original geometric and semantic information with the time dimension. It is very difficult to reflect this original structure with traditional methods and 2D drawing techniques, often resulting in data loss. The loss of data in historical buildings results in incorrect analyses and synthesis, and consequently the wrong decisions for the restoration of buildings. Especially for historical buildings, it will be a great advantage to be able to view the multi-layered heterogeneous information network on a single platform simultaneously. Considering all these requirements, BIM systems, which are used in multiple information management systems for new buildings, have been used as a new documentation method under the name of HBIM in the new digital archive model developed within the scope of this study.

There is a BIM library consisting of holistic and parametric objects developed for use in BIM-based computerised structures. However, historical buildings consist of unique building elements in terms of geometric and semantic properties. For this reason, the building element libraries available in BIM-based computer programs were not used in historical building models. A unique HBIM element library was developed for traditional Gaziantep civil buildings. The library is a database to be used in the creation of HBIM models of Gaziantep historical buildings. The HBIM library is classified according to element type, element geometry, and dimensions within the model. When there is a collapsed or damaged structural element in a historical building to be restored, it can be used as a decision support system to select structural elements according to the classifications in the database. The created building element library was used in HBIM models in the study. HBIM models used in historical buildings are integrated with the time layer, allowing the models to be evaluated in 4D. The fact that the changes of the structures over time can be displayed on a single platform and in 3D will make it easier to understand the structures and prevent data loss.

Conclusion

The study started with a literature review, and firstly the characteristics and original aspects of Gaziantep traditional civil buildings were determined. The importance of documentation systems for the protection and transfer of original features to the next generations is determined. Although there are various studies to document the current status of buildings in historical building studies, no platform was identified to enable different participants to access common data and share the process.

This situation makes historical building research, survey, restitution, and restoration processes difficult. In addition, data losses occur because the survey, restitution, and restoration projects, which are documents, are prepared in 2D, sections and elevations are produced only as much as the restorer or architect draws, and not all details are drawn. Hence, in this study, which is limited to Gaziantep traditional civil buildings, a digital archive model integrated with 3D HBIM is developed for digital storage of data belonging to the historical buildings.

References

Akın, G. (1984). *Doğu ve Güneydoğu Anadoludaki Tarihsel Ev Tiplerinde Anlam*. İstanbul Teknik Üniversitesi.

Anaç, M., & Arun, G. (2024). HBIM supported archive model. *Journal of the Faculty of Engineering and Architecture of Gazi University*, *39*(1), 443–459. https://doi.org/10.17341/gazimmfd.1175113

Arayici, Y. (2008). Towards building information modelling for existing structures. *Structural Survey*, *26*(3), 210–222. https://doi.org/10.1108/02630800810887108

Arayici, Y., Counsell, J., Mahdjoubi, L., Nagy, G., Hawas, S., & Dewidar, K. (2017). *Heritage building information modelling*. Routledge Taylor & Francis Group. https://doi.org/10.4324/9781315628011

Atalar, A. (2004). *Osmanlı Dönemi Antep evleri*. Merinos.

Aygün, H. M. (2011). Kültürel Mirası Korumada Katılımcılık. *Vakıflar Dergisi*, *35*, 191–214. http://acikerisim.fsm.edu.tr:8080/xmlui/bitstream/handle/11352/850/Aygün.pdf?sequence=1&isAllowed=y

Baik, A., Boehm, J., & Robson, S. (2013). Jeddah historical building information modeling "JHBIM" Old Jeddah -Saudi Arabia. ISPRS – International Archives of the Photogrammetry, Remote Sensing and Spatial Information Sciences, *XL-5/W2*(September), 73–78. https://doi.org/10.5194/isprsarchives-xl-5-w2-73-2013

Brusaporci, S., Maiezza, P., Marra, A., Tata, A., & Vespasiano, L. (2023). Scan-to-HBIM reliability. *Drones, 7*(7), 426.

Brusaporci, S., Maiezza, P., & Tata, A. (2018). Computational design for as-built modeling of architectural heritage in HBIM processes. *2018 Metrology for Archaeology and Cultural Heritage (MetroArchaeo)*, 199–203. https://doi.org/10.1109/MetroArchaeo43810.2018.13620

Caetano, I., Santos, L., & Leitão, A. (2020). Computational design in architecture: Defining parametric, generative, and algorithmic design. *Frontiers of Architectural Research*, *9*(2), 287–300. https://doi.org/10.1016/j.foar.2019.12.008

Celli, S., & Ottoni, F. (2023). Managing Information to Improve Conservation: The HBIM of the Wooden Chain of Santa Maria del Fiore. *Sensors,* 23(10), 4860–4887. https://doi.org/10.3390/s23104860

Delis, P., Zacharek, M., Wierzbicki, D., & Grochala, A. (2017). Point cloud derived from video frames: Accuracy assessment in relation to terrestrial laser scanning and digital cameradata. *International Archives of the Photogrammetry, Remote Sensing and Spatial Information Sciences – ISPRS Archives*, *42*(XLII-2/W3), 217–223. https://doi.org/10.5194/isprs-archives-XLII-2-W3-217-2017

Diara, F. (2022). HBIM open source: A review. ISPRS International Journal of Geo-Information, 11(9). https://doi.org/10.3390/ijgi11090472

Gargaro, S., Giudice, M. Del, & Ruffino, P. A. (2018). Towards a multi-functional HBIM model. *Scientific Research and Information Technology Ricerca Scientifica e Tecnologie Dell'Informazio*, *8*(2), 49–58. https://doi.org/10.2423/i22394303v8n2p49

Gül, G. (2005). *Gaziantep İli Bey (Kayacık) Mahallesi Geleneksel Doku Koruma ve Geliştirme Önerisi*. Gazi Üniversitesi.

Kanalıcı, A. K. (2012). *Geleneksel Gaziantep Evleri Yapı Üretimi Analizi*. Lefkoşa Üniversitesi.

Klapa, P., & Gawronek, P. (2023). Synergy of geospatial data from TLS and UAV for heritage building information modeling (HBIM). Remote Sensing, 15(1). https://doi.org/10.3390/rs15010128

Kuban, D. (2001). *Türkiye'de Kentsel Koruma Kent tarihleri ve Koruma Yöntemleri*. In *Tarih Vakfi Yurt Yayınları*.

Lanzara, E., Catello, D., & Casciello, M. (2022). Scan to HBIM for complex reflective metal artefacts. 3D digitisation and restoration. *The International Archives of the Photogrammetry, Remote Sensing and Spatial Information Sciences*, *XLVIII*(December), 15–16.

Lee, Y. C., Eastman, C. M., & Solihin, W. (2016). An ontology-based approach for developing data exchange requirements and model views of building information modeling. *Advanced Engineering Informatics*, *30*(3), 354–367. https://doi.org/10.1016/j.aei.2016.04.008

Li, D., Wang, X., Bai, C., & Wu, C. (2019). Discussion on the problem of regularized reconstruction in HBIM. *The International Archives of the Photogrammetry, Remote Sensing and Spatial Information Sciences*, *XLII-2/W15*(51878440), 657–662. https://doi.org/https://doi.org/10.5194/isprs-archives-XLII-2-W15-657-2019

Martinelli, L., Calcerano, F., & Gigliarelli, E. (2022). Methodology for an HBIM workflow focused on the representation of construction systems of built heritage. *Journal of Cultural Heritage*, *55*, 277–289. https://doi.org/10.1016/j.culher.2022.03.016

Osello, A., Lucibello, G., & Morgagni, F. (2018). HBIM and virtual tools: A new chance to preserve architectural heritage. *Buildings*, *8*(1), 1–12. https://doi.org/10.3390/buildings8010012

Özdemir, E., & Remondino, F. (2019). Classification of aerial point clouds with deep learning. *International Archives of the Photogrammetry, Remote Sensing and Spatial Information Sciences – ISPRS Archives*, *42*(2/W13), 103–110. https://doi.org/10.5194/isprs-archives-XLII-2-W13-103-2019

Pauwels, P., Verstraeten, R., De Meyer, R., & Van Campenhout, J. (2008). Architectural information modelling for virtual heritage application. *Digital Heritage – Proceedings of the 14th International Conference on Virtual Systems and Multimedia*, 18–23. https://biblio.ugent.be/publication/434809

Pavelka, K. (2022). Photogrammetry, laser scanning and HBIM for construction diagnostic. *International Archives of the Photogrammetry, Remote Sensing and Spatial Information Sciences – ISPRS Archives*, *46*(5/W1-2022), 171–176. https://doi.org/10.5194/isprs-archives-XLVI-5-W1-2022-171-2022

Previtali, M., Brumana, R., Stanga, C., & Banfi, F. (2020). An ontology-based representation of vaulted system for HBIM. *Applied Sciences (Switzerland)*, *10*(4). https://doi.org/10.3390/app10041377

Quattrini, R., Malinverni, E. S., Clini, P., Nespeca, R., & Orlietti, E. (2015). From TLS to HBIM. High quality semantically-aware 3D modeling of complex architecture. *The International Archives of the Photogrammetry, Remote Sensing and Spatial Information Sciences*, *XL-5/W4*(February), 25–27. https://doi.org/10.5194/isprsarchives-XL-5-W4-367-2015

Quintero, M. S., Jansen, M., & Oman, S. O. F. (2002). International and multidisciplinary efforts for the conservation of archaeological sites: Introducing sustainable and adapted digital three-dimensional documentation techniques. *World Heritage in the Digital Age, the 30th Anniversary of the World.*

Ünal, Z. (1998). *Bilgisayar Destekli Tarihi Çevre Koruma Bilgi Sistemi Oluşturulması ve Gaziantep Kentsel Sit Alanında Örneklenmesi*. Tıldız Teknik Üniversitesi.

Yıldırım, E. G., & Çağdaş, G. (2017). Gaziantep Geleneksel Mimari Dokusunun Sosyo-Kültürel Bağlamda Mekân Dizimsel Analizi. *Gaziantep University Journal of Social Sciences*, *17*, 508–532.

15 3D Printing Applicability in Housing Projects in Australia

Pedro Miguel Valadas Vicente and Yusuf Arayici

Introduction

In a century of constant changes at a technical and technological level, addressing the issue of how to apply 3D printing across many areas of knowledge and know-how is not only necessary but also a legitimate concern. In construction, it is essential in terms of the current socioeconomic context, where financial concerns are rampant, as are those involving sustainability problems (Kadiri & Shittu, 2013). Hence, it is considered important to take advantage of such an innovative technology to benefit a vast economic sector without neglecting a new way of thinking about resource management: more rigour in production and performance, less waste, and greater civic sense (Perkins & Skitmore, 2015). The exciting thing is that the application of this new technology, in addition to presenting invaluable potential, is within everyone's reach, starting, for example, with a low-cost 3D printer, which fits perfectly on any desk.

3D printing is an automated production technique with layer-by-layer control that has made significant progress over time, notably for decades in the manufacturing sector, but more recently also in the construction sector, for printing bridges, houses, and buildings (Wu et al., 2016).

To many people's surprise, the 3D printer is not such a new technology. American engineer Chuck Hull invented the first 3D printer in 1984, and since its invention, the general concept of how it works has not changed much and remains quite simple. A 3D printer is a piece of equipment that transforms a digital 3D file into a physical part by creating several overlapping layers; what differentiates one 3D printer from another is precisely how each of these layers is constructed in what we call 3D printing technology (Stein, 2017).

A three-dimensional digital model can be built using computer-aided design (CAD) software such as SolidWorks or Draftsight. The file created is translated to stereolithography (STL) in additive manufacturing techniques. This process is responsible for generating a language that the 3D printer will interpret. As soon as the printer receives this program with all the slicing and construction information, the layers of matter begin to be progressively built until the physical 3D model is obtained (Wong & Hernandez, 2012).

DOI: 10.1201/9781003380559-15

Significant developments were achieved due to the use of new materials in 3D printing. Innovations allowed the 3D printing of ceramics and glass, enabling the creation of complex shapes and structures for applications in industries such as electronics, medical devices, and architecture (Jakus, 2019).

In this research study, we explore the different uses and good practices of 3D printing technology, with examples, in the worldwide construction industry based on existing literature and available online information. Additionally, using the same kind of sources, we inquire about this technology's advantages and disadvantages, focusing on concerns such as quality and meeting Australian national construction standards.

We compare housing projects in Australia, in particular traditional construction methods versus 3D printing technology implementation, using data collected from three case studies and interviews with 3D printing companies currently operating in this country. This research intends to provide additional information and a better understanding of the feasibility of 3D printing technology and how it brings value for money.

Justification and Significance

Considering the premises mentioned, it is easy to see that the motivation behind this study is linked mainly to a current and severe concern regarding sustainability issues in the construction world, without neglecting its economic and financial problems. The civil construction industry is recognised as a significant consumer of resources, making it very unfriendly to the environment; even though the industry presents poorly in terms of productivity and innovation, the recent expansion of 3D printing and its undeniable applicability in construction made it a desirable approach to resolving these constraints (Perkins & Skitmore, 2015).

According to a report by McKinsey and Co., the aerospace and automotive industries invest their companies' revenue into innovation by 4.5% and 3.5%, respectively, against less than 1% of construction companies. This difference demonstrates how far behind the construction industry is in implementing new technology (McKinsey & Co., 2016).

3D printing promotes strong development in standardisation, reducing construction time and costs. However, for this technology effectively to be an added value in this production area, it is urgent to understand some of the obstacles to overcome and challenges to face since this technology in construction still assumes a fragmentary nature. It is, therefore, necessary that those responsible for it know the 3D technology, its basic foundations, and its potential so that it is possible to print complete models of buildings and their components on a large scale, meeting the requirements of mass customisation and evaluating the life cycle cost of printed projects.

The digital transformation we are witnessing in a transversal way has not entirely bypassed the civil construction sector. However, there is still much to be done regarding agility, efficiency, and digitisation of processes. The proof is that this sector is one of the least digitised in the world (McKinsey & CO, 2016).

Technology is undoubtedly one of the factors that most contributes to the competitiveness of companies, and civil construction is no different; this is true for large companies and small and medium-sized ones. In fact, it is even more critical for these to grow and become more competitive.

Existing literature focuses on aspects of 3D printing, such as its applicability, advantages, and disadvantages. However, it does not provide information regarding the costs (savings and losses) associated with time, optimisation of resources, challenges, and, overall, the benefit to companies and the final consumer in housing projects. Providing more information regarding this subject will assist in better decision-making regarding innovation investments and incentivisation to use this technology.

This chapter focuses on the research question *How can 3D printing technology bring advancement to housing projects in Australia?*

Accordingly, this chapter elaborates on the following aspects below.

- Explore 3D printing technology views in construction industries.
- Compare and contrast housing projects – traditional versus 3D printing.
- Discuss the feasibility of 3D printing and how it brings value for money.
- Discuss 3D printing strategies for the effective use of this technology.

3D Printing in the Construction Industry

The transformation of outdated 2D paper designs used in building into 3D models utilising CAD provides an illustration of 3D printing in construction. As shown in Craveiro et al. (2019), concrete and other polymetric construction parts can be simultaneously designed and built using 3D printing technology. This is associated with integrating CAD drawings from several trades into the single BIM platform, enabling better project procurement through shared access, clash detection, scheduling, costing, and analytics (Olanipekun et al., 2021).

Several 3D printing systems are used in the construction industry, such as fused deposition modelling (FDM), 3D metal printing, and contour crafting, among others. These 3D printing technologies are further explained.

FDM is a 3D printing/additive manufacturing technology type. It works by melting a thermoplastic filament and depositing it layer by layer through a nozzle to create a 3D object. The nozzle moves according to the design's specifications, laying the material in thin strands that solidify almost immediately upon contact with the build surface or previous layers (Teizer et al., 2016).

According to Rett et al. (2021), FDM is popular due to its accessibility, cost-effectiveness, and the variety of materials available, as it allows architects and engineers to create detailed prototypes and scale models of buildings or structures before actual construction begins. This aids in visualisation, design validation, and identifying potential issues early in the planning phase.

Additionally, FDM can be used to execute custom components and architectural elements, such as unique facades, ornamental details, and complex shapes that might be challenging or expensive to produce with traditional methods.

The use of additively manufactured formwork allows the creation of FDM printed moulds, templates, or frameworks used in construction processes. This technology's simplicity and versatility have made it a cornerstone in the world of 3D printing.

The idea of using FDM technologies in rehabilitation projects for old buildings also came to a construction company when it was awarded a restoration project. It involved the conversion of a historic brewery into an events and office centre, and the biggest challenge was the replacement of segments of five huge stone frames that deteriorated or were destroyed while maintaining their original aesthetics. The conventional technique for manufacturing these frames would include stone masonry, which is a highly skilled activity that generates impressive results but involves high costs and lengthy production time (Bigrep, 2017). The next option would be to use resin-coated foam formwork to mould the concrete, which consists of a lot of waste and toxic waste. Therefore, the construction company contacted a 3D printing company, to explore a third option, 3D printing of formwork, to shape the curved concrete segments of the frames (Figure 15.1).

Using the BigRep ONE printer, they printed the formwork using biodegradable PLA filament from the same company. This printer has the capacity for printing models up to 1 m^3 in volume, making it ideal for this project. After being printed, the various segments were assembled at the location of their implementation, forming the complete frames.

Figure 15.1 Segment of one of the concrete frames after being removed from the 3D printed formwork.

Source: Bigrep, 2023.

Both companies believe that printing the frames, as an alternative to conventional methods, reduced process costs by 50% and production time by 45%, while less labour was used to complete the project. On the other hand, this method also allows architects to control the geometry of the pieces, both on their computers and when manufactured on-site, making the implementation of formwork much more sustainable (Bigrep, 2023).

Currently, there are at least 11 primary metal additive manufacturing (AM) technologies competing for applications in the industrial environment. We focus on the properties of four key technologies (Gibson et al., 2010), as follows:

- Laser powder bed fusion (LPBF): A process where a high-power laser is used as a heat source to selectively melt and fuse metal powder on a bed, building a part layer by layer;
- Direct energy deposition (DED): A process where a focused energy source is used to deposit melted metal – powder or wire – onto a surface where it solidifies, creating a printed part;
- Binder jetting (BJT): A process where a binding agent is deposited onto a bed of metal powder to form a part layer by layer. Afterwards, the part is sintered to remove the binding agent;
- Metal FDM: This is a process in which metal filament with polymer binder is extruded through a heated nozzle, melted, and deposited onto a surface to form a part. Afterwards, the part is sintered to remove the binding material.

Size is also a crucial factor to consider when choosing the technology that works best for you. Overall, metal 3D printing technologies lend themselves to creating smaller, more complex parts. However, DED is the exception to this rule. The DED process means it can be mounted on a robotic arm used to deposit melted material to build a part. Therefore, the size potential of the parts it produces is almost limitless (Steel Printers, 2023).

An excellent example of the use of DED technology can be seen in an implementation in the Netherlands, as shown in Figure 15.2. Dutch robotics company MX3D was given a project for one of the commercial streets in Amsterdam that consisted of a 3D printed metallic pedestrian bridge, designed by Joris Laarman (Arup, 2022).

The structure can be considered a significant technological milestone, as there had never previously been a 3D printed metal part capable of handling pedestrian traffic. Its curved S-shape and balustrades with lattice-like perforations were designed using parametric modelling software. The team behind the project said the technique shows how 3D printing technology can lead to the construction of more efficient and sustainable structures, as shown in Figure 15.3 (MX3D, 2023). "By stepping up our game and willingness to make changes as designers and engineers, we can bring the innovation needed to make a difference to the environment," says engineer Stijn Joosten (Arup, 2022). The bridge has a network of state-of-the-art sensors that allow it to monitor changes such as vibration, temperature, and public interaction.

Figure 15.2 The MX3D Bridge Oudezijds Achterburgwal canal in Amsterdam.
Source: Arup, 2022.

Figure 15.3 Four robots carried out the printing process.
Source: Arup, 2022.

Contour crafting is a large-scale 3D printing technology specifically developed for construction purposes. It aims to automate the construction of buildings and structures by using a massive robotic system to layer concrete or other suitable construction materials, based on a digital model or design.

The process involves a robotic arm or gantry system equipped with a nozzle that extrudes materials (usually concrete) in a controlled manner. The system moves according to the specifications of the digital design, depositing the material precisely to build walls, floors, and other structural elements. Construction is carried out in layers until, in the last phase, a new floor covers the entire structure, or the crane inserts a roof.

One of the main advantages of using the contour crafting technique is the possibility of inserting all types of components at an early stage of construction. Plumbing and electricity pipes/cables are easily laid during the construction of the walls. Another positive aspect of contoured construction is that this method produces minimal debris, facilitating construction management and making the entire process more economical (Khoshnevis, 2004).

In 2023, Europe's largest 3D-printed building was constructed by PERI 3D Construction using the COBOD printer in Heidelberg, Germany. The building contains an IT server hotel with a gross floor area of around 600 square metres (54m × 11m) and 9 metres high. PERI concluded the printing of the walls in just 140 hours, corresponding to around 4 square metres of construction print per hour (PA, 2023).

The exterior facade stands out from other buildings in the city due to the overlapping layers of concrete and the shape of the building being round rather than angular. SSV Architeken and Mense-Korte designed the building, and PERI Group turned the project into reality (Figure 15.4). For Henrik Lund-Nielsen of COBOD,

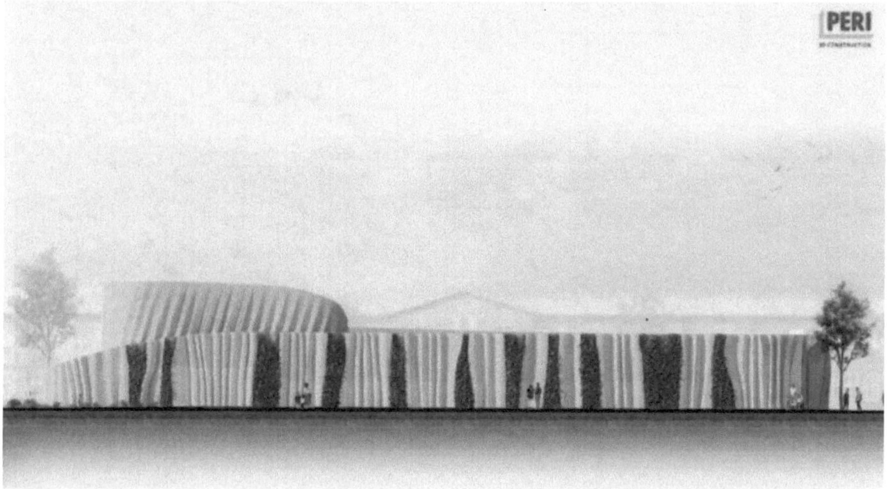

Figure 15.4 3D concrete printed building in Heidelberg, Germany.
Source: PERI, 2023.

Figure 15.5 COBOD 3D printer.
Source: PERI, 2023.

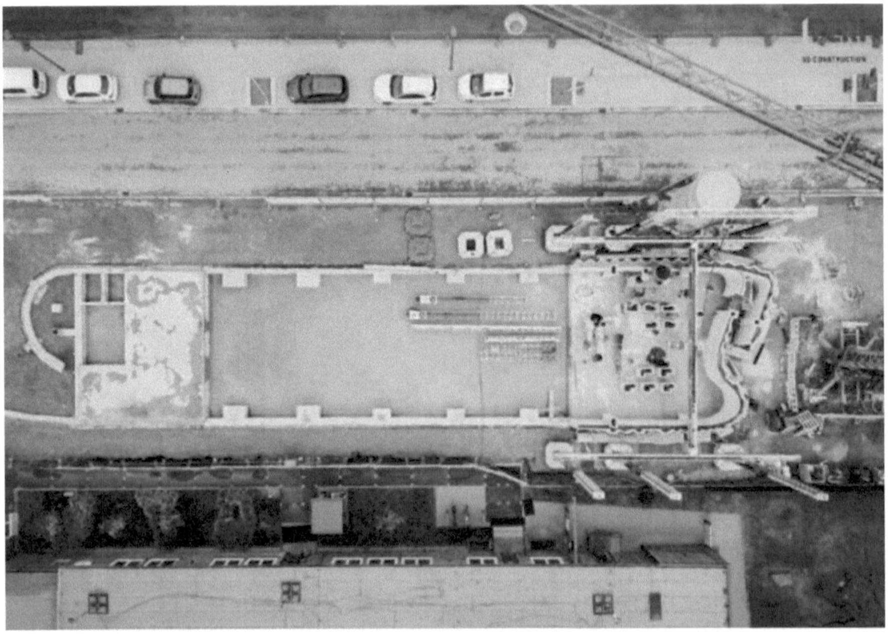

Figure 15.6 3D printing progress.
Source: PERI, 2023.

> In this unique project, PERI is emphasising two of the key benefits of 3D construction printing, speed of execution and design freedom. Because of this, 3D printing technology can carry out everything from materials saving windmill towers over low-cost residential housing in Africa to architectural office-type buildings in Germany. (PA, 2023)

This can be seen in Figures 15.5 and 15.6.

Advantages and Disadvantages of 3D Printing

Extensive customisation options are one factor that encourages the use of 3D printing in construction. Waste reduction and a reduced carbon footprint are two further exciting possibilities. In addition, greater cost-effectiveness is a possibility thanks to increasing automation and a need for less labour. In dangerous environments, automation like 3D printing can reduce risks for human workers (Bergman, 2012).

According to Kothman and Faber (2016), 3D printing can be considered a disruptive technology. It has the potential to impact the eco-performance of entire supply chains. Compared with other technologies, we can see performance improvements that affect the whole supply chain structure by providing more efficient manufacturing methods.

When an organisation adopts disruptive technology, it obtains such a competitive advantage that it is positioned far ahead of its rivals. However, no matter how well these rivals do, using such technology gives them the edge over others who do not employ disruptive technology.

3D printing offers the chance to reconsider the design and, as a result, the associated manufacturing process. This makes it possible to redesign the supply chain. Due to the disruptive nature, there is a much better chance of increasing eco-performance than if modifications were made to just one aspect of the design, manufacturing, or supply chain. Accepting disruptive technologies may hold the key given the rapid growth of technology (Kothman & Faber, 2016).

According to Kothman and Faber (2016), with 3D printing in concrete, there are three key factors (advantages) for using this disruptive technology. Firstly, with concrete 3D printing, the architect not only is responsible for the design and layout of a building but can also now check the structural integrity calculation using simulation techniques on the CAD model. After completing the CAD model, the architect can create the accompanying printing instructions. The construction is then 3D printed. With the 3D printing of concrete, many roles from the traditional design process merge into one. Furthermore, the consumer can use the visual representation of the construction to evaluate the design. Any changes can be visually verified and implemented directly in the design. Role integration allows for a shorter supply chain.

Secondly, element printing on-site eliminates several logistical processes and preparation tasks. For example, a house is typically built using traditional construction concrete, brick-and-mortar, or prefabricated concrete elements. Both methods necessitate extensive planning and logistics. On-site concrete printing involves moulding the raw material into a structure directly rather than using wooden formwork as regular concrete would. In addition, building parts are constantly damaged in transit when prefabricated concrete elements are used. Chipped, cracked, or marked parts are examples of damage that must always be resolved on-site, resulting in additional labour for the contractor.

Moreover, many parts must be over-engineered to withstand transportation, necessitating the use of more reinforcements than required by construction specifications, raising costs. Additionally, for safe transportation and hoisting, parts must have extra features, which necessitate post-assembly finishing work. All of these steps and costs can be avoided by using on-site printing.

Thirdly, the digitisation of a large portion of the manufacturing process results in small batches of complex products with high levels of customisation at minimal additional costs.

Despite all these advantages, it is necessary to consider the social impact of the implementation of such technology. It is estimated that labour is reduced up to 20 times through contour crafting (Kothman & Faber, 2016). This has consequences in terms of employability and the general economy.

Uncertainty surrounding the extent of mass customisation demand and the lack of high-strength printing supplies are two issues that could act as obstacles to implementing this technology. In addition, it is also unclear from the literature whether 3D printing would result in lower or higher construction costs (Wu et al., 2016).

Figure 15.7 Interior of the 3D printer-built poolside cabana.
Source: Contour 3D, 2022.

Currently, available 3D printers have size restrictions, making it challenging to create a whole multi-storey building at once. However, users can print the structural parts separately and assemble them to create a building at the actual scale. Additionally, it is essential to discuss the quality concerns regarding using 3D concrete printing in construction projects. Two 3D printing companies in Australia (Luyten and Contour 3D) have partnered with the University of New South Wales (UNSW) to address the issue. The Centre for Infrastructure Engineering and Safety (CIES) of UNSW developed printable concrete with excellent shape stability after extrusion and high early strength for buildability to support the weight of successive layers as designed. In this concrete mix, they replaced Portland cement with about 30 wt. percentage of supplementary cementitious materials (SCM) alkali-activated materials (geopolymers). With this process change, they could lower the embodied carbon of the printed concrete walls (UNSW, 2023). Group Architects (design) and Contour 3D tested this concrete mix in a 3D printed building. The 60 $m^2 \times 3$ m high structure was built as a poolside cabana featuring a living area, dining area, bathroom, and bar (Figure 15.7). This project was executed in the Macedon Ranges in Victoria, Australia, using a modular gantry system and an extrusion-based printer, and it was the first printed house in the Southern Hemisphere (CIES, 2022), as shown in Figure 15.8.

The research of this developed material won the prestigious National Award of "Excellence in Concrete" in Technology and Innovation from the Concrete Institute of Australia (CIA) at the Concrete 2023 biennial conference in Perth in September (UNSW, 2023).

Regulatory issues about 3D printed housing in Australia are another area that raises concerns about adopting this technology. Australia is lagging behind regulatory and standards jurisdictions in other countries concerning formulating codes to address such new technology. Organisations such as Underwriters Laboratories

Figure 15.8 Exterior of the 3D printer-built poolside cabana.
Source: Contour 3D, 2022.

(UL) and the International Codes Council in the United States of America have published UL 3401 *Outline for Investigation for 3D Building Construction* and AC509 *Acceptance Criteria for 3D Automated Construction for 3-D Concrete Walls*, respectively. It is also understood that codes have been produced and are under development in Europe (DRC, 2022).

Since there are no specific prescriptive code requirements for 3D printed concrete wall construction, approval authorities must consider each individual project under the relevant performance solution provisions for their evaluation and approval. A report produced by Contour 3D and the University of NSW for the Dubbo Council clarifies the status of regulating this technology according to the National Construction Code (NCC).

As mentioned in the DRC Report (2022), the body responsible for the National Construction Code is the Australian Building Codes Board (ABCB). The NCC currently comprises three codes:

- Building Code of Australia Volume 1 (relates primarily to Class 2 to 9 buildings);
- Building Code of Australia Volume 2 (relates primarily to Class 1 and 10 buildings); and
- Plumbing Code of Australia (relates primarily to plumbing and drainage associated with buildings).

The NCC is produced at a national level, but it allows for state and territory variations on its technical provisions. The NCC has no legislative weight but must be adopted and referenced by legislation in each State/Territory to have legal effect. The Building Code of Australia (BCA), which comprises Volumes 1 and 2, contains the technical requirements for designing and constructing new buildings. A performance-based code allows building design and construction to be undertaken for alternative requirements to its deemed-to-satisfy (Dts) specifications. This allows for innovation and variation from the Dts requirements while maintaining the principal levels of health, safety, and amenity of building occupants.

Other forms of wall construction such as straw bales, earth (adobe, pise, etc.), and timber logs are not listed as Dts methods under the current BCA and, consequently, would have to be the subject of a performance solution under a Construction Certificate/Complying Development Certificate application. 3D printed concrete walls, usually unreinforced, would fall into this latter category and similarly necessitate a performance solution.

The BCA currently does not address 3D printed building construction as a Dts construction method, including its referenced documents. However, BCA has provided a compliance assessment undertaken for a detached single storey Class 1a dwelling utilising 3D printed concrete walls, revealing that such construction is achievable but would necessitate performance solutions to address certain aspects. Current Australian standards and deemed-to-satisfy provisions of the BCA do not specifically address 3D printing technology, necessitating the preparation of

Table 15.1 Expected Performance Solutions

BCA Provision	
P2.1.1	Structural stability and resistance and Part 3.0 structural provisions
P2.2.2	Weatherproofing
3.10.6	Attachment of decks and balconies to external walls of buildings

performance solutions in the application process to address those areas where there is a shortfall/inadequate coverage. Table 15.1 summarises those aspects where a performance solution would be expected to be required (DRC, 2022).

Value for Money of 3D Printing

The long- and short-term benefits of digital technologies are revealed via their business cases. The business case for digital technologies with high starting costs, like 3D printing, should outline the value generated over time. When used over time, such technologies are more likely to provide greater value (Craveiro et al., 2019). Moreover, applying this technology can be expensive regarding resources, complexity, and time. As a result, the business case needs to be structured for a long-term investment. The justification of investment in digital technologies for the long and short term in the construction industry is a strategy that needs to be included in a company's overall strategy, not only at the project level (Olanipekun & Sutrisna, 2021).

Measuring the performance of traditional construction projects against projects that use new construction methods is paramount for providing a better understanding of value for money. Performance measurement is an essential point of management activity, providing managers with the information necessary to make decisions, monitor and coordinate performance, and appropriately apply available resources (Russo, 2006). According to Russo (2006), there are several reasons to measure the performance of organisations or projects, pointing out the following as the most relevant:

- To align the mission, strategy, values, and behaviours.
- To continue improving processes, products, and results.
- To quantify the successes or failures achieved.

According to Toor and Ogunlana (2010), the lack of information about the behaviour and performance of the project throughout its execution is highlighted as a cause of problems, such as exceeded budgets, delays in fulfilling work schedules, low quality, and low productivity in the construction process, given that performance measurement in projects has been dominated by conventional measures of time, cost, and quality, which are seen as the three traditional performance indicators used in the UK construction industry.

Lantelme (1994) states that construction industry projects were initially analysed only based on these three traditional indicators (time, cost, and quality). Still, aspects related to safety, health, and training of resources, among others, can also be identified as essential factors in a company's performance. However, according to this author, construction companies have considerable difficulties applying performance indicators due to their specificities. Therefore, many companies in this sector choose to use more straightforward performance indicators that result in performance measurement more focused on the project and a shorter term instead of focusing on the company's total needs in the medium and long term.

The key performance indicators (KPIs) emerged from the "Rethinking Construction Report" prepared by John Egan in 1998, with the purpose of measuring performance and improvements in the United Kingdom's construction industry. In this report, Egan highlighted that the sector had low profitability and invested very little in research, development, and training and that, in addition, many customers were dissatisfied with its overall performance. Thus, specific goals were identified for improvement in terms of the enterprises' productivity, profits, quality, safety, and performance (Egan, 1998).

The most widely used is the Constructing Excellence approach from the UK, which is used in Australia equally. The UK system of KPI assessment currently in practice is tabulated Table 15.2.

Several authors (Ahmad & Dhafr, 2002; Neely et al., 1996) were concerned with defining criteria for the design and selection of efficient and effective KPIs. These authors suggest that KPIs should be transparent, practical, easy to implement, simple to understand, have a visual impact, focus on improvement, have low cost, and be related to the organisation's strategy and its objectives.

Table 15.2 UK KPIs (Constructing Excellence, 2006)

Economic KPIs	*Social KPIs*	*Environment KPIs*
Client satisfaction – product	*Employee satisfaction*	*Environmental impact*
Client satisfaction – service	*Staff turnover*	*Energy use - product*
Defects	*Sickness absence*	*Energy use –process*
Predictability – cost	*Safety*	*Water use – product*
Predictability – time	*Working hours*	*Water use – process*
Safety	*Qualifications and skills*	*Waste removed from site*
Productivity	*Equality and diversity*	*Commercial vehicle movements*
Profitability	*Training*	*Impact on biodiversity*
Construction cost	*Pay*	*Area of habitat created / retained*
Construction Time	*Investors in people*	*Whole of life performance product*

Case Study Analysis

KPI Selection

One of the objectives of this research is to perform content and contrasting analyses of housing projects in Australia that included 3D printing technology. Therefore, defining a set of KPIs allows this analysis and provide more detailed knowledge of how the various aspects contributing to the work's efficiency are developing. Many KPIs could be chosen, but due to this study's time scale and information available, they were narrowed to seven. They are as follows:

1. Costs
    - Cost variance: Baseline vs actual
    - Prelims and management: Baseline vs actual
2. Quality
    - Reworks: Actual cost % of total baseline cost
    - Defects correction: Actual cost % of total baseline cost
3. Schedule
    - Duration variance: Baseline vs actual
    - Defects correction: Duration % of baseline total duration
4. Environmental Impact
    - Material waste: Actual cost % of total baseline cost

Case Studies

Case Study 1: Traditional Housing Project, Three Bedroom

- Contractor: Glenway Homes
- Construction GFA (gross floor area): 200 sqm
- Foundations: Concrete slab-on-ground
- External walls: Double brick, brick veneer and cladding
- Internal: Lightweight framing
- Roof: Metal sheeting with steel support structure
- Project duration: 11 months

Case Study 2: 3D Printing Project

- The Dubbo Regional Council appointed Contour3D to execute a 3D printed amenities block in West Dubbo, NSW, Australia

Figure 15.9 Three-bedroom house.
Source: AG, 2016.

- Construction footprint: 148 m^2
- Foundations: Concrete slab 18.5 m × 8 m × 0.5 m
- Internal and external walls: 3D printed concrete with the addition of oxide colour to the mix to eliminate the need for painting
- Roof: Steel support structure and insulation panels
- Internal fit-out was completed by local tradespeople
- Project duration: 2.5 months (25 hours 3D print only)

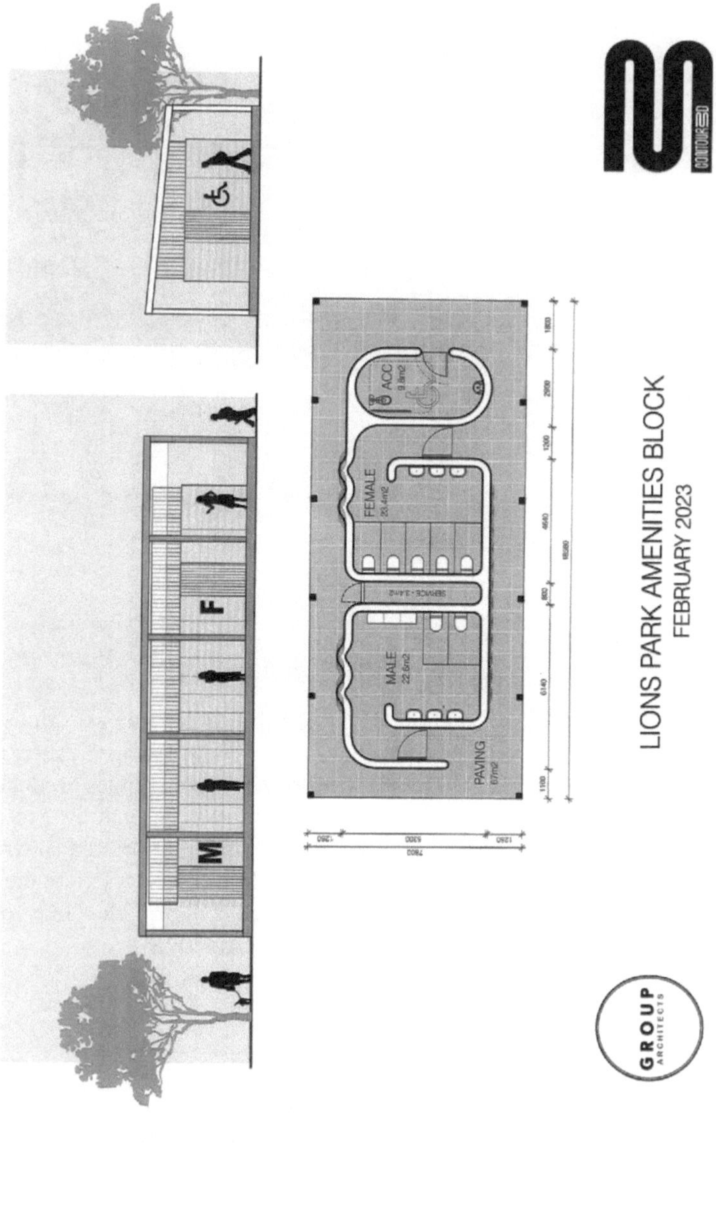

Figure 15.10 Lions Park amenities block.

Source: DRC, 2023.

Figure 15.11 Contour 3D printing: Day 2.
Source: DRC, 2023.

Case Study 3: Three-Bedroom House Project with 3D-Printed Walls

From the information available and collected from the other two case studies, the author of this research has produced a third case study with the application of 3D printing technology in external and internal walls only. A take-off was produced for all the external and internal walls of the house presented in case study 1 to determine the quantities of concrete required to replicate them in 3D-printed concrete. It was considered for this research that all other materials and services were the same as those used in case study 1. This additional case study will provide a like-for-like comparison of two three-bedroom housing projects in Australia.

Conceptual Analysis of Case Studies

The information collected for the case studies was obtained from Glenway Homes and Contour 3D. These companies used this data to track the project's progress from start to finish. Therefore, the case studies to be explored are based on real projects. To obtain the KPIs charts, Microsoft Excel was used to prepare the existing data

Figure 15.12 Project completion.
Source: DRC, 2023.

into data that could be used to compare the case studies. There are many ways to report KPI data, with the main options being line charts, bar charts, Venn diagrams, tables, and representations. For our analysis, we are using bar charts to compare the performance of three case studies on each KPI. The advantage of the bar chart is that a comparison with other states can be undertaken to determine if the targets are being met or are not being met. Additionally, we are including a Venn diagram to illustrate the performance against multiple KPIs at once. This will provide a better visual representation of the overall performance against all KPIs. KPI reference numbers are tabulated Table 15.3:

Costs

These KPIs were based on the information received for each project's budget and actual costs. The total cost is the sum of all activities of the project and uses the following formula:

Table 15.3 KPIs list

KPI 1	**Cost**
KPI 1.1	Cost variance – Baseline vs Actual
KPI 1.2	Prelims & Management – Baseline vs Actual
KPI 2	**Quality**
KPI 2.1	Reworks – Actual cost % of Total Baseline cost
KPI 2.2	Defects correction – Actual cost % of Total Baseline cost
KPI 3	**Time**
KPI 3.1	Duration variance – Baseline vs Actual
KPI 3.2	Defects correction – Duration % of Baseline Total duration
KPI 4	**Environmental Impact**
KPI 4.1	Material waste – Actual cost % of Total Baseline cost

$$Total\ costs = \sum task\ costs \qquad \text{Eq. 15.1}$$

Additionally, to calculate the percentage, the following formula was used:

$$Percentage = \frac{Costs}{Total\ Costs} \times 100 \qquad \text{Eq. 15.2}$$

The following formula was used to determine the cost variance of each project:

$$Cost\ variance = \frac{Actual - Baseline}{Baseline} \times 100 \qquad \text{Eq. 15.3}$$

Moreover, an additional calculation was included to show the delta between the budget costs and actual costs and the percentage increase with the use of the following formulas:

$$\Delta\ Cost = Actual\ cost - Baseline\ cost \qquad \text{Eq. 15.4}$$

$$\Delta\ Percentage = \frac{Actual\ Cost}{Baseline\ Cost} \qquad \text{Eq. 15.5}$$

Tables 15.4 to 15.6 show the total costs of these projects using equations 1 to 5. These tables are paramount to understanding the costs associated with project completion and can be compared with other projects of similar characteristics. They serve as a benchmark for allowing cost verification of savings or overspends.

For Tables 15.4, 15.5, and 15.6 and Equations 15.1 to 15.5, we calculated the KPIs and achieved the results seen in Figure 15.13.

In case study 1, changes to the project requested by the client during the construction stage, such as the kitchen layout, resulted in additional costs not only to the kitchen supply and installation but also in the services installation works (hydraulic, electrical, HVAC), resulting in an additional $9.8k in costs. These additional works have a direct cost implication for the preliminaries and project management due to the extension of time required to complete them. For the same reasons, reworks ($11.8K) and defects correction ($4.7K) incur additional costs. The overall project costs for case study 1 are $458,232.58.

In case study 2, we can see that the project performed much better. After completion, the client requested additional work (painting the walls) despite the concrete mix, including oxide colour. These additional works ($4.5K) represented a

Table 15.4 Case Study 1: Total Costs

Case Study 1: Three-Bedroom House		*Baseline*		*Actual*		*Delta*	
		Cost	*%*	*Cost*	*%*	*Cost*	*%*
1	Prelims and management	$42,800.00	10.0%	$46,610.00	10.2%	$3,810.00	108.9%
2	Earthmoving/ reinforced concrete	$77,040.00	18.0%	$84,744.00	18.5%	$7,704.00	110.0%
3	Framing	$36,808.00	8.6%	$39,016.48	8.5%	$2,208.48	106.0%
4	Roofing	$23,540.00	5.5%	$24,717.00	5.4%	$1,177.00	105.0%
5	Windows and external doors	$29,104.00	6.8%	$29,831.60	6.5%	$727.60	102.5%
6	External cladding / brickwork	$38,520.00	9.0%	$40,446.00	8.8%	$1,926.00	105.0%
7	Internal walls and ceiling finishes	$29,960.00	7.0%	$31,008.60	6.8%	$1,048.60	103.5%
8	Floor finishes	$27,392.00	6.4%	$27,939.84	6.1%	$547.84	102.0%
9	Internal doors and wardrobes	$12,840.00	3.0%	$13,482.00	2.9%	$642.00	105.0%
10	Sanitary ware and accessories	$11,128.00	2.6%	$11,684.40	2.5%	$556.40	105.0%
11	Kitchen	$30,816.00	7.2%	$35,438.40	7.7%	$4,622.40	115.0%
12	Hydraulic works	$26,536.00	6.2%	$28,791.56	6.3%	$2,255.56	108.5%
13	Electrical works	$20,116.00	4.7%	$21,624.70	4.7%	$1,508.70	107.5%
14	HVAC-solar	$21,400.00	5.0%	$22,898.00	5.0%	$1,498.00	107.0%
	Additional works					$9,884.66	
	Reworks					$11,817.08	
	Defects correction					$4,720.84	
	Total Cost	**$428,000.00**	100.0%	**$458,232.58**		**$30,232.58**	107.1%

Table 15.5 Case Study 2: Total Costs

Case Study 2: 3D Printed Amenities	*Baseline*		*Actual*		*Delta*	
	Cost	*%*	*Cost*	*%*	*Cost*	*%*
1 Prelims and management	$24,153.08	10.0%	$25,353.08	10.3%	$1,200.00	105.0%
2 Earthmoving / reinforced concrete slab	$19,350.00	8.0%	$19,350.00	7.9%	–	100.0%
3 Roofing	$33,170.22	13.8%	$33,170.22	13.5%	–	100.0%
4 External doors	$12,881.64	5.4%	$12,881.64	5.2%	–	100.0%
5 External and internal walls (3D print)	$53,550.00	22.2%	$53,550.00	21.7%	–	100.0%
6 Floor finishes	$20,610.62	8.6%	$20,610.62	8.4%	–	100.0%
7 Internal doors and partitions	$22,542.87	9.4%	$22,542.87	9.1%	–	100.0%
8 Sanitary ware and accessories	$19,322.46	8.0%	$19,322.46	7.8%	–	100.0%
9 Hydraulic works	$19,966.54	8.3%	$19,966.54	8.1%	–	100.0%
10 Electrical works	$15,135.93	6.3%	$15,135.93	6.1%	–	100.0%
Additional works - painting	–	0.0%	$4,500.00	1.8%	$4,500.00	101.4%
Total Cost	**$240,683.36**	100.0%	**$246,383.36**	100.0%	**$5,700.00**	102.4%

direct cost increase to the preliminaries and project management due to the extension of time required to complete them ($1.2K). The overall project costs for case study 2 are $246.383.36.

Case study 3 performed better than case study 1, even though we considered the same costs associated with the additional works ($9.8K), reworks ($11.8K), and defects correction ($4.7K) of case study 1. The overall project costs for case study 3 are calculated to be $353,617.50.

Implementing 3D printing concrete technology on an average three-bedroom housing project can save $104K in project costs. We believe that the savings could be much higher if case study 3 were a real project, because a 3D printing project will have BIM integration (3D modelling) that will avoid service clashes and construction methods optimisations that would result in a substantial reduction of reworks, defects correction, and materials and time savings.

Table 15.6 Case Study 3: Total Costs

Case Study 3: Three-Bedroom House with 3D Print	*Baseline*		*Actual*		*Delta*	
	Cost	*%*	*Cost*	*%*	*Cost*	*%*
1 Prelims and management	$34,000.00	10.2%	$37,810.00	10.7%	$3,810.00	111.2%
2 Earthmoving/ reinforced concrete slab	$29,000.00	8.7%	$31,900.00	9.0%	$2,900.00	110.0%
3 3D print (external and internal walls)	$67,500.00	20.2%	$67,500.00	19.1%	–	100.0%
4 Roofing	$23,540.00	7.1%	$24,717.00	7.0%	$1,177.00	105.0%
5 Windows and external doors	$29,104.00	8.7%	$29,831.60	8.4%	$727.60	102.5%
6 Floor and ceiling finishes	$27,392.00	8.2%	$27,939.84	7.9%	$547.84	102.0%
7 Internal doors and wardrobes	$12,840.00	3.9%	$13,482.00	3.8%	$642.00	105.0%
8 Sanitary ware and accessories	$11,128.00	3.3%	$11,684.40	3.3%	$556.40	105.0%
9 Kitchen	$30,816.00	9.2%	$35,438.40	10.0%	$4,622.40	115.0%
10 Hydraulic works	$26,536.00	8.0%	$28,791.56	8.1%	$2,255.56	108.5%
11 Electrical works	$20,116.00	6.0%	$21,624.70	6.1%	$1,508.70	107.5%
12 HVAC – solar	$21,400.00	6.4%	$22,898.00	6.5%	$1,498.00	107.0%
Additional works					$9,884.66	
Reworks					$4,804.60	
Defects correction					$1,746.24	
Total Cost	**$333,372.00**	100.0%	**$353,617.50**		**$20,245.50**	106.1%

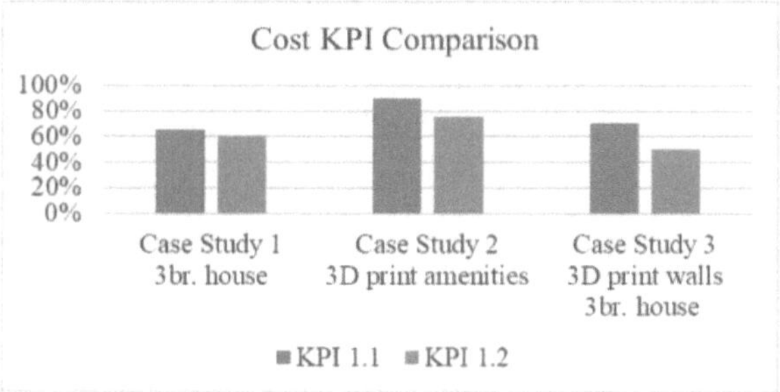

Figure 15.13 Cost KPI comparison chart.

Quality

This indicator is usually used in the construction industry to determine a defect rate by measuring the number of defects or issues found in completed work against the total work completed. This could include issues identified during inspections or client walkthroughs.

In this exercise, the KPIs reference defects correction and rework costs. Therefore, an analysis was conducted to determine the impact of these additional costs against the total budget by using the following formulas:

$$Reworks\ cost = \frac{Actual\ Cost}{Total\ Baseline\ Cost} \times 100 \quad \text{Eq. 15.6}$$

$$Defects\ correction\ cost = \frac{Actual\ Cost}{Total\ Baseline\ Cost} \times 100 \quad \text{Eq. 15.7}$$

This analysis will provide insights into the overall quality, allowing for adjustments and improvements for future projects. KPIs were calculated using equations 15.6 and 15.7 and costs from Tables 15.4 to 15.6, achieving the results seen in Figure 15.14.

Case study 1 was the project with the worst results. Changes mentioned in the cost analysis directly affected some reworks ($11.8K). Defects correction ($4.7K), unfortunately, is something that exists in the construction industry and always has an impact on time and cost.

Case study 2 had the best results. 3D printing technology does not (in theory) rely on excellent planning of all activities. 3D modulation is produced in computer software to identify service clashes and service installation elements, such as pipes and encasing. This substantially reduces errors in the delivery stage.

Case study 3 performed better than case study 1, even though we considered the same costs associated with reworks ($11.8K) and defects correction ($4.7K). This project would have performed much better in normal circumstances for the reasons stated in case study 2. The production of a 3D modulation would substantially reduce the costs for reworks and defects.

Schedule / Time

The data available was used to produce a high-level program for each case study with a combination of headings from the original programs. Business days were used to determine the duration of each heading.

For this KPI, the following formula is used to determine the duration variation:

$$Duration\ variance = \frac{Actual - Baseline}{Baseline} \times 100 \quad \text{Eq. 15.8}$$

Table 15.7 Cost KPI Data

KPI	*Case Study 1*	*Case Study 2*	*Case Study 3*
KPI 1.1 Cost variance: Baseline vs actual	65%	90.0%	70.0%
KPI 1.2 Prelims and management: Baseline vs actual	60%	75.0%	50.0%

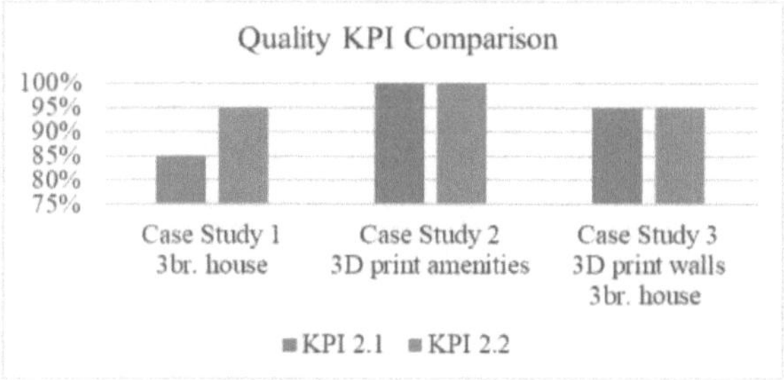

Figure 15.14 Quality KPI comparison chart.

Table 15.8 Quality KPIs data

KPI	*Case Study 1*	*Case Study 2*	*Case Study 3*
KPI 2.1 Reworks: Actual cost % of total baseline cost	85%	100.0%	95.0%
KPI 2.2 Defects correction: Actual cost % of total baseline cost	95%	100.0%	95.0%

Additionally, it is essential to discuss the defects correction duration impact on the overall duration of a project. Therefore, a KPI was included with the use of the following formula:

$$\textit{Defects correction duration} = \frac{\textit{Actual Cost}}{\textit{Total Baseline Cost}} \times 100 \qquad \text{Eq. 15.9}$$

Moreover, an additional calculation was included to show the delta between the baseline and actual duration, with the use of the following formula:

$$\Delta\ \textit{Duration} = \textit{Actual duration} - \textit{Baseline duration} \qquad \text{Eq. 15.10}$$

Tables 15.9 to 15.11 compare the actual duration against the baseline duration for each case study.

KPIs were calculated using equations 15.8 to 15.10 and Tables 15.9 to 15.11, and the following results were achieved in Figure 15.15.

In case study 1, changes to the project requested by the client during the construction stage, as mentioned in the costs analysis section, including defects correction,

Table 15.9 Case Study 1: Programme

Case Study 1: Three-Bedroom House	*Baseline*			*Actual*			*Delta*
	Duration	*Start*	*Finish*	*Duration*	*Start*	*Finish*	*Duration*
Construction	*173*	*2/02/2022*	*2/10/2022*	*203*	*2/02/2022*	*12/11/2022*	*30*
1 Prelims and management	173	2/02/2022	2/10/2022	203	2/02/2022	12/11/2022	30
2 Earthmoving/reinforced concrete	70	2/02/2022	10/05/2022	77	2/02/2022	19/05/2022	7
3 Framing	47	12/05/2022	17/07/2022	50	21/05/2022	29/07/2022	3
4 Roofing	24	19/07/2022	20/08/2022	25	31/07/2022	2/09/2022	1
5 Windows and external doors	21	17/07/2022	15/08/2022	23	29/07/2022	30/08/2022	2
6 External cladding / brickwork	21	17/07/2022	15/08/2022	23	29/07/2022	30/08/2022	2
7 Internal walls and ceiling finishes	40	16/06/2022	10/08/2022	42	27/06/2022	23/08/2022	2
8 Floor finishes	18	10/08/2022	2/09/2022	18	23/08/2022	15/09/2022	0
9 Internal doors and wardrobes	13	30/08/2022	15/09/2022	14	13/09/2022	2/10/2022	1
10 Sanitary ware and accessories	6	10/09/2022	19/09/2022	9	23/09/2022	5/10/2022	3
11 Kitchen	7	18/09/2022	27/09/2022	10	4/10/2022	17/10/2022	3
12 Hydraulic works	62	16/06/2022	10/09/2022	80	27/06/2022	15/10/2022	18
13 Electrical works	62	16/06/2022	10/09/2022	80	27/06/2022	15/10/2022	18
14 HVAC – solar	40	17/07/2022	10/09/2022	43	27/07/2022	24/09/2022	3
Defects correction				25	10/10/2022	12/11/2022	25

resulted in an extension of an additional 30 business days for project completion. The overall project duration for case study 1 is 203 business days.

In case study 2, we can see that the project performed much better. The additional work for painting the walls represents an extension of four business days. It is worth mentioning that printing all of the concrete walls only took two business days (25 hours). The other two days were for site mobilisation and demobilisation. The overall project duration for case study 2 is 52 business days.

Case study 3 performed much better than case study 1. Printing the external and internal 3D concrete walls was calculated to take five business days. The overall project duration for case study 3 is calculated to be 127 business days.

Implementing 3D concrete technology on an average three-bedroom housing project can thus save 76 business days in project duration. Again, as mentioned in the cost KPIs section, it is considered that these time savings could be much higher if case study 3 were a real project because a 3D printing project will have BIM integration (3D modelling).

Table 15.10 Case Study 2: Programme

Case Study 2: 3D Printed Amenities	*Baseline*			*Actual*			*Delta*
	Duration	*Start*	*Finish*	*Duration*	*Start*	*Finish*	*Duration*
Construction	*52*	*15/06/2023*	*27/08/2023*	*56*	*15/06/2023*	*31/08/2023*	*4*
1 Prelims and management	52	15/06/2023	27/08/2023	56	15/06/2023	31/08/2023	4
2 Earthmoving/reinforced concrete slab	16	15/06/2023	6/07/2023	16	15/06/2023	6/07/2023	0
3 Roofing	5	14/07/2023	20/07/2023	5	14/07/2023	20/07/2023	0
4 External doors	2	20/07/2023	22/07/2023	2	20/07/2023	22/07/2023	0
5 External and internal walls (3D print)	4	8/07/2023	13/07/2023	4	8/07/2023	13/07/2023	0
6 Floor finishes	8	20/07/2023	31/07/2023	8	20/07/2023	31/07/2023	0
7 Internal doors and partitions	9	31/07/2023	10/08/2023	9	31/07/2023	10/08/2023	0
8 Sanitary ware and accessories	7	10/08/2023	20/08/2023	7	10/08/2023	20/08/2023	0
9 Hydraulic works	25	14/07/2023	17/08/2023	25	14/07/2023	17/08/2023	0
10 Electrical works	25	14/07/2023	17/08/2023	25	14/07/2023	17/08/2023	0
Additional works: painting				4	27/08/2023	31/08/2023	4

Table 15.11 Case Study 3: Programme

Case Study 3-3bedroom house with 3D Printed Walls	*Baseline*			*Actual*			*Delta*
	Duration	*Start*	*Finish*	*Duration*	*Start*	*Finish*	*Duration*
Construction	*117*	*2/02/2023*	*15/07/2023*	*127*	*2/02/2023*	*30/07/2023*	*10*
1 Prelims and management	117	2/02/2023	15/07/2023	127	2/02/2023	30/07/2023	10
2 Earthmoving/reinforced concrete slab	22	2/02/2023	3/03/2023	22	2/02/2023	3/03/2023	0
3 3D print (external and internal walls)	5	5/03/2023	10/03/2023	5	5/03/2023	10/03/2023	0
4 Roofing	23	15/03/2023	15/04/2023	24	15/03/2023	17/04/2023	1
5 Windows and external doors	22	15/04/2023	16/05/2023	23	18/04/2023	18/05/2023	1
6 Floor finishes	18	16/05/2023	8/06/2023	19	18/05/2023	13/06/2023	1
7 Internal doors	13	6/06/2023	22/06/2023	14	8/06/2023	27/06/2023	1
8 Sanitary ware and accessories	7	18/06/2023	27/06/2023	9	20/06/2023	30/06/2023	2
9 Kitchen	8	26/06/2023	5/07/2023	10	10/07/2023	22/07/2023	2
10 Hydraulic works	61	15/04/2023	10/07/2023	79	15/04/2023	3/08/2023	18
11 Electrical works	61	15/04/2023	10/07/2023	79	15/04/2023	3/08/2023	18
12 HVAC–solar	40	16/05/2023	10/07/2023	42	15/05/2023	11/07/2023	2
Defects correction				10	15/07/2023	30/07/2023	10

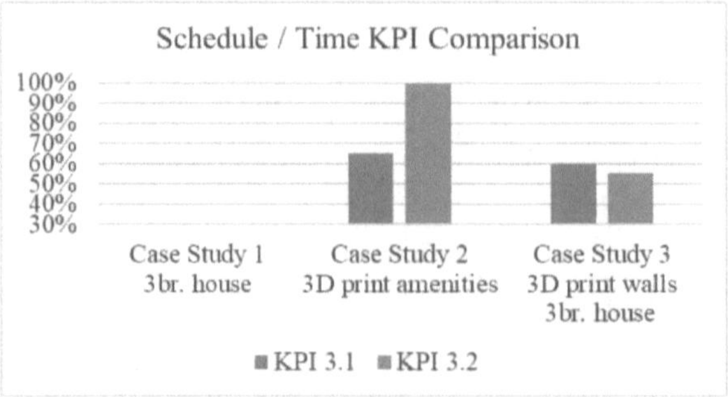

Figure 15.15 Schedule KPI comparison chart.

Table 15.12 Schedule KPI Data

KPI	*Case Study 1*	*Case Study 2*	*Case Study 3*
KPI 3.1 Duration variance: Baseline vs actual	30.0%	65.0%	60.0%
KPI 3.2 Defects correction: Duration % of baseline total duration	30.0%	100.0%	55.0%

Table 15.13 Case Study 1: Waste Calculations

Material Waste Calculations: Three-Bedroom House		*Costs*
2	Concrete	$4,237.20
3	Framing	$1,950.82
4	Roofing	$1,235.85
6	External cladding / brickwork	$2,022.30
7	Internal walls	$1,550.43
8	Floor finishes	$1,396.99
12	Hydraulic works	$1,439.58
13	Electrical works	$1,081.24
14	HVAC – solar	$1,144.90
	Total	**$16,059.31**

Environmental Impact

The data available that could be used for this KPI is related to material waste costs, which are considered 10% of the total materials cost for each project's main activities, as shown below.

The KPI calculation compared the actual cost against the overall budget, as considered in the following formula:

$$Material\ waste\ cost = \frac{Actual\ Cost}{Total\ Baseline\ Cost} \times 100 \qquad \text{Eq. 15.11}$$

This analysis will provide great insight into the importance of waste reduction for project cost savings. KPIs were calculated using Tables 15.13 to 15.15 and Equation 15.11, and the following results are shown in Figure 15.16.

Case study 1 was the project with the worst results. The volume of work is higher than the other two projects. The conventional construction methods produce lots of waste; in this project, they were calculated to be $16K.

Case studies 2 and 3 projects produced similar results. 3D printing technology does not (in theory) produce any waste. As mentioned in this research's environmental impact analysis, the 3D modulation and printing processes substantially reduce the wasted materials in the delivery stage.

Case studies 2 and 3 have included waste for all other activities not included in the 3D process during delivery. Therefore, the costs are calculated at $5.4K and $7.9K, respectively. Implementing 3D concrete technology on an average three-bedroom housing project can save $8.1K in material waste costs.

Table 15.14 Case Study 2: Waste Calculations

Material Waste Calculations: 3D Print Dubbo		*Costs*
2	Earthmoving / reinforced concrete slab	$967.50
3	Roofing	$1,658.51
5	External and internal walls (3D print)	–
6	Floor finishes	$1,030.53
9	Hydraulic works	$998.33
10	Electrical works	$756.80
	Total	**$5,411.67**

Table 15.15 Case Study 3: Waste Calculations

Material Waste Calculations: Three-Bedroom with 3D Print		*Costs*
2	Concrete	$1,595.00
4	Roofing	$1,235.85
6	Floor finishes	$1,396.99
10	Hydraulic works	$1,439.58
11	Electrical works	$1,081.24
12	HVAC – solar	$1,144.90
	Total	**$7,893.56**

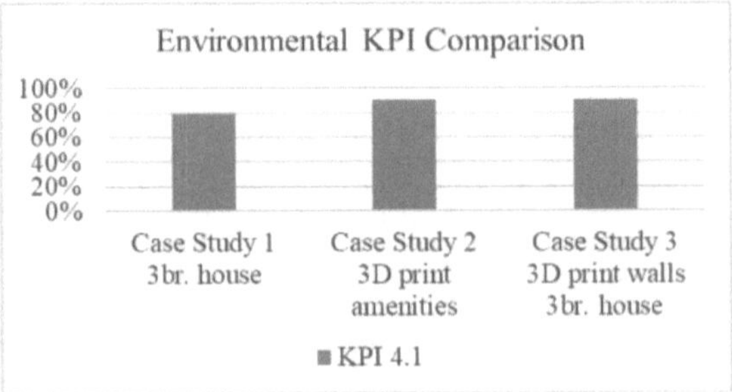

Figure 15.16 Environmental KPI comparison chart.

Table 15.16 Environmental KPI data

KPI	*Case Study 1*	*Case Study 2*	*Case Study 3*
KPI 4.1 Material waste: Actual cost % of total baseline cost	80.0%	90.0%	90.0%

Overall KPIs Performance Comparison

The Venn chart shown in Figure 15.17 visually represents all KPI performance for the three case studies against the target of achieving the cost and time baseline (100%).

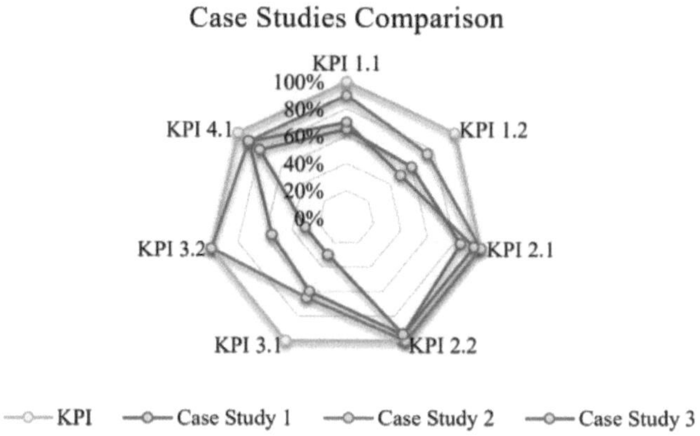

Figure 15.17 Case Studies performance comparison.

Table 15.17 Case Studies: KPI Comparison Data

KPI	*Compare Case Study 1 vs Case Study 2 vs Case Study 3*	*Performance*			
		Target	*C. Study 1*	*C. Study 2*	*C. Study 3*
KPI 1.1	Cost variance: Baseline vs actual	100%	65.0%	90.0%	70.0%
KPI 1.2	Prelims and management: Baseline vs actual	100%	60.0%	75.0%	50.0%
KPI 2.1	Reworks: Actual cost % of total baseline cost	100%	85.0%	100.0%	95.0%
KPI 2.2	Defects correction: Actual cost % of total baseline cost	100%	95.0%	100.0%	95.0%
KPI 3.1	Duration variance: Baseline vs actual	100%	30.0%	65.0%	60.0%
KPI 3.2	Defects correction: Duration % of baseline total duration	100%	30.0%	100.0%	55.0%
KPI 4.1	Material waste: Actual cost % of total baseline cost	100%	80.0%	90.0%	90.0%

Average Cost Building a House in Australia

It is essential to analyse the case studies selected for this research against the construction costs of similar projects in Australia. According to BMT Quantity Surveyors (2023), the current cost to build a three-bedroom house in Australia per square metre is shown in Table 15.18.

In Table 15.19, we calculated the actual costs per square metre of the projects included in case studies 1 and 2 and compared the results with the average square metre market costs.

The results show that the square metre cost of the project in case study 1 ($2,291/sqm) is 8% higher when compared to the square metre cost average for a three-bedroom whole brick project home with the same characteristics ($2,114/sqm),. On the other hand, the project in case study 3, which replaced the walls of case study 1 with 3D-printed concrete internal and external walls, shows that the square metre cost of this project is $1,768, which represents a reduction of 16% when compared to the square metre cost average for a three-bedroom whole brick project home ($2,114/sqm),.

INTERVIEWS DATA ANALYSIS

The information collected from the interviews with 3D printing companies provided a good understanding of the status of concrete 3D printing in the Australian construction industry. Figure 15.22 illustrates the views of 3D printing experts on the subject and shows a concept map that accumulatively and pictorially presents the interview data analysis. We can identify that the common main themes of discussion when interviewing these companies about using this technology were value for money, quality, opportunities, and challenges.

Table 15.18 Average Cost to Build a Three-Bedroom House in Australia

Construction Type	*Level of Finish (Quality)*		
	Low	*Medium*	*High*
3BR weatherboard project home	$1,786.00	$1,999.00	$2,473.00
3BR brick veneer project home	$1,904.00	$2,120.00	$2,532.00
3BR full brick project home	$1,895.00	$2,114.00	$2,626.00

Table 15.19 Case Studies 1 and 3: Sqm Build Costs Calculations for a Three-Bedroom House

Case Study	*GFA (SQM)*	*Actual Cost*	*Actual Cost / SQM*	*Medium Cost/ SQM*	*Delta*
Case Study 1	200	$458,232.58	$2,291	$2,114	1.08
Case Study 3	200	$353,617.50	$1,768	$2,114	0.84

Value for Money

According to Ahmed Mahil (2023), CEO of Luyten, the cost of building or buying a house is too high, not only in Australia but also all over the world. Producing houses with 3D printing offers an opportunity for first-time buyers to get more affordable homes. Builders can produce more houses faster with the use of 3D printing technology.

In Mahil's view, this technology can be used in the construction industry like a factory:

> you can bring 100 workers into the factory, do things manually, and make some profit, but your margins will be extremely thin. You automate most of the processes that don't need to be super complex, and then you make some more money, and you end up producing more, so the overall cost for the end user drops significantly.

For Holden (2022), CEO of Contour 3D, the concept of 3D printing in housing projects requires some economy of scale to be profitable. They are not planning to do one-off houses in different locations for individual clients. As he observes, "It's not about the money in the first iteration of the business model. We want to get it right." The intent is to have five to six structures on the ground outdoors and then review their business model – that is, print around ten houses on a particular site to gain the efficiencies and volume efficiencies that come with that. As Holden (2022) notes, "we've got one control system operating, you know, two machines, and we have two people just monitoring between two houses. That sounds like something that would be sort of attractive."

Additionally, the 3D printing technology saves time compared to traditional construction methods. Holden (2023) states, "at the moment when the slab is down already to print a single story four-bedroom home, internal, external walls, there's around about three to four days' worth of printing in that."

These experts' commentary aligns with this research's findings regarding the performance of case studies 1 and 3 (KPIs), whereby implementing 3D printing concrete technology on a three-bedroom housing project can save $104K in project costs and around 76 business days in time. Printing the external and internal 3D concrete walls was calculated to take only five business days, including mobilisation and demobilisation.

Quality

Regarding the quality concerns of implementing this technology, Mahil (2023) mentions that you need to have an excellent product to be in this industry. His company has printed a 3D house using Ultimatecrete, "which is an eco-friendly cementitious mixture. And by ecofriendly, we mean, like, we reduced the amount of cement that you would need to make concrete behave in that way." This was to

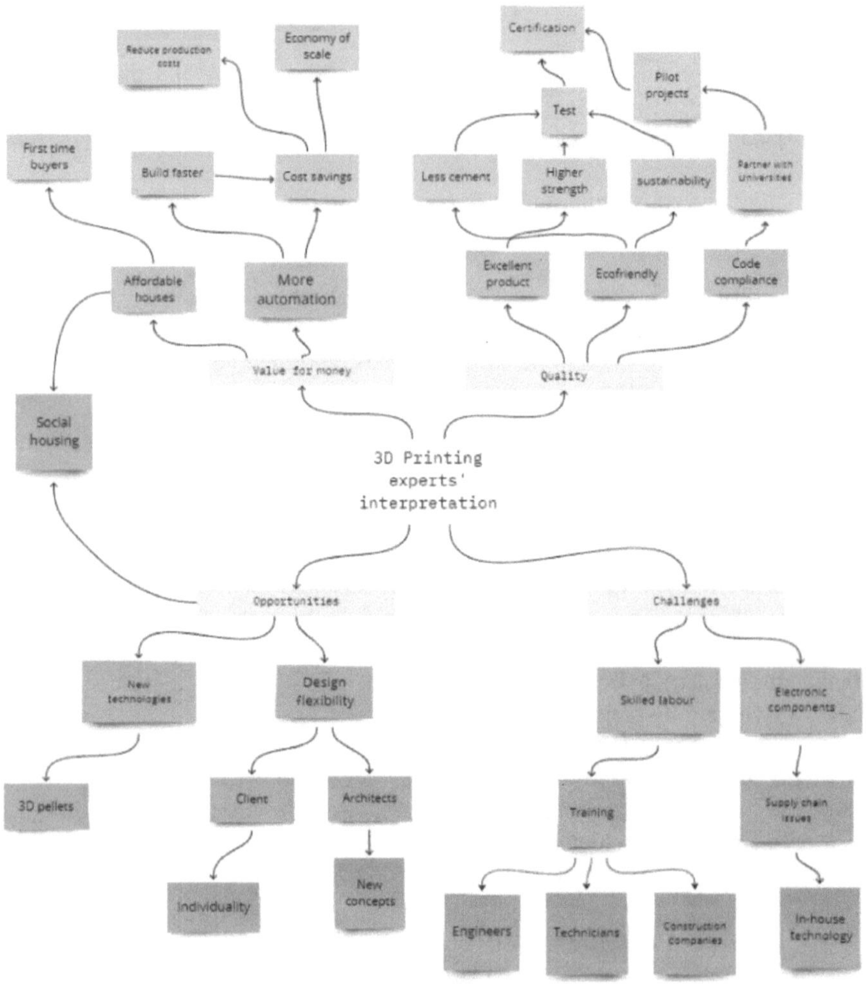

Figure 15.18 Interview data holistic classification for the thematic analysis.

showcase the technology and code compliance. It "surpassed the code four times, let's say, in some of the metrics."

According to Holden (2022), the current standards in Australia do not include 3D printing. Therefore, any 3D-printed building must be certified independently for the permit to be issued. However, Contour 3D is working with the regulators to complete the required standards for the construction industry.

The quality of the concrete mix is a key fixture of 3D print technology. Contour 3D has a full-time material scientist working with the company. They have partnered with the UNSW for the material development.

> At the moment, we've got 40% recycled materials in our mix that we call Contour Crete. By the end of this year, we will get that towards 75%. So, it's a very sustainable option, very quick and obviously super robust in a very high-performance concrete. (Holden, 2023)

In their project for the Dubbo Council (amenities block), the concrete structure has achieved a strength of 50Mpa, which far exceeds the strength requirements for a typical concrete slab of a house (20Mpa) (Holden, 2023).

As mentioned in the literature review section of this research, these companies have partnered with universities in Australia to create and test the concrete mixers necessary to extrude the product on their printers. Additionally, they are using the same partnership to develop the required standards with the Australian authorities to achieve certification in housing projects.

New Opportunities

For the final consumer, this technology gives "that choice to change the outlay or the design of their house, which gives them that individuality." The best response about using 3D printing in the construction industry comes from the architects. According to Mahil (2023), they were suffering for years in their designs and their concepts. "They have to deal with this statement and try to minimise curves. It costs more in the final project." Architects can now use different designs. "Whatever computer-aided draught they made; we can make it into a 3D printable code" (Mahil, 2023).

3D printing in construction is not just about executing projects for developers or construction companies. It can also tackle social issues such as homelessness. For Holden (2022),

> there's a lot of talk about 3D printing solving some affordability crisis and social housing, there's lots of social housing that needs to be built in Australia, and the government are on board with that. I fortunately know some of the right people, so that won't be a difficult market, but I think it's more speed than cost.

For Holden (2022), there is a potential to use thermoplastic systems.

> As part of an overall 3D business, I would like to incorporate some thermoplastics within that, and there are companies in Australia that are developing or that by the end of this year, they will have a machine that can generate or manufacture the 3D pellets here in Australia so we're not having to import that, and you can get almost 100 recycle components.

As mentioned earlier and discussed in the literature review, 3D printing technology gives architects creative freedom when building what was designed. Through 3D printer technology, details or finishes can be easily made. This represents a notable paradigm break as well as greater professional independence concerning industry standards. Moreover, design, balance, or engineering flaws can be easily identified and resolved during the 3D printing process.

Challenges

One of the biggest challenges for 3D printing technology is the lack of skill set required to work with this kind of technology. For Mahil (2023), Australia lost much of the automotive industry, and as such, the skilled labour necessary to work in automation is required in the construction industry.

> We've got some very skilful technicians who used to work in the automotive industry, and they used to use robots, make certain circuitry. I believe the supply of a well-versed workforce would go a long way, and it's a problem right now because you get engineers, but it's not all about the engineers. We need technicians as well.

Another challenge relates to the electronics supply and manufacture. There's a shortage of available products (chips and sensors) in Australia due to the lack of production. Therefore, they need to be imported, which causes delays and sometimes "certain model is not produced anymore. So, it affects the quality of the consistency of your products using the same sensors or et cetera, software wise or even electronic wise" (Mahil, 2023).

Holden (2022) intends to produce all in-house control systems and machinery to tackle the issue of electronic components, such as servo motors for control sensors and componentry, to reduce supply chain risks. In his view, a challenge for 3D printing technology is providing the correct training and skill set to the construction companies to allow them to operate the equipment and print concrete efficiently – in a sense, how to implement a simple-to-use and "bulletproof" system.

These challenges are pertinent considering that one of the advantages of using 3D printing is less labour required to execute construction housing projects, which can tackle the current issue of skill shortage, as mentioned in the literature review of this research. The 3D printing companies seem to face the same issues as the construction companies in these terms, but on a different scale.

Discussion and Conclusion

This research-based study is conducted to provide additional information about implementing 3D printing technologies in Australian housing construction projects. The existing literature could not provide the required information to evaluate and measure the performance of these kinds of projects against traditional construction projects, motivating the researchers to perform this analysis.

In this research, via literature review, the researchers explored the use of 3D printing technology in the worldwide construction industry, providing examples of good practices and progressing to consider the advantages and disadvantages of using this disruptive technology, focusing on current concerns such as quality and standards.

These concerns were addressed and discussed in the analysis of the interviews with 3D printing companies currently operating in Australia. These companies are working with Australian universities and Australian regulatory authorities to achieve the required certifications for materials and build according to national standards. The findings suggest that, although these companies are trying to execute pilot tests to showcase the technology to the construction industry and population, a considerable amount of work still needs to be done.

Furthermore, the researchers compared three housing projects in Australia, comparing traditional construction methods versus the implementation of 3D printing technology, using data collected from three case studies. An analysis of these projects' cost, time, quality, and environmental performance (KPIs) was conducted. From the findings, the researcher concluded that the 3D printing projects performed substantially better than traditional construction housing projects. Also, it is concluded that implementing this technology in housing projects could represent a cost reduction of 16% when compared to benchmark data for square meterage average costs for a three-bedroom house in Australia. This did not consider any economy of scale that could be achieved if more houses were constructed simultaneously, which, as mentioned in the interviews analysis, could result in even more savings.

Limitations

The results and outcomes of this study are based on case studies of construction projects executed in Australia, with case study 3 being created by the researcher to provide a like-for-like comparison of a traditional housing project versus a 3D concrete printing technology project. As such, it cannot be considered representative of all 3D printing projects executed in the country. However, the data collected and the corresponding findings provided a good understanding of the benefits of using this technology.

The study was conducted across a limited timeframe, and despite several attempts to approach key players in the 3D printing construction industries in Australia, the researchers were unable to get the desired interviews. Due to these limitations, the researchers used instead available online interviews with the CEOs of the two 3D printing companies operating in Australia and performed their analysis based on this secondary data. Despite those restrictions, this information complemented the data analysis performed in the case studies, offering valuable insights into how these companies see the current 3D printing technology in terms of value for money, challenges, quality, and opportunities.

Recommendations

Although this research provided insightful information regarding the use of 3D printing technology in construction projects, it is recommended that future research be conducted using more case studies to validate this research. Analysing further successful case studies and real-world applications of 3D printing in construction will allow a better understanding of the challenges, successes, and areas for improvement.

Emerging from the findings of this research, this research would be complemented through further explorations of economy of scale by researching ways to make this technology more scalable and economically viable for large-scale projects aligned with design optimisation to reduce material wastage and improve efficiency.

Another topic that arises from this research is that of regulatory and legal frameworks. Research on the regulatory and legal challenges associated with implementing 3D printing in construction would complement this research and provide a better understanding for addressing these challenges, which is crucial for broader adoption in the construction industry.

Ultimately, the primary questions are about the effective use of this technology in housing projects and its value for money. Hopefully, the research area will continue to attract further interest.

References

Ahmad, M., & Dhafr, N. (2002). *Establishing and improving manufacturing performance measures. Robotics and Computer-Integrated Manufacturing*, 18(3–4), 171–176.

The Arup Journal Vol.57 No.1 (2022). Available at: www.arup.com/perspectives/publications/the-arup-journal/section/the-arup-journal-2022-issue-1 (Accessed: 9 September 2023)

Bigrep (2017). *Geiger 3D Printing Concrete Casting Molds*. Available at: www.youtube.com/watch?v=hQK5nGXc0UA&t=35s (Accessed: 5 November 2023)

Bigrep (2023). *Advanced Applications with 3D Printed Formwork*. Available at: https://bigrep.com/applications/concrete-formwork/ (Accessed: 11 October 2023)

BMT Quantity Surveyors (2023). *How Much Does It Cost to Build a House in 2024?* Available at: www.bmtqs.com.au/construction-cost-table (Accessed: 10 December 2023)

Contour 3D. (n.d.). *The Block Tree Change. Contour 3D*. Available at: https://contour3d.com.au/the-block-tree-change/ (Accessed: 05 December 2023).

Constructing Excellence (2006). *UK Construction Industry: Key Performance Indicators*. Constructing Excellence.

Craveiro, F., Duarte, JP, Bartolo, H., & Da Silva Bartolo, P. J. (2019). *Additive manufacturing as an enabling technology for digital construction: A perspective on Construction 4.0. Automation in Construction*, 103, 251–267. https://doi.org/10.1016/j.autcon.2019.03.011

Crossan, F. (2003). *Research philosophy: Towards an understanding. Nurse Researcher*, 11(1), 46–55. https://doi.org/10.7748/nr2003.10.11.1.46.c5914

Daymon, C., & Holloway, I. (2002) Interviews. In: *Qualitative Research Methods in Public Relations and Marketing Communications*. Routledge, 166–185.

Dubbo Regional Council. (n.d.). Macquarie Lions West Amenity Block. Dubbo Regional Council. Available at: https://yoursay.dubbo.nsw.gov.au/Macquarie-Lions-West-Amenity-Block (Accessed: 05 December 2023)

Dubbo Regional Council (DRC) – Ordinary Council Meeting 21 September 2022, Report: *Regulatory Issues Pertaining to 3D Printed Housing in NSW and Australia (2022)*. Available at: www.dubbo.nsw.gov.au/ArticleDocuments/278/Business%20Paper%20-%20Ordinary%20Council%20Meeting%20-%2021%20September%202022.PDF.aspx (Accessed: 15 December 2023)

Egan, Sir John (1998). *Rethinking Construction*. Report of the Construction Task Force to the Deputy Prime Minister, John Prescott, on the scope for improving the quality and efficiency of UK construction. Available at: https://constructingexcellence.org.uk/wp-content/uploads/2014/10/rethinking_construction_report.pdf (Accessed: 10 September 2023)

GAO (1990). *Case Study Evaluations*. United States General Accounting Office.

Gibson, I., Rosen, D. W., Stucker, B. (2010). *Additive Manufacturing Technologies*. Springer.

Holden, N. (2022). *Interview for Automate Construction Podcast*. Available at: www.youtube.com/watch?v=cH4XbvH1U_A (Accessed: 10 September 2023)

Holden, N. (2023). *Interview for Triple M Radio*. Available at: https://omny.fm/shows/mmm-dubbo-breakfast/matho-with-nick-holden-contour-3d-full-lisntr-exte (Accessed: 10 September 2023)

Holden, M. T., & Lynch, P. (2004). *Choosing the appropriate methodology: Understanding research philosophy. The Marketing Review*, 4(4), 397–409. https://doi.org/10.1362/1469347042772428

Hyde, K. (2000). *Recognising deductive processes in qualitative research. Qualitative Market Research: An International Journal*, 3(2), 82–89.

Jakus, A. E. (2019). *An Introduction to 3D Printing d Past, Present, and Future Promise, 3D Printing in Orthopaedic Surgery*. Elsevier Inc. https://doi.org/10.1016/B978-0-323-58118-9.00001-4

Kadiri, D. S., & Shittu, A.A. (2013). *Causes of time overrun in building projects in Nigeria: Contracting and consulting perspectives. Journal of Chemical Information and Modeling*, 53(9), 1689–1699.

Khoshnevis, B. (2004). *Automated construction by contour crafting – Related robotics and information technologies. Automation in Construction*, 13(1), 5–19. https://doi.org/10.1016/j.autcon.2003.08.012

Kothman, I., & Faber, N. (2016). *How 3D printing technology changes the rules of the game. Insights from the construction sector. Journal of Manufacturing Technology Management*, 27(7), 932–943. https://doi.org/10.1108/JMTM-01-2016-0010

Lancaster, G. (2005). *Research Methods in Management: A Concise Introduction to Research in Management and Business Consultancy*. Butterworth-Heinemann.

Langley, A. (1999). *Strategies for theorising from process data. Academy of Management Review*, 24(4), 691–710.

Lantelme, E. (1994). *Proposal for a Quality and Productivity Indicator System for the Construction Industry*. Porto Alegre, Brazil, Federal University of Rio Grande do Sul. M.Eng. Dissertation (in Portuguese).

Mahil, A. (2023). *Interview for AuManufacturing*. Available at: www.aumanufacturing.com.au/self-belief-cementitious-3d-printing-and-sensors (Accessed: 08 September 2023).

McKinsey & Co. (2016). *Imagining Construction's Digital Future*. Available at: www.mckinsey.com/capabilities/operations/our-insights/imagining-constructions-digital-future (Accessed: 20 September 2023).

MX3D (2023). *A Smarter Bridge*. Available at: https://mx3d.com/industries/design/smart-bridge/ (Accessed: 09 September 2023)

Neely, A., Mills, J., Platts, K., Gregory, M., & Richards, H. (1996). *Performance measurement system design: Should process based approaches be adopted? International Journal of Production Economics*, 46–47, 423–431. https://doi.org/10.1016/S0925-5273(96)00080-1

Olanipekun, A. O., & Sutrisna, M. (2021). *Facilitating digital transformation in construction – A systematic review of the current state of the art. Frontiers in Built Environment*, 7, 660758. https://doi.org/10.3389/fbuil.2021.660758

PERI 3D Construction. (n.d.). *Europe's Largest 3D Printed Building. PERI.* Available at: https://cobod.com/europe-largest-3d-printed-building/#:~:text=PERI%203D%20Construction%2C%20a%20pioneer,meter%20of%20building%20per%20hour.&text=Henrik%20Lund%2DNielsen%2C%20Founder%20&,Thank%20you. (Accessed: 12 December 2023).

Perkins, I., & Skitmore, M. (2015). Three-dimensional printing in the construction industry: A review. *International Journal of Construction Management*, 15(1), pp. 1–9. https://doi.org/10.1080/15623599.2015.1012136

Rett, J. P., Traore, Y. L., & Ho, E. A. (2021). *Sustainable materials for fused deposition modeling 3D printing applications. Advanced Engineering Materials*, 23(7), 1–8. https://doi.org/10.1002/adem.202001472

Russo, J. (2006). *Balanced Scorecard para PME e Pequenas e Médias Instituições*. (L. Lidel, Ed.) (6th ed.). Lisboa.

Saunders, M., Lewis, P., & Thornhill, A. (2016). *Research Methods for Business Students.* 7th ed. Pearson Education Limited.

Steel Printers (2023). *Metal 3D Printing: The Benefits and Drawbacks of 4 Technologies.* Available at: www.thesteelprinters.com/news/metal-3d-printing-the-benefits-and-drawbacks (Accessed: 16 August 2023)

Stein, J. A. (2017). *The political imaginaries of 3D Printing: Prompting mainstream awareness of design and making the political imaginaries of 3D printing: prompting mainstream awareness of design and making. Design and Culture*, 7075(February), 1–25. https://doi.org/10.1080/17547075.2017.1279941

Teizer, J., Blickle, A., King, T., Leitzbach, O., & Guenther, D. (2016). *Large scale 3D printing of complex geometric shapes in construction.* In: *33rd International Symposium on Automation and Robotics in Construction.* ISARC Publications. https://doi.org/10.22260/ISARC2016/0114

Toor, S. R., & Ogunlana, S. O. (2010). *Beyond the "iron triangle": Stakeholder perception of key performance indicators (KPIs) for largescale public sector development projects. International Journal of Project Management*, 28(3), 228–236.

Wong, K. V., & Hernandez, A. (2012). *A review of additive manufacturing. ISRN Mechanical Engineering*, 2012, 1–10. https://doi.org/10.5402/2012/208760

Wu, P., Wang, J., & Wang, X. (2016). *A critical review of the use of 3-D printing in the construction industry. Automation in Construction*, (68), 21–31. https://doi.org/10.1016/j.autcon.2016.04.005

Yin, R. K. (2009) *Case Study Research: Design and Methods.* 4th ed. SAGE Publications.

Žukauskas, P., Vveinhardt, J., & Andriukaitienė, R. (2018). *Philosophy and Paradigm of Scientific, Management Culture and Corporate Social Responsibility*, IntechOpen. https://doi.org/10.5772/intechopen.70628

16 Commercial Property Vacancy for Building Adaptation and Re-use

Kevin Muldoon-Smith, Jonathan Pearson, Paul Greenhalgh, Jane Loxley, and Leo Moreton

Introduction

Research into property market vacancy is not new. The changing consumer demand of occupiers regularly renders property assets redundant, obsolete, and vacant – exhibiting the creative destruction outlined by Joseph Schumpeter (1950). Moreover, some commercial property vacancy is a 'necessary' attribute (referred to as natural vacancy in this chapter) of property markets. This type of vacancy can be understood as that part of property stock that is taken up in response to the needs of occupier demand. This process of vacancy is generally referred to as initial, frictional, or cyclical in nature (Kerris and Koppell, 2006) in contrast to structural vacancy, which is more persistent, demanding some kind of adaptation into improved or new use. However, what is new is the increased incidence and volatility of commercial vacancy in some sectors and the media attention given to ghost towns, zombie high streets, and moribund buildings following Covid-19. More recently, this led to suggestions in some quarters that society faces a new economic crisis due to overvalued commercial properties (Hammond, 2022).

This turbulence is due in large part to the international property sector experiencing a set of structural growing pains in response to dynamic changes in business practices, for example, the appetite for smaller commercial floor plates in the office sector due to changing working practices and the advancement of technology and the increase in vacant retail units due to online purchasing patterns. However, a counterpoint to this narrative is the parallel reduction in small industrial units due to demand from the logistics sector for last mile deliveries and the Covid-19 related demand for more floor space to accommodate distancing expectations. The result is that vacancy in buildings and locations is now more frequent (Henneberry, 2017) and more difficult to anticipate in contrast to the relatively regular occurrence of vacancy following long lease expiry or business cycle activity seen through built environment history. In response to this more volatile nature of vacancy, there is a pre-emptive need for a conceptual framework that captures the nature of vacancy across the building life cycle.

This analytical aperture directs the primary aim for this chapter. For researchers and practitioners to reflect on commercial property vacancy, and in particular to consider when, where, and how to adapt buildings, they need to have a conceptual

DOI: 10.1201/9781003380559-16

framework that can be used to reflect on the material reality of vacant properties – one that moves beyond the binary distinction of natural (Lausberg, 2008) and structural vacancy (Remoy, 2010). Consequently, the primary objective of this chapter is to develop a framework that can be used to examine vacancy throughout the building life cycle – starting with the initial construction phase and ending with demolition and redevelopment; and then to test this framework in a sample of six commercial locations.

Originality rests in conceptual utility as the first known holistic examination of commercial property vacancy beyond that of an abstract economic factor. While vacancy is considered as a singular construct (e.g., vacant, or not vacant) or as an econometric simplification when considering rental adjustment over time, it is grounded in the contemporary situation of regular property adaptation. In addition, in grounding the conceptual framework in a vacant property data set (across six major locations in the UK), the research is the first known holistic analysis of vacancy within the contemporary market context. Furthermore, in comparing office, retail, and industrial property markets, the research is the first known comparative asset analysis of commercial vacancy. Novelty is found in the conceptual development of the vacancy typology that considers the finer grain detail and segmentation of vacancy, the positioning of this in the market context, and the processes that construct vacancy. In addition, novelty is found in the detailed creation of a vacant building database that has adapted information originally created for property tax purposes by the British Government.

Research significance is found in the basis that this research provides for governments and property managers to evaluate ideas for vacant building management and adaptive re-use in the commercial property sector. For those property professionals involved in the day-to-day management of commercial property assets in the developed world, the chapter provides an approach to understanding the wider significance of vacant buildings and how they adapt through the building life cycle, which the authors hope will contribute to more knowledgeable and effective practice in relation to vacant building strategies and considerations of adaptive re-use. The expectation is that expanding knowledge in this area will help city leaders and asset managers in mature urban areas deal with the challenges of adapting an ageing and poorly performing property stock.

Theoretical Context

In recent decades researchers have studied vacancy through a variety of perspectives: those interested in obsolescence and depreciation (Baum, 1991, 1993; Baum and McElhinney, 1997; Dunse and Jones, 2005; Andrew and Pitt, 2006; Crosby and Devaney 2006; Crosby et al., 2011); those interested in the adaptation of vacant properties and the wider built environment (Gann and Barlow, 1996; Beauregard, 2005; Kincaid, 2002; Heath 2001; Geraedts and van der Voordt, 2003; Agre, 2005; Langston et al., 2008; Remoy, 2010; Remoy and Wilkinson, 2012; Wilkinson and Read, 2011; Armstrong et al., 2021; Lanz and Pendlebury, 2022; Armstrong et al., 2023) those who want to map the characteristics of vacancy (Myers and Wyatt, 2004; Katyoka and Wyatt, 2008; Remoy H. and Koppels, 2009); those who model the cyclical behaviour of the economy and property (Barras, 2009; Wheaton 1999); and those who reflect on the medium to long-term rental adjustment process (Blank and Winnick, 1953; Wincott, 1997; Voith and Crone, 1988; Crone, 1989; Grenadier, 1995; Pissarides, 2000; Sanderson et al., 2006; Miceli and Sirmans, 2013). Concurrently, professional practices regularly use relative vacancy levels (alongside absorption and take-up, rent and yield) to monitor the performance of local markets (see quarterly updates from international commercial real estate companies, CBRE, Savills, Colliers, Avison Young, Cushman and Wakefield, BNP Paribas, Jones Lang LaSalle). However, broadly speaking, in this perspective functional real estate assets grow old, become less productive, and must then be improved or replaced. Through this process, loss of value occurs gradually in a typically linear fashion related to the original function of the building rather than under external conditions of sudden market disruption (Christensen, 1997).

In contrast, the recent turn towards issues of transience and permanence (Henneberry, 2017) associate vacant land and premises in the post-industrial city (Burkholder, 2012), an engagement with DIY, guerrilla, and tactical urbanism (Deslandes, 2013), an emphasis on temporary and informal uses (Columb, 2012; Bishop and Williams, 2012; Oswalt et al., 2013), and the pragmatic steps involved in transferring a temporary activity into a mainstream process (Andres, 2013; Crosby and Henneberry, 2015) has highlighted a more dynamic and volatile property evolution process. Rather than gradual depreciation leading to vacancy over a period, sudden change and transience are the order of the day in this literature – reflecting shorter lease lengths, the disruption of Covid-19, and the general need for occupiers to expand and contract in response to changing circumstances.

Transience is a long-held natural characteristic of the real estate development process. Buildings are produced in response to socio-economic circumstances to meet extant demand. As that demand evolves through economic restructuring, technical innovation, social change, and so on, existing buildings and uses become obsolete and new buildings and uses are required to replace them – a 'natural' building development cycle (Barras, 2009; Henneberry, 2017). Transience within this conceptual framework might be considered to occur at two levels relating to the temporality of 'permanent' buildings / uses and, when their redevelopment is stalled, of the passing uses that are made of derelict / vacant land and buildings in the interim.

If we examine this broad framework in more detail in the commercial property market, we can see that transience can relate to the relative permanence of physical buildings before they are altered or redeveloped in response to physical obsolescence or external stimuli such as new occupier requirements and technological change (Drane, 2013; Muldoon-Smith and Greenhalgh, 2016). Transience can also be illustrated by the parallel use and movement of business between physical premises as they make new location decisions. For instance, 'filtering' describes the movement of businesses between properties as they filter up the property ladder into better quality premises or down the property ladder into lower quality premises.

'Take up' or 'absorption' describes the rate at which businesses occupy property within a specific period. In addition, displacement is often related in the classic gentrification literature to the push factors of new, wealthier businesses that increase local property prices and consequently price-out the original business community (see Smith, 1979; Marcuse, 1986; Lees et al., 2010). In commercial property markets, displacement can also be related to the mismatch between the buildings that firms occupy and their actual needs (see Fothergill et al., 1987; Harris, 2002; Greenhalgh and King, 2013). In this instance, the pull factors of new premises and attractive socio-economic conditions elsewhere can provide an incentive to move from existing locations. Viewed in this way, commercial property is not a rigid construction set in stone; rather, it is a 'transient manifestation of human activity' (Barras, 2009, 2).

In certain locations urban development also responds faster to occupier needs, due to buoyant socio-economic conditions that assist commercial viability. In others, that response may be slower, due to adverse economic conditions – the consequence is variegated market performance and degrees of vacancy between locations. This is because commercial real estate markets and their locations are not uniform. Instead, they have their own distinctive traits, rhythms, and cycles of change (Bryson 1997; Barras 2009). Finally, transience can also relate to the temporary use of commercial stock, as meanwhile and interim uses move into vacant premises to exploit advantageous rental conditions or to minimise the holding costs associated with vacancy and dereliction on behalf of the landlord. This occurs because of the frequent interval between one building use and the next (Henneberry, 2017).

However, the contemporary consideration of transience rarely foregrounds research with a consideration of vacancy across the building life cycle, the container for the transient practices. In addition, some of the wider societal imperatives that are related and coterminous to the consideration of transience, for example the adaptive re-use of buildings; sustainable retrofit of properties in response to the climate crisis; and considerations of the circular economy (the re-use of obsolescent products) rarely consider the manifestation of property vacancy – even though vacancy is the clearest indicator of the need for adapting obsolescent buildings. This gap in knowledge can be extended within the commercial property development industry itself, where designers, financiers, and agents rarely consider the market reality of vacancy before proposing a new building – the advent of new construction activity can regularly be cheek by jowl with concentrations of vacancy in town and city centre locations around the world.

Developing an Understanding of Vacancy

Historically, it is research into rental adjustment and professional practice that has given most attention to commercial property vacancy. Much of this traditional research specifically focused on the natural rate of vacancy and the prime markets (those properties considered most efficient and most lucrative for investors). Typically, this language borrowed from neo-classical economics, particularly its cyclical nature, and surveys of the labour market. This is most clearly seen in the parallel utilisation of the natural rate of unemployment and property vacancy and the utilisation of initial, cyclical, and frictional categories of unemployment and property vacancy (outlined by Kerris and Kopells, 2006).

In the study of employment, initial vacancy is taken to mean those potential employees who are recently qualified but yet to find employment. The parallel example in commercial real estate are those commercial properties that were constructed but were not filled yet. Cyclical unemployment occurs in parallel with the economic cycle; for example, when the economy is in decline unemployment will rise and vice versa. A similar process takes place, although lagged, in commercial real estate as the property cycle oscillates over time. Furthermore, frictional unemployment is a result of the movement of employees between firms and the consequent time taken to hire and refill vacant positions. This same process takes place in commercial real estate as businesses expand and retract. While structural unemployment is the consequence of a permanent change in the composition of the economy that leads to mismatches between the requirements of business and the available employee skills and training base to fill these positions (for a rare discussion of structural property vacancy, see Remoy, 2010, and of the impact of vacancy, see Armstrong et al., 2021, 2023). However, although the employment market analogy is useful, there are key differences between employment and commercial property markets. For example, new entrants into the employment market typically enter at the bottom rung of the employment ladder and work their way up. In contrast, new property stock typically enters at the top rung of the property ladder and then descends as it depreciates over time.

It is worth noting that initial, cyclical, and frictional concepts of vacancy implicitly assume that the market process will correct itself over time as the market clears. It is only structural vacancy that considers the other side of this situation, those properties that do not clear the market and fall off the business cycle. On a certain level, this existing set of terminology covers both sides of the commercial market, those properties that are temporarily vacant and those that are permanently vacant. However, under closer scrutiny this argument starts to fall apart when we consider that the natural rate of vacancy, which by most estimates only accounts for 4–10% of stock, received the most academic attention. The rest of the vacant commercial stock, considered structurally vacant, remains relatively unexplored (Lausberg, 2008).

This chapter responds to this situation by setting out a conceptual framework that delves under this condition, particularly the transition from natural to structural vacancy, and reveals the operation of perceived sub-optimal variants of vacancy that have received less attention in academia and practice. It achieves this aim by introducing two other commercial property ingredients into the discussion, the commercial property descriptions 'prime' and 'secondary.' In this chapter, prime property is taken to mean the most recent additions to, and most desirable segments of, commercial stock. In contrast, secondary property is taken to mean older stock in relation to the traditionally more desirable prime stock. The secondary focus is vindicated in the vacancy typology when it becomes increasingly apparent that a simple bifurcation between natural and structural vacancy does not exist. Secondary vacancy transcends both positions, indicating the ambiguous and dynamic nature of commercial vacancy. It is important to note at this juncture that important research took place into market segmentation, for example, the work of Jones (2013), Dunse and Jones (2002), Dunse et al. (2001), and Dunse et al. (2002). However, this research largely investigated sub-markets through a spatial land markets perspective. The conceptual framework outlined in this chapter does interact with sub-markets, for example prime and secondary segments; however, it does so to consider vacancy across the building life cycle, rather than across space.

Methodology

The research in this article is informed by a five-year research project into commercial market obsolescence, depreciation, and vacancy in the United Kingdom. While conducting this research, based on an ongoing interview process with more than 100 industry professionals, it quickly became apparent that the traditional language used in academia and practice to describe commercial vacancy was not adequate to express or explain the various manifestations of vacancy present in the commercial market, nor its variability and adaptation. This learning was taken forward to develop an original commercial vacancy typology that better describes the dynamic and transient processes taking place in contemporary towns and cities. This typology was then sense checked with the respondents via a consensus building Delphi technique to ensure that the framework reflected the insights and judgements of those working in the sector.

The elite/expert practitioner strategy described by Temenos and McCann (2013) and McGuinness et al. (2015) was used by the researcher team to identify respondents. There are fewer than ten major commercial real estate organisations in England due to amalgamations over time; the sample is therefore considered to reflect well the practices of commercial real estate in England with a strategy of quality over quantity adopted to capture insight from within the commercial market. This sample of respondents was also supplemented by those working in the planning and economic development profession – while not involved in the day-to-day management or transaction of commercial properties, these professionals do have an important perspective of the market due to their role in land, infrastructure, business, and investment planning and management.

A comparative sample of 18,284 vacant properties (in six cities) was then developed by utilising a snapshot of vacant commercial properties in England during October 2021. Secondary data sets created by local government for vacant property taxation and valuation purposes form the basis of this stage of enquiry, primarily the National Non-Domestic Rate Returns (NNDR) – similar methods were explored and used by Katyoka and Wyatt (2008) and Myers and Wyatt (2004). In the United Kingdom, all vacant properties must pay empty property rates (EPR); those that do not must prove exemption (information related to exemption is also recorded in the baseline NNDR information). The six locations chosen were:

- Birmingham
- Leeds
- Manchester
- Bristol
- Liverpool
- Newcastle upon-Tyne

These locations were chosen because they are six of the largest regional commercial real estate locations in the United Kingdom and demonstrate commercial real estate markets in operation. Each property is referred to in tax valuation as a hereditament – that is, a taxable real estate entity that is generating rent. For example, a multi-storey office building may be subdivided into ten office suites for lease. In this case, rather than one building, there would be ten hereditaments. All landlords must pay tax on their vacant properties in the United Kingdom or prove suitability for exemption. Whether valid for empty property taxation, or not, this information is monitored and, in most cases, published. The result is a consistent appreciation of commercial vacancy in each location. The data set provides a relatively raw depiction of vacancy within a locality – for example, a given building is either vacant or non-vacant.

It is important to note that this data is not the most up-to-date property information for these locations (it dates from 2021); more recent information is available on various commercial property information systems like Co-star and EGI Radius. However, these data sets are often inconsistent between location and do not offer finer grain property detail. Instead, the authors elected to use the slightly older treatment of NNDR data as this is more consistent between locations and offers more data fields for analysis. In this case, the research objective was not up-to-date market data but rather consistent data to test the conceptual framework. For example, the data set provides information in relation to duration of vacancy and is considered relatively exhaustive in coverage, depicting all vacancies within a local government administrative area. Importantly, the NNDR returns hold information in relation to duration of vacancy. This allowed the authors to start applying the conceptual framework into the empirical data set by using duration of vacancy as a baseline parameter for the segmentation.

Property Vacancy Typology

Orthodox econometric thought suggests that commercial vacancy can be separated into two broadly distinct tiers, that of natural vacancy and that of structural vacancy. However, this then interacts with the realities of commercial property practice, which is separated into the prime market and the secondary market. However, these parallel bifurcations do not run contiguously. Each vacancy tier, natural and

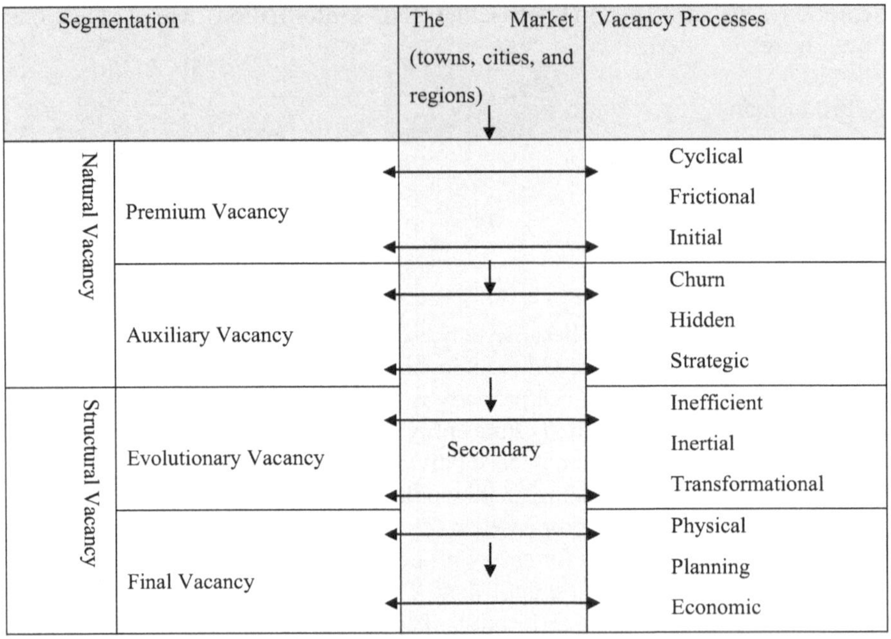

Figure 16.1 Typological model of vacancy segmentation.

structural, has its own characteristics, and although both are part of the same commercial market, they operate and manifest themselves quite differently.

Figure 16.1, the Typological Model of Commercial Property Vacancy, and the proceeding narrative explain this situation. Figure 16.1 should be read from left to right and top to bottom. The horizontal dimension describes the operational variation inherent in commercial property vacancy, running from the macro scale to the micro scale. This is denoted by the horizontal arrows that pass through Column 3. The vertical dimension represents the property ladder, the filtering process of tenants as they move between commercial buildings, and the building life cycle. The best properties are added to the top in a funnel-like system and the worst ones eventually drop out of the bottom depending on their contingent circumstance (following the vertical arrows in Column 3).

The first column describes the commonly held tiers of commercial property vacancy, natural vacancy, and structural vacancy. Natural vacancy describes those properties that efficiently clear through the classic supply and demand mechanism, while structural vacancy describes those properties that no longer clear through the supply and demand mechanism. In this formulation, Column 1 describes the macro-level description of the vacancy process commonly held in econometric literature. To understand this situation in greater detail, this bifurcation can then be subdivided to reflect real market segmentation. The natural rate can be subdivided into premium and auxiliary vacancy. Premium vacancy, as the name suggests, represents in this typology the very best buildings that are on the market and is associated with the familiar initial, frictional, and cyclical vacancy (Kerris and Koppells, 2006; Lausberg, 2008).

Auxiliary vacancy describes those vacant secondary properties that still have a role to play in the commercial real estate market. Auxiliary vacancy describes non-prime secondary properties that are held in reserve to 'fill in' prime supply shortages. The concept of 'filling in' is, by its very nature, temporary and transient. This is because it presumes that once new prime buildings are constructed, tenants will move to higher specification accommodation. Filling in is most likely to take place in buoyant areas with constrained supply conditions and during and following times of recession when speculative construction has abated, resulting in lagged development. Auxiliary vacancy is more permanent in those areas with adverse economic conditions, where it is difficult to justify the cost of development. In these locations it is important to safeguard this type of space to fulfil the requirements of occupier demand and economic development (in such areas auxiliary vacancy is closer to premium vacancy). In turn, structural vacancy can then be subdivided into evolutionary vacancy and final vacancy. Evolutionary vacancy describes those properties that could still have a future in alternative use if adapted – relevant to current debates of adaptive re-use and conversion from one use into another (Armstrong et al., 2023; Clifford et al., 2018; Fern et al., 2020). Final vacancy, as the name suggests, describes those properties that no longer have a future in either their present or alternative use and should therefore be removed from property supply altogether.

The first two columns can then be related to the overall commercial market (Column 3), which, for simplicity, is divided into prime property and secondary property. It is important to note that the prime market only intersects with premium vacancy, while secondary vacancy accounts for auxiliary, evolutionary, and final vacancy. It is this part of the model that lays out the disparity and non-alignment between natural and structural vacancy, and the prime and secondary market (they are not one and the same). Demonstrating the influence of the secondary market, this model indicates that it is, in part, included in both tiers of vacancy, natural and structural, as it also forms part of the auxiliary layer of vacancy. It is this non-alignment that exposes the myth that all secondary vacancy is bad and that the natural rate of vacancy only contains prime property. The third column, representing the property market (and its contingent location), forms the spinal structure of the model. The left-hand side (of which) considers the segmentation of vacancy in market locations, while the final column to the right considers the micro processes of vacancy that take place in these locations. It is these processes that reflect and make sense of the dynamic change and movement that take place within and between the respective segments of commercial property vacancy – and that is currently seen in contemporary society.

'Cyclical,' 'frictional,' and 'initial' vacancy are relatively well known in international literature (Kerris and Koppells, 2006; Lausberg, 2008; Remoy, 2010). These concepts are typically associated with the 'natural' rate of vacancy, market clearing, and concepts of equilibrium and, in this typology, premium vacancy. By themselves these processes are an efficient means of describing premium vacancy as it oscillates around equilibrium (cyclical), as it facilitates the movement of firms (frictional) in the economy, and as new property enters the market to create (initial) vacancy. These three types of vacancy process are helpful in the conventional economic sense as they 1) facilitate the efficient operation of the property market and 2) help us understand how this happens – all are presumed to be temporary in nature.

Moving down Column 4, churn, hidden, and strategic vacancy describe those types of commercial vacancy that take place within auxiliary vacancy. Churn vacancy is a variation of frictional vacancy, describing this concept after it has begun to filter down the property ladder. Churn vacancy takes place when the push and pull factors of new development at higher specification are constructed and cause existing tenants to filter up the property ladder through a 'flight to quality.' It is different to frictional vacancy because it leads to a downward revision in rent, capital value, and yield (without significant property improvement) and takes place more regularly. It is not a negative attribute of vacancy (this type of filtering and absorption is directly related to new start-ups and small businesses); however, it is a signal that such property is no longer a prime investment. Hidden vacancy describes that portion of vacancy that is difficult to detect, often consciously so. It includes those properties that are taking shelter from empty property taxation (but are vacant for all intents and purposes) and those properties often considered grey space by practitioners – those that are leased but surplus to tenant requirements (Norman, 2023).

Strategic vacancy is a potentially negative attribute of the commercial market. It describes those instances when landlords forcibly evict or coerce tenants to leave their buildings in pursuit of higher values associated with alternative building use even though they are still relatively viable in their present use (hence why it sits in the auxiliary segment). Strategic vacancy is particularly prevalent in England, following planning changes that have incentivised landlords in certain locations to target more profitable use (the advent of relaxed planning regulation, through permitted development rights, has been seen to favour office to residential conversion due to the higher economic value of the latter). All three of these concepts are still part of natural vacancy but are also associated with degrading performance and an increase in void space.

Inefficient vacancy, transformational vacancy, and inertial vacancy take place in the evolutionary vacancy layer. These types of vacancy can be considered on a progressive redevelopment spectrum and chart the transition of commercial property into potential new use. Inefficient vacancy describes those properties that are inefficient in terms of operational cost, holding cost, and embodied carbon. These properties are functionally and economically obsolete and ready to transition into alternative use (or potentially within use following major improvement). Inertial vacancy describes the regular impasse between operational use (in original form) and transformation (into new use). It does not happen in all cases but can be a consequence of restrictive tenancy covenants, planning negotiations, and financial due diligence. As the name suggests, transformational vacancy describes those properties going through new development, and details the final transition between inefficient use and such properties leaving supply altogether.

Physical, planning, and economic (often interrelated rather than separate categories) vacancy processes make up final vacancy. Planning vacancy includes those properties that cannot be adapted into alternative use (but are no longer viable in their present use) because they are constrained by planning regulation that places restrictions on alternative use. Physical vacancy describes those properties that have either depreciated beyond repair or have restrictive designs that do not lend themselves to re-use. Economic vacancy describes those properties that are not supported by viable local rental levels. In other words, the underlying rental levels that underpin such buildings do not cover existing running cost or the cost of new development. The only way these buildings can be re-used is through the introduction of subsidies.

The segmentation is not a static model. There is a great deal of transference between the fuzzy boundaries of the four segments, especially between auxiliary and evolutionary vacancy (and increasingly between market segments as the boundaries between use dissolve). It is also important to note that the model will also vary between towns, cities, and international locations depending on the prevailing market conditions in those locations.

Commercial Vacancy in the Market Context

The following section grounds the conceptual framework in the vacant building data model. In turn the quantity of vacancy and then its segmentation is presented. Table 16.1 describes how the different types of vacancy were segmented using duration of vacancy. Only the first three columns of the conceptual framework were grounded in the market data – this is because the micro processes of vacancy are very difficult to isolate in terms of duration of vacancy. However, the ground of the first three columns does present a foreground for analysis into these micro processes, while the conclusion presents a discussion of the limitations to this approach and opportunities for further research on refinement and the use of additional indicators.

Figure 16.2 describes the amount of space (in m^2) accounted for by vacant buildings in the six locations. While this is not an extrapolation from the conceptual

Table 16.1 Basis of segmentation breakdown using duration of vacancy

Natural Vacancy	0–24 months
Structural Vacancy	24+ months
Premium Vacancy	0–12 months
Auxiliary Vacancy	12–24 months
Evolutionary Vacancy	24–48 months
Final Vacancy	48+ months
Prime Vacancy	0–12 months
Secondary Vacancy	12+ months

Figure 16.2 Quantity of commercial property vacancy (m^2).

framework, it does give an initial foregrounding in relation to the scope of the vacant building issue; collectively there are 7.329 million m^2 of vacant commercial office space. In Birmingham, 433,258 m^2 of office space, 951,127 m^2 industrial space, and 387,767 m^2 retail space is vacant. Liverpool has the least vacant retail space (134,915 m^2), while Newcastle upon-Tyne has the least vacant industrial (136,245 m^2) and office space (149,787 m^2). It is important to note that these are nominal numbers and should not be used for comparison purposes. This is because Birmingham has a bigger real estate market than Newcastle upon-Tyne and indeed all the other locations. Instead, the numbers are presented to show the magnitude in each location.

Figure 16.3 describes the value of vacant property in the six locations, using rateable value as a proxy of the commercial value of stock. In total, the value of vacant properties across the asset sectors was £577 million, while the empty property tax liability was £295 million per year (based on the empty property rate tax multiplier).

Figure 16.3 indicates that the value of vacant industrial properties in Birmingham is highest (£29.9m), while the value of office (£97m) and retail (£61m) properties is highest in Manchester. Reflecting Figure 16.2, the lowest value of vacant office (£14.9m) and industrial (£4.4m) is in Newcastle, while the lowest values of retail (£11.3m) are in Liverpool. Figure 16.4 begins to break down the vacant building model into types of vacancy, in this case natural and structural vacancy represented in terms of metres squared (m^2).

While natural and structural vacancy are relatively common descriptions of property vacancy, they are rarely grounded in actual vacancy data – market reports in industry typically quantify vacancy in gross terms without any sub-division. In the office market, in most cases, natural vacancy accounts for the most vacancy, ranging from 54% (Bristol) to 68% (Newcastle). The exception to this rule was Liverpool and Manchester, where structural vacancy accounted for the most vacant space (66%) and (65%). The situation in the retail market followed a similar trend: natural vacancy accounted for the most vacant space in Bristol (57%) and Newcastle (59%). In Liverpool and Manchester, structural vacancy accounted for the most vacant space (ranging from 65% to 74%).

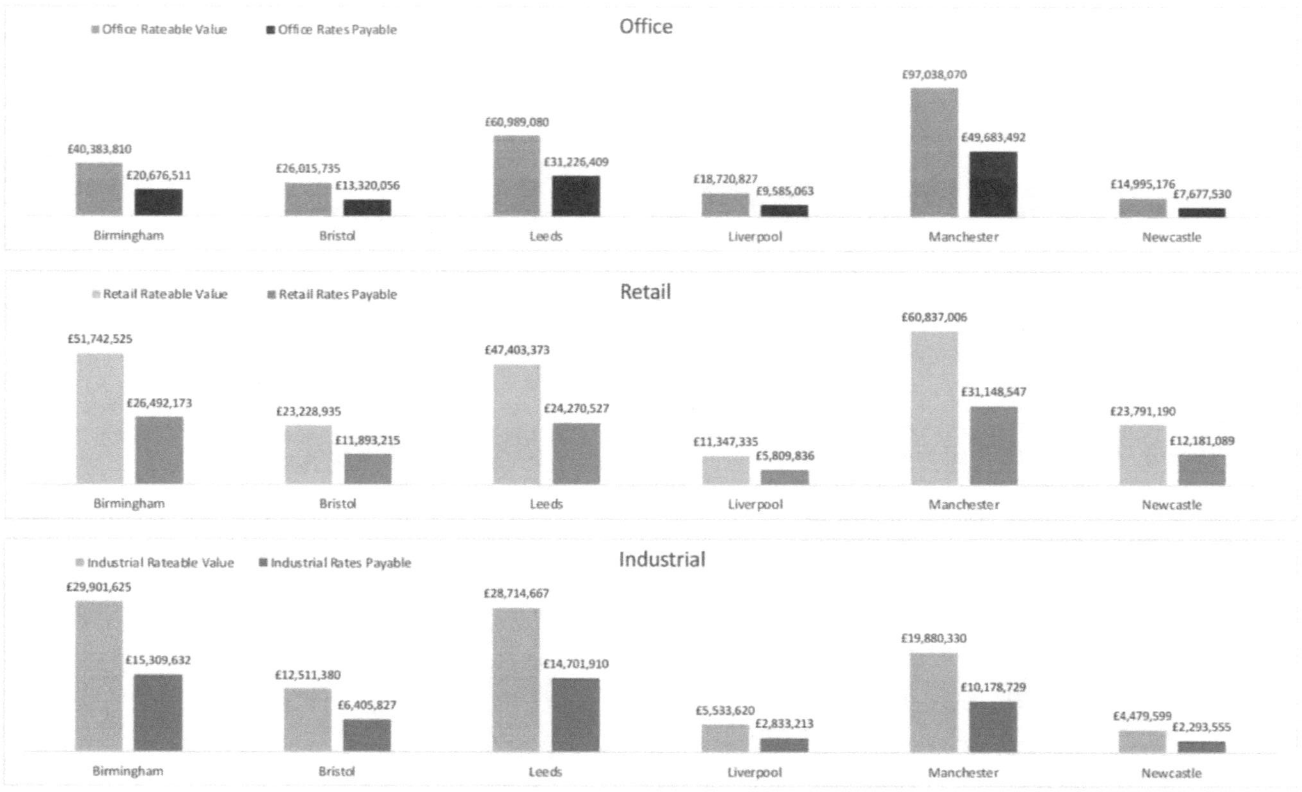

Figure 16.3 Total rateable value and tax representation of vacancy.

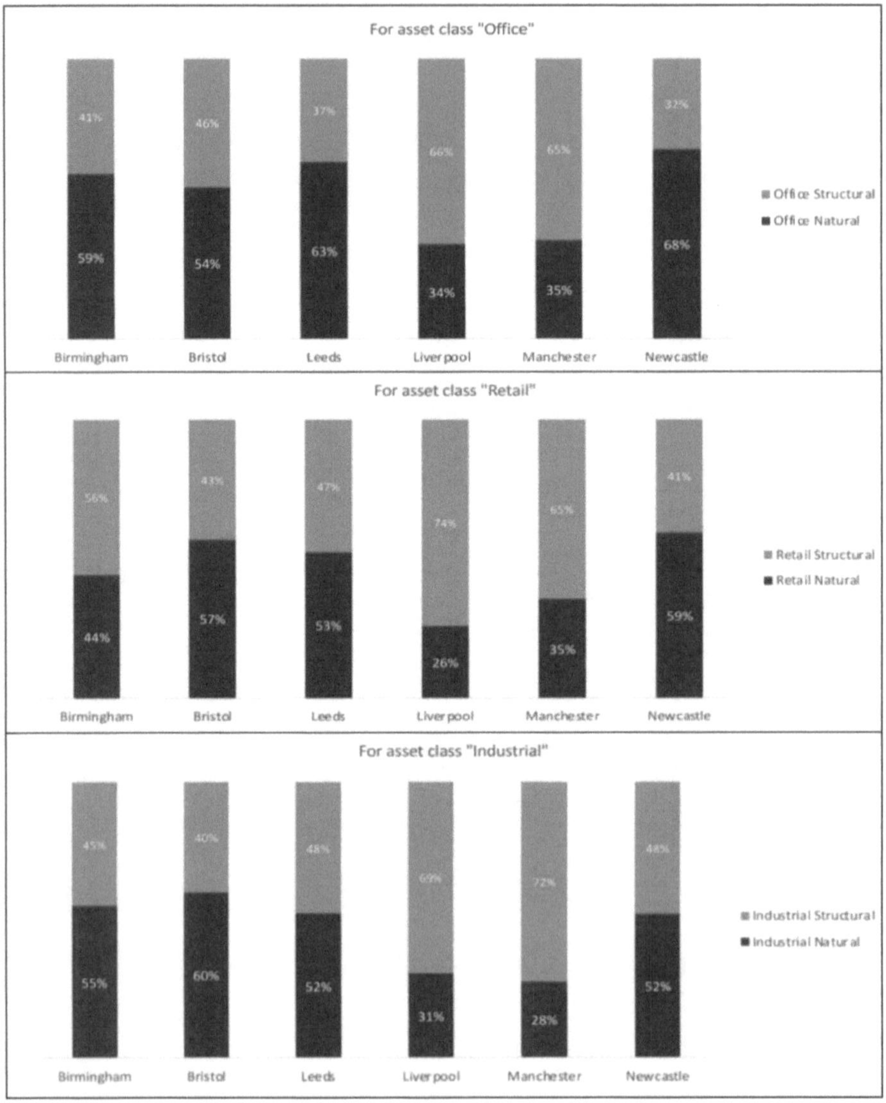

Figure 16.4 Natural and structural vacancy represented by m^2 (% distributed by m^2).

This is an unusual finding because most historical estimates suggest that natural vacancy accounts for only a small proportion of vacant buildings. There are several reasons for this finding, such as a large degree of new development (initial vacancy) or firms moving into new space (frictional vacancy), but perhaps more realistically it is because of the large number of tenants in the office and retail sector that are suffering the economic strain of Covid-19 and due to changing consumer and working practices in these sectors. The industrial sector shows a similar pattern to the office and industrial sector. In all locations, excluding Liverpool and Manchester, natural vacancy accounts for the most vacant space, ranging from 52% (Newcastle) to 60% (Bristol). Structural vacancy accounted for 69% in Liverpool and 72% in Manchester. Taken together, these findings signal a potential shift in the

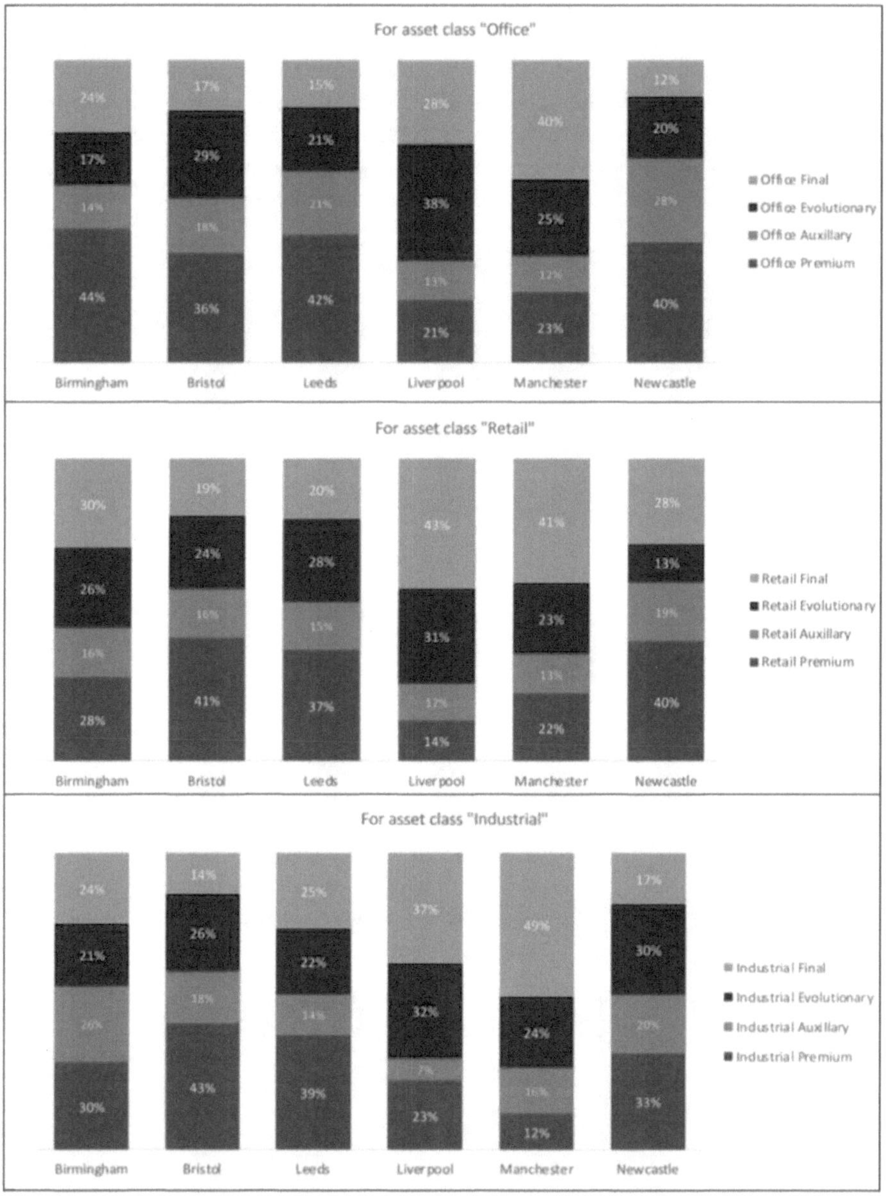

Figure 16.5 Vacant building segmentation.

long-held dynamics of commercial property vacancy, where the greatest quantity of vacancy was typically held in those properties vacant for the longest.

Figure 16.5 breaks this situation down into different segments of vacancy. In the office sector, evolutionary vacancy typically accounted for most vacant space in Liverpool (38%), whereas final vacancy accounted for most space in Manchester (40%). Both locations had a comparatively lower premium vacancy, 21% (Liverpool) and 23% (Manchester), than other locations, which range from 36% (Bristol) to 40% (Newcastle). In Bristol, Leeds, and Newcastle, auxiliary and evolution vacancy account for 42% to 47% of vacancy, as opposed to Birmingham, which has 31% with a higher comparative level of final vacancy. In the retail sector, final vacancy again accounted for most vacant space in Liverpool (14%) and Manchester (22%), while premium vacancy accounted for the least vacant space in these locations. Final and premium vacancy were most prevalent in Newcastle and

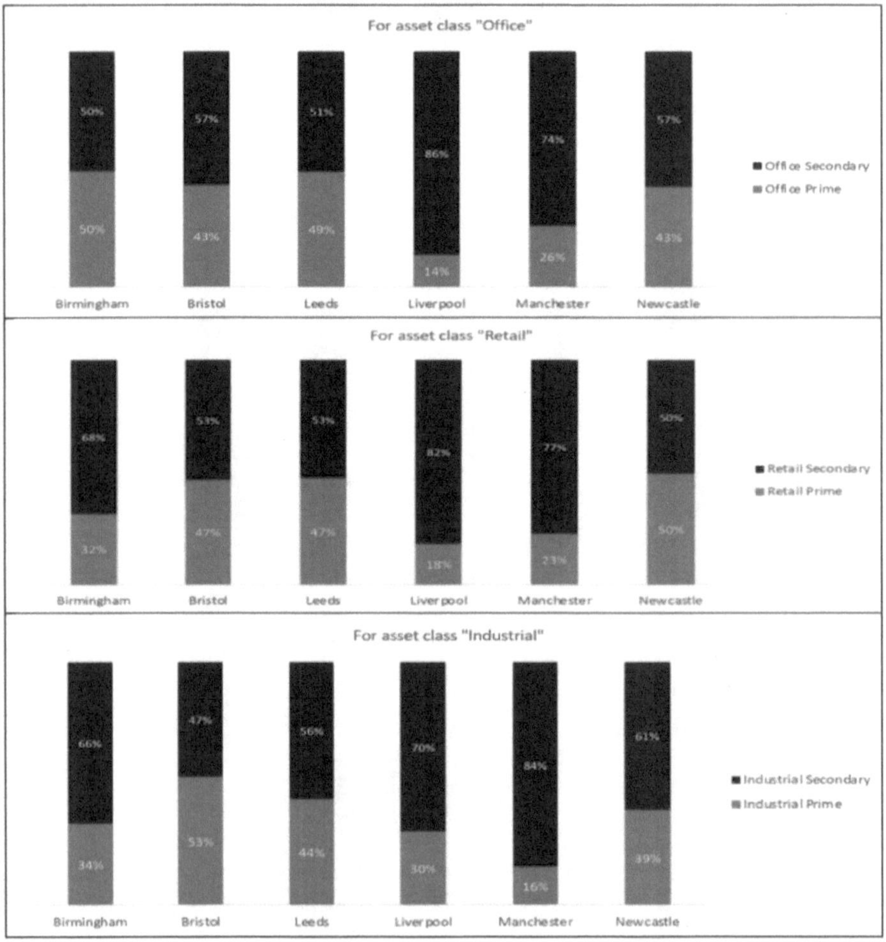

Figure 16.6 Prime and secondary vacancy.

Birmingham, while premium and auxiliary vacancy were most prevalent in Bristol and Leeds. In the industrial sector, Liverpool (37%) and Manchester (49%) had the highest level of structural vacancy. However, premium vacancy was most prevalent in Birmingham, Bristol, Leeds, and Newcastle (ranging from 30% to 43%).

Figure 16.6 offers a subsequent layer of analysis, separating vacancy into prime and secondary components. In the office sector, secondary vacancy accounts for the most vacant space in all but one location (ranging from 51% to 86%). The exception is Birmingham, where prime vacancy accounted for 50% of vacant space. In the retail sector, secondary vacancy accounts for the most vacant space again in all but one location (ranging from 53% to 82%), with Newcastle the exception with 50% prime space. In the industrial sector, secondary vacancy accounted for the most vacancy in five out of six locations (ranging from 56% to 84%). The exception was Bristol, where prime vacancy accounted for 53% of vacancy. Across all sectors, Liverpool and Manchester have the lowest proportion of premium vacancy across all asset classes. The third lowest proportion in the industrial (34%) and retail (32%) asset classes is Birmingham, and the office class is Newcastle and Bristol (43%).

Conclusion

This chapter explicated a conceptual framework for commercial vacancy that moves beyond the positive facets of vacancy, such as initial, frictional, and cyclical vacancy types (Keeris and Koppels, 2006) and the general approximation of structural vacancy. To enhance this understanding, the first challenge for global commercial property market stakeholders, their professional bodies, and academics relates to the recognition of vacancy beyond traditional definitions of natural and structural vacancy. The conceptual framework introduced in this chapter, and testing in six major commercial locations, goes towards this aim. However, new empirical research and theorisation needs to take place to create the informational baselines that reflect the existence and nature of the different types of vacancy outlined in this article. We intend for this research to be a staging post for new research into vacancy and validation in different market contexts. Although we maintain that the typology is a useful way of understanding vacancy, we expect each market context to be different. This is because the dynamic process of transience and permanence is now a key feature of the real estate market dynamic. Furthermore, we concede that considerations of the respective types of vacancy are inherently colloquial – defined by the unique market tendencies in each location but also the subjective appraisal by market actors. For example, many of those properties considered secondary in London, Frankfurt, and New York are likely to be considered prime in regional cities. In the developed world this may be possible using market transaction data and estimations of vacancy and building quality. However, in less developed locations this may not be straightforward due to issues of information transparency. Therefore, other situational specific methodologies could potentially be more useful, such as small, location-specific studies.

In addition, the authors argue that further blending of multi-disciplinary conceptual domains will be necessary to understand and interpret the variable contexts of commercial property vacancy. For example, in growing cities, in both developed (through agglomeration) and less developed countries (through urbanisation), there is a need to accommodate rapidly increasing levels of population through more efficient building use. Understanding the evolution of vacancy across the building life cycle in developed countries could provide a potential opportunity to minimise vacancy in less developed countries. This is possible because large amounts of the built environment in less developing locations have not been constructed yet. However, it would be inaccurate to claim that all towns, cities, and regions are growing. There are countless examples of shrinking cities in the international context (for example the rust belt cities in North America, the Ruhrgebiet in Germany) and increasingly ghost cities (for example Ordos in China) that have never been inhabited to their full potential. Understanding vacancy in these contexts can help counter and potentially take advantage of urban blight. Each location is very different and will have its own characteristics and rhythms.

However, to begin to understand the nature of vacancy, it is necessary to qualify the research findings in this chapter. First, the UK focus of the research reveals the need for some cautionary words in relation to the context and content of the findings and conclusions in this chapter. We must be careful of overgeneralisation and simplification. Each location in the world contains a variety of comparable but highly specific real estate markets that are contingent on and socially produced in each context. It is therefore likely that the operation of vacancy will be different in alternative market contexts. Moreover, grounding the conceptual framework in the six locations made use of vacancy duration as the primary means of separating vacancy. Duration of vacancy is only an approximation of the various types of vacancy introduced in this chapter. It is recognised that there will be additional indicators that can be brought to bear through further methodological development of each vacancy type, particularly to bring into the foreground the micro-level process of vacancy in the final column of the conceptual framework. Indeed, the authors consider that this is the main opportunity for further research. It is hoped that the conceptual framework set out in this chapter is used as a framework for discussion and methodological evolution, rather than rigid structure.

References

Andres, L. (2013) Differential spaces, power hierarchy and collaborative planning: A critique of the 471 role of temporary uses in shaping and making places. *Urban Studies*, 50(4), 759–775.

Andrew, A. and Pitt, M. (2006) Property depreciation in government. *Journal of Property Investment 474 & Finance*, 24(3), 259–263.

Armstrong, G., Soebarto, V. and Zuo, J. (2021) Vacancy visual analytics method: Evaluating adaptive reuse as an urban regeneration strategy through understanding vacancy. *Cities*, 115, 103220.

Armstrong, G., Wilkinson, S. and Cilliers, E.J. (2023) A framework for sustainable adaptive reuse: Understanding vacancy and underuse in existing urban buildings. *Frontiers*, 5. https://doi.org/10.3389/frsc.2023.985656

Barras, R. (2009) *Building cycles: Growth and instability*. Chichester: Wiley-Blackwell.
Baum, A. (1991) *Property investment depreciation and obsolescence*. London: Routledge.
Baum, A. (1993) Quality, depreciation, and property performance. *Journal of Real Estate Research*, 8(4), 541–565.
Baum, A. and McElhinney, A. (1997) *The causes and effects of depreciation in office buildings: A ten-year update*. London: RICS – Royal Institution of Chartered Surveyors.
Beauregard, R.A. (2005) The textures of property markets: Downtown housing and office conversions in New York City. *Urban Studies*, 42(13), 2431–2445.
Bishop, P. and Williams, L. (2012) *The temporary city*. London: Routledge.
Blank, D.M. and Winnick, L. (1953) The structure of the housing market. *Quarterly Journal of Economics*, 67(2), 181–203.
Bryson, J.R. (1997) Obsolescence and the process of creative reconstruction. *Urban Studies*, 34(9), 1439–1458.
Christensen, C. (1997) *The innovators dilemma*. Cambridge, MA: Harvard Business Review Press.
Clifford, B.P., Ferm, J., Livingstone, N. and Canelas, P. (2018) *Assessing the impacts of extending permitted development rights to office-to-residential change of use in England*. London, UK: Royal Institution of Chartered Surveyors.
Columb, C. (2012) Pushing the urban frontier: Temporary uses of space, city marketing, and the creative city discourse in 2000s Berlin. *Journal of Urban Affairs*, 34(2), 131–152.
Crone, T. (1989) *Office vacancy rates: How should we interpret them?* Pennsylvania: Business Review, Federal Reserve Bank of Philadelphia.
Crosby, N. and Devaney, S. (2006*) Depreciation and its impact on the total return of UK commercial real estate 1994–2003*. Reading, UK: University of Reading working papers in Real Estate and Planning.
Crosby, N., Devaney, S. and Law, V. (2011) Benchmarking and valuation issues in measuring depreciation for European office markets. *Journal of European Real Estate Research*, 4(1), 7–28.
Crosby, N. and Henneberry, J. (2015) Financialisation, the valuation of investment property and the 510 urban built environment in the UK. *Urban Studies*, 53(7), 1424–1441.
Deslandes, A. (2013) Exemplary amateurism: Thoughts on DIY urbanism. *Cultural Studies Review*, *513*, 19(1), 216–227.
Drane, J. (2013) *The state of contemporary property development theory*. 19th Annual Pacific-Rim Real Estate Society Conference, Melbourne, Australia, 13–16 January 2013.
Dunse, N. and Jones, C.A. (2002) The existence of office sub-markets in cities. *Journal of Property Research*, 19(2), 159–182.
Dunse, N. and Jones, C.A. (2005) Rental depreciation, obsolescence, and location: The case of industrial properties. *Journal of Property Research*, 22(2/3), 205–223.
Dunse, N., Leishman, C. and Watkins, C. (2001) Classifying submarkets. *Journal of Property Investment and Finance*, 19(3), 236–250.
Dunse, N., Leishman, C. and Watkins, C. (2002) Testing for the existence of office sub-markets: Comparison of evidence from two cities. *Urban Studies*, 39(3), 483–506.
Fern, J., Clifford, B., Canelas, P. and Livingstone, N. (2020) Emerging problematics of deregulating the urban: The case of permitted development in England. *Urban Studies*, 58(10), 2040–2058. https://doi.org/10.1177/0042098020936966
Fothergill, S., Monk, S. and Perry, M. (1987) *Property and industrial development*. London: Hutchinson.

Gann, D. M., & Barlow, J. (1996) Flexibility in building use: the technical feasibility of converting redundant offices into flats. *Construction Management and Economics, 14*(1), 55–66. https://doi.org/10.1080/014461996000000007

Geraedts, R. and Van der Voordt, T. (2003) Offices for living in: An instrument for measuring the potential for transforming offices into homes. *Open House International*, 28(3), 80–90.

Greenhalgh, P. and King, H. (2013) Developing an indicator of property market resilience – investigating the potential of GIS to analyse business occupier displacement and property market filtering: A case study of Tyne and Wear. *Urban Studies*, 50(2), 372–390.

Grenadier, S.R. (1995) Local and national determinants of office vacancies. *Journal of Urban Economics*, 37(1), 57–71.

Hammond, G. (2022) Goldman Sachs sounds alarm on UK commercial property. *Financial Times*, October 23.

Harris, R. (2002) Evolution in the supply of commercial real estate; the emergence of a new relationship between suppliers and occupiers of real estate, in Guy, S. and Henneberry, J. (Eds.) *Development and developers: Perspectives on property*. Oxford: Blackwell.

Heath, T. (2001) Adaptive re-use of offices for residential use: The experiences of London and Toronto. *Cities*, 18(3), 173–184.

Henneberry, J. (2017) *Urban transience*. London, UK: Wiley-Blackwell.

Jones, C. (2013) *Office markets and public policy*. London, UK: Wiley-Blackwell.

Katyoka, M. and Wyatt, P. (2008) An investigation of the nature of vacant commercial and industrial 550 property. *Planning Practice and Research*, 23(1), 125–145.

Keeris, W. and Koppels, P.W. (2006) *Uncertainty about the vacancy rate in the Dutch office market due to the different vacancy types and stratified structure*. ERES. Annual European Real Estate Society Conference, Weimar, Germany, Bauhaus-Universität Weimar.

Kincaid, D. (2002) *Adapting buildings for changing uses: Guidelines for change of use refurbishment*. London: Spon Press.

Langston, C., Wong, F.K.W., Hui, E.C.M. and Shen, L.Y. (2008) Strategic assessment of building adaptive reuse opportunities in Hong Kong. *Building and Environment*, 43(10), 1709–1718.

Lanz, F. and Pendlebury, J. (2022) Adaptive reuse: A critical review. *Journal of Architecture*, 27(2–3), 441–462.

Lausberg, C. (2008) *The rising natural vacancy rate in the German office market and its effects on real estate finance*. 15th Annual European Real Estate Society Conference in Kraków, Poland.

Lees, L., Slater, T. and Wyly, E. (Eds.). (2010) *The gentrification reader* (1st ed.). London: Routledge.

Marcuse, P. (1986) Abandonment, gentrification and displacement: The linkages in New York City, in Smith, N. and Williams, P. (Eds.) *Gentrification of the city*. London: Unwin Hyman, pp. 153–177.

McGuinness, D., Greenhalgh, P. and Pugalis, L. (2015) Is the grass always greener? *Geographical Journal*, 181(1), 26–37.

Miceli, T. and Sirmans, C. (2013) Efficiency rents: A new theory of the natural vacancy rate for rental housing. *Journal of Housing Economics*, 22(1), 20–24.

Muldoon-Smith, K. and Greenhalgh, P. (2016) Greasing the wheels or a spanner in the works: Permitting the adaptive re-use of redundant office buildings in residential use in England. *Planning Theory and Practice*, 17(2), 175–191.

Myers, D. and Wyatt, P. (2004) Rethinking urban capacity: Identifying and appraising vacant buildings. *573 Building Research & Information*, 32(4), 285–292.

Norman, P. (2023) London office take-up rebounds but 'grey Space' at record high. *Co-star*. February 6.
Oswalt, P., Overmeyer, K. and Misselwitz, P. (2013) *Urban catalyst: The power of temporary use*. Berlin: DOM Publishers.
Pissarides, C. (2000) *Equilibrium unemployment theory* (2nd ed.). Cambridge, MA: MIT Press.
Remoy, H. (2010) *Out of office*. Amsterdam: University of Delft Publishing.
Remoy, H. and Koppels, P. (2009) Structural vacancy of office buildings; the influence of building and location. 16th Annual European Real Estate Society Conference – Stockholm, Sweden 2009.
Remoy, H.T. and Wilkinson, S.J. (2012). Office building conversion and sustainable adaptation: A comparative study. *Property Management*, 30(3), 218–231.
Sanderson, B., Farrelly, K. and Thoday, C. (2006) Natural vacancy rates in global office markets. *Journal of Property Investment and Finance*, 24(6), 490–520.
Schumpeter, J. (1950) *Capitalism, Socialism and Democracy* (3rd ed.). New York: Harper.
Smith, N. (1979) Toward a theory of gentrification a back to the city movement by capital, not people. *Journal of the American Planning Association*, 45(4), 538–548.
Temenos, C. and McCann, E. (2013) Geographies of policy mobilities. *Geography Compass*, 7(5), 344–357.
Voith, R. and Crone, T. (1988) National vacancy rates and the persistence of shocks in U.S. office 595 markets. *Real Estate Economics*, 16(4), 437–458.
Wheaton, W. (1999) Real estate 'cycles': some fundamentals. *Real Estate Economics*, 27(2), 209–230.
Wilkinson, S. and Reed, R. (2011) Examining and quantifying the drivers behind alterations and 598 extensions to commercial buildings in a central business district. *Construction Management & 599 Economics*, 29(7), 725–735.
Wincott, D.R. (1997) Vacancy rates and reasonableness. *Appraisal Journal*, 65(4), 361–370.

17 BIM Implementation in Construction Companies in the United Kingdom

David Laurance and Yusuf Arayici

Introduction

The UK government recognised the importance of innovation within the construction industry, agreeing a Construction Sector Deal in 2018 bringing together construction, digital technology, manufacturing, materials, and energy sectors. Building information modelling (BIM) is a transformative technology, one that is becoming a necessity on construction projects. The amount of data generated by BIM is tenfold that of projects using 2D plans and specifications (Reddy, 2012, p. 6).

The adoption of BIM within construction companies has been treated with hesitancy for numerous reasons. Resistance to change; the steep learning curve of tools required, and their real-life applications; and belief among groups that existing projects will suffer have been highlighted as factors in reluctance to integration (UK Construction Media, 2018). Academic studies have been conducted regarding the uptake of BIM within different prescribed conditions. Gledson et al. (2012) investigated perspectives towards BIM within differing size UK construction companies. Results found that SMEs were more reluctant to implement BIM than larger companies. However, research was limited, and the sampling method could have been more purposively targeted.

Research exploring benefits and obstacles of BIM implementation by Migilinskas et al. (2013) was conducted through case studies of Lithuanian construction contractors. The study highlighted differing levels of BIM maturity within companies, emphasising the need for interoperability and standardisation of work methods among project participants for successful implementation. Summarised research conducted by Plan Radar (Construction Digital, 2021) indicates that the United Kingdom has the largest number of construction companies using BIM at level 2 and above. It also highlights that only 62% of UK small businesses actively use BIM, compared to 80% of large companies. This disparity reinforces the theory that BIM is not being implemented across the United Kingdom to its full potential and raises the question as to why uptake is significantly lower within SMEs.

A key research paper by Khosrowshahi and Arayici (2012) explored BIM implementation in Finland. Through developing concept mapping they generated a roadmap for effective and informed methods of implementing BIM within the United Kingdom. The study evaluated the state of readiness of organisations,

DOI: 10.1201/9781003380559-17

analysing the culture, education, training, and information management. It explored the understanding and awareness of BIM and assessed barriers and challenges to its implementation. The research summarised that effective implementation needed to cover technology, processes, and people, through soft system methodology, information engineering, and process innovation.

From the past studies on BIM implementation, it can be determined that there is a lack of recent exploration on the integration of the latest BIM regulations. While research has been undertaken on BIM maturity levels across differing size companies, analysis seems limited. Comparison of BIM adoption through case studies at the international level, and the varying responses for reasons of reluctance, highlights the need for further exploratory academic study within UK companies. To develop an overarching vision of BIM implementation within the UK construction industry, there is a need to examine adoption across each UK region, with companies of differing sizes, and within varying construction professions.

The roadmap developed by Khosrowshahi and Arayici (2012) is now over a decade old, and there is a need to analyse how this potential implementation procedure has been adhered to within the UK construction industry. For the reasons outlined above, it is necessary to evaluate and compare the uptake across different regions within the United Kingdom, and to assess specific factors resulting in differences between locations. This established the research question 'to what extent has BIM been successfully implemented within the construction industry comparatively across the regions of the United Kingdom?'

To identify opportunities and hindrances that different size construction companies have encountered with BIM implementation, evaluating analogous factors that could influence integration across national locations, the chapter will endeavour to address the following objectives:

- Explore the development of BIM within the United Kingdom and analyse reasons for reluctance of implementation
- Explain current BIM guidelines and assess currently available BIM systems to establish maturity stages at which construction companies may be positioned
- Determine the proportional representation of construction firm locations across the United Kingdom to provide a data set to validate research results against
- Quantify advantages and pitfalls highlighted when reviewing the application of BIM software and hardware within construction firms
- Compare and evaluate this information to assess factors that may influence BIM adoption, cross-analysing both size and location of companies with their maturity stage
- Provide recommendations to enable more prosperous fulfilment of BIM operations at a national level.

Development of BIM in the UK

The initial UK Government Construction Strategy, published in 2011, resulted in the establishment of a BIM Task Group. The report mandated fully collaborative

3D BIM in place within all contractors by 2016 (Cabinet Office, 2011, p. 14). The rapid development of BIM led to a global strategy being recognised, highlighting the need for engagement with EU partners to assist in a consistent adoption of BIM (HM Government, 2012, pp. 10–12). Construction 2025 (HM Government, 2013, pp. 19–61) set a vision for the UK construction industry to lead the world in research and innovation, transformed by digital design, embracing the transition to a digital economy and smart construction.

The Construction Sector Deal (HM Government, 2018, p. 7) builds on Construction 2025 by providing a framework to achieve cost and time reductions in construction by focusing on three strategic areas: digital techniques at all phases of design through BIM to optimise performance; offsite manufacturing technologies to minimise waste and inefficiencies; and whole life asset performance to shift focus from costs of construction to cost of the building whole life cycle. The perceived benefits of BIM from a political standpoint demonstrates the value placed on successful implementation for an enhanced future of the UK construction industry. This furthers the goal of the report, to gain greater objective and subjective views of the advantages and hindrances found with BIM integration.

There is a common misconception that BIM is simply CAD. Reddy (2012, pp. 1–7) outlines that BIM is a process improvement methodology leveraging data to analyse and predict outcomes throughout phases of the building life cycle. Reddy (2012) expresses that BIM is a subset of CAD software, but the processes are very different. He examines how BIM can be broken into sets based on the information's author and user: design intent (designer); build intent (contractor); fabrication intent (subcontractor); and facility management (owner). This is furthered by Kensek and Noble (2014), who posit that BIM is a natural progression in the evolution of computer-supported practice, revolutionising how building partners practice and document work, describing it as a landmark technology and how opting out is not a choice. By understanding this progression of BIM, we can recognise how the industry has an enormous task in encouraging a cultural shift, and the inherent resistance that companies may feel towards this digital transformation.

Reluctance of BIM Implementation

BIM is a paradigm shift, which usually happens at pace; however, the architecture, engineering, and construction (AEC) sector is moving at a much slower speed due to the nature of the system at large. The current AEC processes and end state of the paradigm shift are well recognised; however, the adjustment from the current position to this end state is tumultuous (Reddy, 2012, p. 2). This argument is furthered by Eynon (2016, p. 65), who correlates the expression 'collaborative working' alongside BIM. He states that collaborative working seems to be both a panacea and equally the biggest headache for the AEC industry.

Smith and Tardif (2009, pp. 51–53) highlight that key reasons cited by business leaders for implementing BIM include maintaining or enhancing their competitiveness and streamlining operations. They stress the importance for companies to perform due diligence, ensuring they are selecting the most beneficial software, and necessary training required for an effective BIM implementation strategy. Kensek (2014, pp. 116–119) explores seven steps companies should examine to address implementation concerns: motivation, goals, plan, investment, software, tools, and hardware.

A study by Siebelink et al. (2020, pp. 2–4) identifies barriers that hinder implementation of BIM, classifying them as: management support; people and culture; technological aspects; and use of defined processes and standards. These factors were also highlighted by Gu and London (2010, pp. 992–993), who raised key issues such as: work practice and process-related issues – version management, data integrity, and data organisation; technical issues – standards, security, and registry of communication; and other issues – training support, roles, and responsibilities.

Eastman et al. (2018, pp. 797c–797g) raised common misconceptions for delaying BIM integration, categorising them as process barriers and technology barriers. These include the misbelief that the market is not ready; potential negative ROI; training costs and learning curve being too high; costly legal barriers; and lack of willingness to participate. This final factor was expanded by Harty, Kouider, and Paterson (2015, p. 72) with relevance to SMEs. They determine a factor affecting BIM adoption is if one or more stakeholders do not fully engage in BIM. They stress that for BIM to be successfully implemented it is very important for information inside the software to be processed and properly integrated, as it will result in errors if not administered effectively.

The reasons for reluctance of BIM implementation that have been established are manifold. There is a need to initiate a defined classification of these factors from which questions can be assigned to companies. This will allow for an enhanced view as to the hierarchy within which these elements have played a part in the issues encountered with BIM integration and help generate potential solutions for how these can be resolved.

BIM Implementation Roadmap

The research paper by Khosrowshahi and Arayici (2012, pp. 610–611) acknowledges challenges met by construction companies regarding knowledge sharing and states that 'technology alone cannot influence the required changes'. It highlights that for the benefits of BIM to be realised, it is essential for precise implementation at an organisational level alongside full integration at the industry level. The paper informs that, at the time of writing, Finland was seen as a leader in BIM implementation. The researchers conducted face-to-face interviews with Finnish construction companies that had unique experiences of BIM. Results from these interviews enabled the generation of a concept map of BIM implementation, allowing the researchers to develop a questionnaire to be distributed among construction companies.

This research project will follow similar steps and processes to Khosrowshahi and Arayici, observing their study as an essential launch point. The questions generated within their study are grouped into five categories: BIM understanding and awareness; barriers and challenges; problems solved with BIM; future estimations for BIM implementation; and personal experience of respondents. These queries closely align with the proposed study and will therefore help form the basis of initial online interviews. The questions will be adapted to achieve responses that could highlight potential benefits or hindrances companies find with BIM implementation, specifically within the UK construction industry.

Current BIM Guidelines

The introduction of BS EN ISO 19650 in 2019 aimed to benefit contractors by gaining efficiencies through better specification and delivery, concerning design, construction, operation, and maintenance of buildings and infrastructure, using appropriate technologies (UK BIM Framework, 2021). Further transitional guidance in 2020 resulted in a set of internationally agreed standards to generate information in a systematic and consistent way (BSI Group, 2022, p. 6).

RICS (2020, p. 19) examines how BS EN ISO 19650 represents a shift towards a standardised framework to internationally standardise principles and terminology used rather than being UK centred. The report explores the transition from a single collaborative model to a federated information model implementation.

By examining the current BIM guidelines within the UK construction industry, we can gain a greater understanding of the maturity stages within which companies may operate. Due to the interoperable nature of BIM, for businesses to embrace its full potential it is essential for stakeholders of projects to have capabilities for operating at the same level as one another. To establish the maturity stage that companies align, their understanding of BIM needs to be determined, along with the technology they use to position themselves at this level. Comprehending these aspects will build a further understanding of BIM implementation within the UK construction industry at a holistic level.

The adoption of BIM upends conventions drawn over decades regarding collaboration between architects, engineers, contractors, and clients (Holzer, 2015, pp. 4–6). Holzer demonstrates that, in the early 2000s, implementation of BIM remained limited to architectural designers and structural engineers; however, in 2010 there were major developments allowing BIM integration to increase pace. Within the construction industry there is a large range of BIM software and hardware available for use across specific sectors. These vary in complexity and interoperability, as highlighted within Figure 17.1, and thus different sectors have variable BIM options that will determine the maturity stage where they sit. Eastman et al. (2018, pp. 1a–2b) establish diverse ways BIM platforms may be used: by architects for design modelling; by engineers for structural data; by surveyors for financial and logistical data; and by contractors for building construction coordination models, detailing, and Facilities Management (FM).

It is essential to analyse the various BIM applications related to specific sectors, to effectively assess the BIM maturity stage at which different companies sit comparatively. Industry sectors have been assessed to ensure suitability for research questioning. Due to the interoperable, collaborative nature of BIM, many professions will overlap with their use of different software and hardware, and therefore the profession most relevant to their usage has been evaluated.

BIM Software for Architects

Revit for Architecture (Autodesk, 2023c) allows for generation of floor plans, elevations, 2D and 3D views, and accurate renderings. It enables building performance optimisation through cost estimates and performance monitoring over the lifetime of the project and building. Yori, Kim, and Kirkby (2019, p. 289) demonstrate the interoperability of Revit with other software through use of industry foundation classes (IFC). IFC exchange allows for effective data exchange with applications, including linking AutoCAD files into Revit and coordination between Revit and Allplan, Vectorworks, or ArchiCAD.

Bentley Systems (2023a) has architectural systems including OpenBuildings Designer and Microstation. OpenBuildings Designer allows development and design of buildings or facilities to produce plans, framing layouts sections and elevations. It produces 3D images, animations, and photo-realistic renderings and allows building documentation and report generation. Microstation builds the foundation of infrastructure design projects, allowing for the design of geospatially enabled 2D and 3D drawings.

Graphisoft ArchiCAD was an originator of BIM with its virtual building concept and is still one of the most used architectural programs (Graphisoft, 2023). Epstein (2011, pp. 15–45) outlines how ArchiCAD was designed to follow how architects work. She explains how, like Revit and Microstation, ArchiCAD uses parametric objects allowing for IFC descriptors to enable interoperable data exchange. This interoperability between software is an essential benefit for BIM implementation, allowing for companies to collaborate, even when not using the same software.

RIBA Plan of Work 2020

The Centre for Digital Built Britain highlighted potential technologies to evolve BIM such as artificial intelligence (AI), digital twins, and offsite manufacturing initiatives, highlighting that the use of virtual reality is reducing the need for 2D information (RIBA, 2020, pp. 108–110). It theorises that change will be focused on:

- Rules-based design – design automation for repeat clients
- Generative design – generation of multiple designs through complex coding
- 4D – 3D design plus time measurement tools, unlocking the most efficient way of manufacturing and constructing
- 5D – addition of cost reporting to 4D principles
- Integrated system analysis – providing real-time feedback from models
- Manufacturing workflow – allowing for faster, more effective ways to build
- Artificial intelligence – seen as unlikely to cause major disruption due to enormous amounts of data required but creates new ways of designing.

While in theory these potential developments of BIM seem like they could lead to a digital revolution, the research conducted regarding reluctance to change, and the paradigm shift required, causes further scepticism as to how BIM will be integrated to this level across the industry.

BIM Software for Structural Engineers and Civil Engineers

In addition to the architectural modules within their software packages, Autodesk and Bentley Systems both offer systems designed for engineering. Autodesk (2023a) offers both Robot Structural Analysis for structural engineers and Civil 3D for use by civil engineers to create plans for infrastructure projects, including site design, road and motorway design, and rail design. Similarly, Bentley Systems (2023b) created numerous modules for structural analysis, including RAM Structural, ProStructures, and STAAD, and civil engineering modules OpenSite, OpenRoads, OpenRail, and OpenBridge (Bentley Systems, 2023b).

Trimble offers a similar suite of programs for structural and civil engineering, including Tekla Structures, Tekla Structural Designer and Tekla Tedds (Trimble Solutions Corporation, 2023). All Tekla programs allow collaboration with other software through IFC operability. Eastman et al. (2018, p. 18) highlight the 4D capabilities of Tekla Structures with Gantt chart scheduling, allowing the model to be played between dates with allowance for time-dependent attributes. This time scheduling efficiency of BIM software is key for companies with their ability to operate at higher BIM maturity stages.

Clash Analysis and Deficiency Detection

3D model review software is used within the AEC industry to complement design packages. Autodesk Navisworks is a collaboration software that can open multiple 3D files and combine them into a single workspace. Similar software includes Tekla BIMsight, Bentley Navigator, and Solibri Model Checker, which all provide likewise functionality (Hardin and McCool, 2015, p. 196). Navisworks can run clash detection to check for objects inhabiting the same space and is often seen as essential during the design phase. Rules-based checking systems like Solibri contain automated tasks to detect for deficiencies and inconsistencies in groups of objects (Epstein, 2011, pp. 50–56). Clash detection is seen as a key benefit for companies operating at higher BIM levels, in their ability to save time and money at the front line of construction sites.

BIM Software for Quantity Surveyors and Building Surveyors

While most of the software for the AEC industry is made for design and structural processes used by architects and engineers, there are features within some software and separate surveying specific software available. In addition to being used by engineers for clash detection, Autodesk Navisworks has the feature *quick take-off for quantity surveying*. This allows surveyors to measure lines, areas, and counts with 2D and 3D take-off rather than carrying out manual calculations (Autodesk, 2023b). Similarly, Bluebeam has specific quantity estimation features allowing for measurements to be taken and running cost totals to be tracked in real-life (Bluebeam, 2023).

3D Laser Scanning

There are numerous technological hardware systems available, providing key data for further BIM implementation. Reed and Sims (2014, p. 246) indicate that, alongside BIM technology, 3D laser scanning offers the ability to capture, map, and analyse building information. They highlight that while this tends to focus on design, benefits are also seen by surveyors. High precision, detailed photographic output allows the surveyor to see building characteristics, providing more intricate collection of data than conventional methods. They note that while it may be cost prohibitive for smaller companies, the benefits of accurate 3D data replicas to use within BIM software are enormous.

Whyte (2002, p. 17) explains how use of 3D laser scanning and geometry capture allows models to be built using as-built data, rather than geometrical CAD data, which is especially beneficial for use within virtual reality programming. As previously outlined, these digital developments within the construction industry do not seem to be scalable to the industry. Alternatively, highly developed technology seems tailored for companies looking to gain traction at a tendering position from their bespoke ability to operate at projects requiring greater BIM necessities.

BIM Hardware for Contractors

While some companies may use the BIM software previously outlined, more common use of BIM by independent contractors comes from use of information collection and modelling hardware. Tal and Altschuld (2021, pp. 23–25) explain how drones are full cycle tools for AEC industry projects due to the ability to be used at startup, through development and construction, and post-completion. Data collected at each phase can be used alongside BIM software with high interoperability. They emphasise the key advantages of drone technology being affordability; flexibility; range and access; integrable nature; and safety.

Aghimien et al. (2021, p. 52) express how the term 'fourth industrial revolution' (4IR) is used to cover digitisation of products and services, as well as innovative digital business models. In addition to including virtual reality and AI, it can be used to describe systems such as 3D printing. Emmitt (2018, pp. 38–39) describes how rapid development within the 3D print sector has made it possible for building components to be produced. Typically, this is carried out offsite in factories. However, it is possible to use large-scale printers on-site with print heads travelling along scaffolds to print layers.

BIM and Facilities Management (FM)

Atkin and Brooks (2015, p. 34) summarise that BIM is expected to grow rapidly within the FM sector, particularly in relation to new and refurbished facilities. It allows for a current data model of assets to be maintained, which can be used for both facility planning and facility management. They raise the issue that the adoption of BIM for FM with existing buildings is less certain due to the cost of creating a new model, with the term 'retrospective BIM' being used. IFMA (2013, pp. 1–9) outline issues with current FM practice in existing buildings and express how BIM integration can address these issues. They highlight that the answer is for integration of data systems for the life cycle of a facility, notably: the need for graphics and data varies over the life cycle; the need for interoperability between systems; and owner benefits of BIM FM integration.

Locality of UK Construction Firms

Research conducted by the Office for National Statistics (2020) examined the structure of the UK construction industry. The report highlights proportional representation of construction firms and employees across Great Britain. Key takeaways include the main concentration of firms across Southern England, particularly London and the Southeast. As expected, this proliferation is also evident in the location of employees.

A notable standout is the larger number of employees in the Northwest and Scotland than the Southwest, which has a higher proportion of firms, as shown in Figure 17.4. The lack of firms and employees in the Northeast and Wales stands out as an indicator of potential lack of political and economic investment within those regions. It should be noted that Northern Ireland has not been included within the research, so it is unclear what the proportion across this region is.

By determining the locality of construction companies within the United Kingdom, a presumed data origin point can be set against which results obtained within the research can be evaluated. Results can be validated based on the percentage of responses received in relation to the expected level of responses from each region.

Research Design

The study has followed an ontological approach of both objectivist and subjectivist perspectives. Bourdieu argues there is a necessity to go beyond mutual antagonism of subjectivism and objectivism, preserving gains from each (Grenfell, 2014, p. 43). The objectivist hypothesis underlies the research of existing factual historical disparity between UK regions, alongside the subjective interpretive experience of both researcher and proposed interviewees, and how relevant this is to their reality.

In line with the ontological approach, it has followed a critical realist epistemological philosophy, where the researcher acknowledges historical reality of regional inequality but observes there are external societal systems with inherent causality. Matthews and Ross (2010, p. 29) describe how critical realist perspectives focus on identification of knowledge that is real but, however, unobservable other than in effects it has. They highlight how this approach uncovers power relations through revealing hidden structures and mechanisms.

The research undertook a relative abductive approach, combining a deductive hypothesis that regional inequality has implications on implementation of BIM across locations, with an exploratory inductive research system expanding upon this existing theory. Dubois and Gadde (2002, pp. 553–560) propose a specific approach of 'systematic combining' grounded in abductive logic. They indicate an abductive approach is advantageous if the objective is to discover new theories, whereas systematic combining builds on refining existing theories rather than inventing new ones.

Methodological Choice and Strategy

The research followed a sequential exploratory mixed method research design, with qualitative research being followed by quantitative study, as recommended by Saunders, Lewis, and Thornhill (2019, p. 182). Qualitative online interviews will be sent to selected contractors across predetermined regions. The United Kingdom is split into 12 different regions, which are then further split into 66 individual counties (Table 17.1).

To create a database of relevant companies to approach, based on the informed literature review, the research has focused on online searches within each county for architects, structural engineers, civil engineers, quantity surveyors, building surveyors, contractors, and facilities management. This extensive research resulted in a database of 3,319 companies for the quantitative survey. From this list, two companies from each region, with background knowledge of BIM, were selected to participate with initial qualitative online interviews.

Denscombe (2014, pp. 197–199) addresses benefits of conducting online interviews: negligible costs; written responses, requiring no need to transcribe; and remoteness, reducing culture and gender effect of interaction. Drawbacks include time delay in receiving responses, the need to ensure private communication, potential psychological remoteness, and differences in how people express themselves verbally and in writing. These consequences were considered and influenced the questions used. Due to time constraints, the horizon of the study is cross-sectional. Fellows and Liu (2015, p. 106) explain that 'longitudinal studies are superior to cross-sectional studies for making causal assertions'. In an ideal scenario the project would be replicated frequently at future dates as a form of repeated measure design and to produce more accurate results.

Following interview responses, a quantitative online survey was created, drawing from subjective feedback given by interviewees. Data collected was analysed, concluding in the generation of recommendations for more prosperous fulfilment of BIM operations nationally.

Table 17.1 Counties within the regions of the UK

Southwest			**Southeast**		
Bristol	Cornwall	Devon	Berkshire	Buckinghamshire	East Sussex
Dorset	Gloucestershire	Somerset	Hampshire	Isle Of Wight	Kent
Wiltshire			Oxfordshire	Surrey	West Sussex
East of England			**London**		
Bedfordshire	Cambridgeshire	Essex	City of London	Greater London	
Hertfordshire	Norfolk	Suffolk			
East Midlands			**West Midlands**		
Derbyshire	Leicestershire	Lincolnshire	Herefordshire	Shropshire	Staffordshire
Northamptonshire	Nottinghamshire	Rutland	Warwickshire	Worcestershire	
Yorkshire and The Humber			**Northeast**		
East Riding of Yorkshire	North Lincolnshire	North Yorkshire	County Durham	Northumberland	Tyne and Wear
South Yorkshire	West Yorkshire				
Northwest			**Scotland**		
Cheshire	Cumbria	Greater Manchester	Central Scotland	Glasgow	Highlands and Islands
Lancashire	Merseyside		Lothian	Mid Scotland and Fife	Northeast Scotland
			South Scotland		
Wales			**Northern Ireland**		
Central Wales	East Wales	North Wales	County Antrim	County Armagh	County Down
South Wales	West Wales		County Fermanagh	County Londonderry	County Tyrone

Data Collection and Analysis

Using information obtained through secondary data collection, questions were formalised into a structured online interview and sent to predetermined construction contractors across each region. Questions were made largely open-ended, ensuring honest responses were returned, giving a holistic view of interviewees' opinions on BIM. Davies and Hughes (2014, pp. 183–185) emphasise the need to undertake an exploratory and preparatory phase, stressing risks of researcher bias. The researcher has ensured that questions are not leading and remain informal. Data collected from interviews has been categorised and interpreted. Mason (2017, p. 204) describes how interpretive readings of data involve the researcher constructing a version of what the data signifies, and what phenomena or concepts it can lead to. This data interpretation has formed questions in the quantitative online survey that was sent to a large sample of contractors across the United Kingdom.

Results of the questionnaire have been triangulated with responses from online interviews and have undergone bivariate analysis to examine any relationship between company size and its BIM maturity stage, as well as exploring any causality between the location of the firm and their position on BIM. Rowley (2014, p. 325) describes how bivariate analysis focuses on relationships between two variables, through options including contingency tables, correlation, and regression. This analysis will help determine whether any outlying external factors may impact upon implementation of BIM across different companies.

The quantitative survey has a larger research population of over 3,000 participants. This sample has been selected through probability sampling using a stratified method. Candidates were selected through random identification within identified subsets, selecting a set number of respondents from each of the geolocations chosen. Cohen, Manion, and Morrison (2007, pp. 112–113) identify that organising a stratified random sample involves two-stages: identifying characteristics appearing in the wider population that must therefore appear in the sample; and randomly sampling within these groups, with the group size being determined by the researcher. They highlight that a stratified random sample is a useful blend of randomisation and categorisation.

Online Interviews

Using data collected through secondary research, adapting question categories highlighted by Khosrowshahi and Arayici (2012), structured online interview questions were developed. To keep responses as relevant to the research objective as possible, interview questions were in a structured format, while still being open-ended, allowing for honest responses. The following questions based on the outlined categories were generated:

1. BIM understanding and awareness

    - Within your company's profession(s), what examples of BIM software and/or hardware can you give for each ISO 19650 BIM maturity stage?

2. Personal experience of the respondents
   - What BIM maturity stage would you place your company at?
   - Which BIM software and/or hardware does your company currently use?
   - What BIM software and/or hardware would your company be looking to use in the future to increase your BIM capabilities?
3. Problems solved with BIM use
   - What benefits has your company achieved from the BIM software and/or hardware that you use?
4. Barriers and challenges
   - What has hindered your company from implementing the BIM software and/or hardware that you would like to use in the future?
5. Future estimations for BIM implementation
   - What potential reasons (e.g., financial, political, educational) do you think might be affecting companies within the UK from implementing BIM to their full capabilities?

To gain greater understanding of responses in relation to the size and location of companies approached, online interviews initiated with:

- How many staff does your company employ?
- In which region(s) within the UK is your company based?
- Within the construction industry, in which profession(s) does your company operate?

It was determined that for more accurate responses, locational choice would be set for respondents to the 12 UK regions, with the option for multi-locational firms, through use of tick boxes. It was observed that professions within which companies operate should also be set as choices through use of tick boxes, ensuring questions were as simple to answer as possible, as recommended by Saunders, Lewis, and Thornhill (2019). Due to the diverse nature of the construction industry, the decision was made to expand upon grouped sectors to achieve a more holistic view of the industry. Therefore, disciplines such as architecture, structural engineering, civil engineering, quantity surveying, building surveying, contractor, facilities management, project management, and building services are included in the interviews.

Upon completion of the pilot test and amendments to final online interview questions, 24 companies were selected from the bulk data set. These companies were selected across the 12 regions, two chosen from each, to collate a diverse range of responses, with different size companies selected from various sectors of the AEC.

Interview Data Analysis

Of the 24 respondents contacted, 17 responses were received, resulting in a 71% response rate. Multiple attempts were made to contact each company with no initial response. Sapsford and Jupp (2006, p. 51) outline that even after second copies of questionnaires are sent out and chased up, response rates above 80% are rarely obtained. Stoop et al. (2010, p. 5) indicate that there is no response rate that guarantees high quality results, and that surveys with high response rate but poor questions will never be a good study. The response rate to online interviews conducted is therefore considered to be acceptable. Responses came from companies across a range of business sizes. The Foreign, Commonwealth and Development Office (2023) differentiate the size of SME companies by either turnover or headcount. As turnover of respondents is unknown, the number of employees was used to determine business size: micro – fewer than ten staff; small – fewer than 50 staff; medium – fewer than 250 staff; large – more than 250 staff. Company size and brief descriptions of respondents are found in Table 17.2, with data anonymised for ethical purposes.

Results were returned from companies across each region barring Wales, with four companies operating across multiple locations and 13 from single regions (Table 17.3). As expected from the proportion of firms across the United Kingdom, many companies operating in multiple locations had bases within London and the Southeast.

Results showed that respondents operate within a range of industry sectors. Respondents of different sizes included four micro, eight small, three medium, and two large sized companies, of which six operate within single sectors and 11 in multiple sectors. The results showed that respondents operate within a range of industry sectors. Figure 17.1 shows the size of the companies and the sectors

Table 17.2 Online interview respondents company size and description

	Company Size	*Description*
Company A	Micro	Building surveying company based across two regions
Company B	Medium	Multi-sector company working in London
Company C	Small	Engineering firm operating in Scotland
Company D	Small	Architectural company based in the Southeast
Company E	Small	Engineering business located in the Southeast
Company F	Large	Multi-sector firm operating within two regions
Company G	Large	Multi-sector company operating across numerous regions
Company H	Small	Multi-sector business located in the Southwest
Company I	Small	Engineering company operating in Scotland
Company J	Micro	Engineering business located in the East of England
Company K	Micro	Engineering firm based within numerous regions
Company L	Small	Architectural business working in the East Midlands
Company M	Small	Architectural firm operating in the West Midlands
Company N	Small	Multi-sector company operating in Northern Ireland
Company O	Micro	Multi-sector business based in Northern Ireland
Company P	Medium	Contractor operating in the Southeast
Company Q	Medium	Building Services company working in numerous regions

Table 17.3 Online interview respondent locations

Location	*Respondents*
Southeast	5
London	5
Southwest	2
East of England	2
East Midlands	1
West Midlands	2
Yorkshire and The Humber	3
Northeast	1
Northwest	1
Wales	0
Scotland	2
Northern Ireland	2
Multiple Locations	**4**
Single Location	**13**

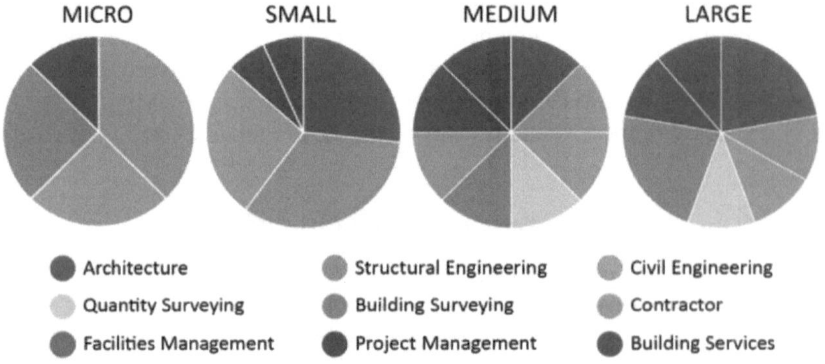

	MICRO	SMALL	MEDIUM	LARGE
Multiple Industry Sectors	3 respondents	5 respondents	1 respondent	2 respondents
Single Industry Sector	1 respondent	3 respondents	2 respondents	0 respondents

Figure 17.1 Online interview respondent company size and industry sector.

in which they operate. The breakdown of respondents between the different sizes included four micro, eight small, three medium, and two large companies, of which six operate within single sectors and 11 in multiple sectors.

Online interviews conducted generated a large amount of qualitative data, providing further information regarding BIM implementation, benefits and hindrances companies had found, and potential reasons why existing companies may have issues integrating BIM. To help create the questionnaire survey to be sent out in bulk, the process of concept mapping was used to generate more specific questions,

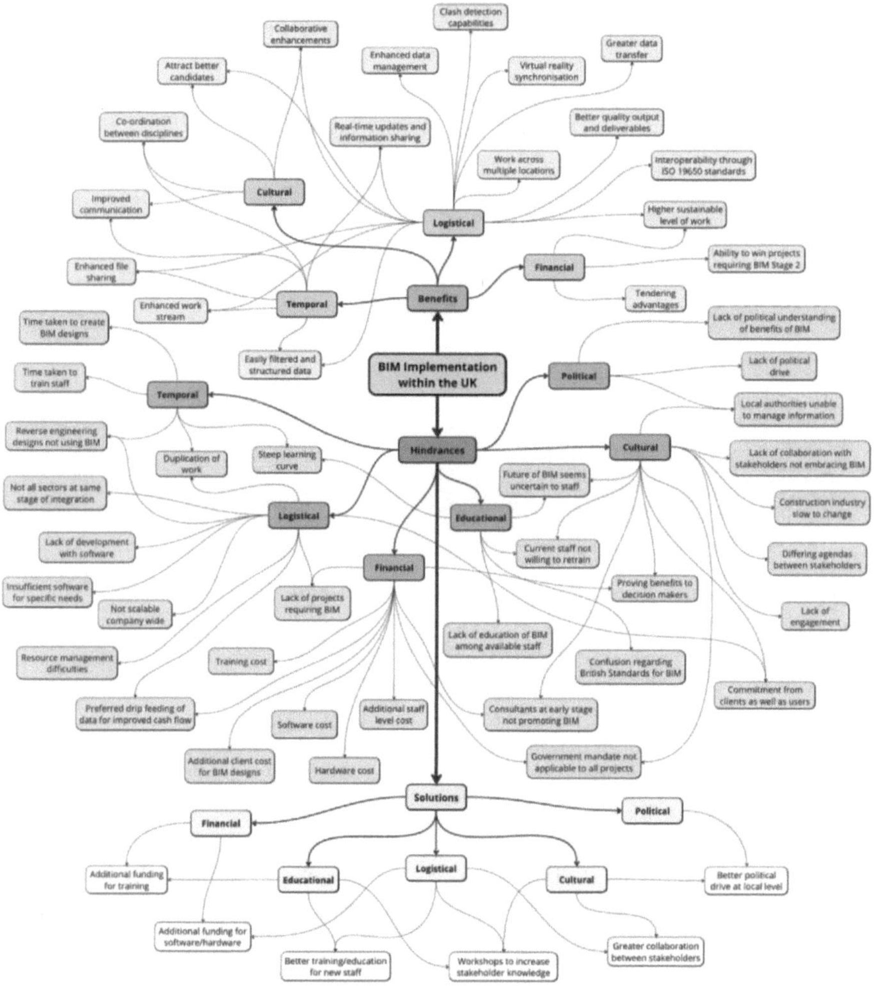

Figure 17.2 Concept map of interview responses for benefits and hindrances of BIM implementation and potential solutions.

to enable quantitative data analysis of survey results. Torres and de Cássia Veiga Marriott (2009, p. 79) explain how concept maps are hierarchical trees, with concepts connected via labelled, graphical links, providing a more visual representation to examine results. The map displays how benefits and hindrances were allocated to core themes:

- Temporal
- Financial

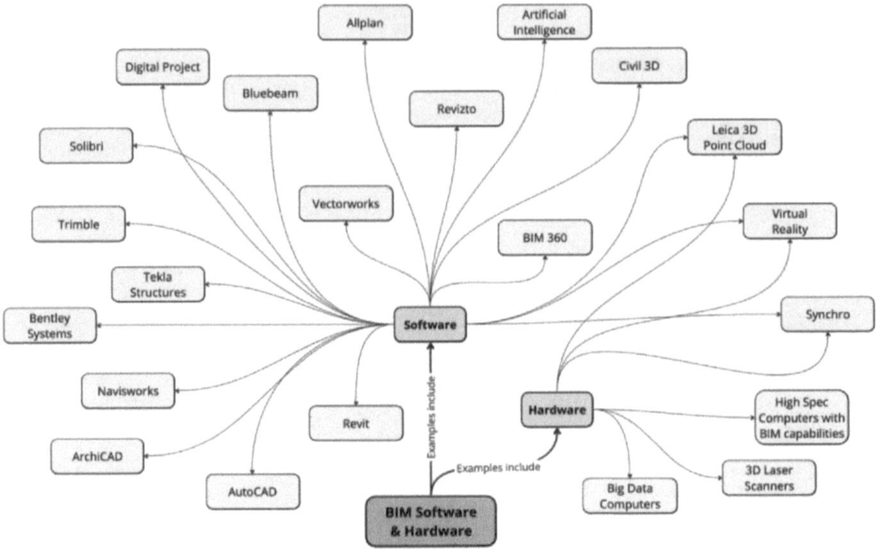

Figure 17.3 BIM software and hardware used by online interview respondents.

- Logistical
- Educational
- Political
- Cultural

By associating responses to these themes, potential solutions to the derived hindrances could be generated. A concept map (Figure 17.2) was also created to demonstrate the software used by respondents, giving a greater picture of the BIM maturity stage where each company sat.

A concept map (Figure 17.3) has also been created to show the different responses as to the software used by the respondents, which gave a greater picture as to the BIM maturity stage where each company sat.

Questionnaire Survey Data Analysis

Creation and analysis of concept maps allowed multiple-choice survey questions to be produced and distributed to the bulk data set. To ensure questions created were specifically focused on answering the set objectives, the following headings were created as topics to assess:

Objective 1: Explore the development of BIM within the UK and analyse reasons for reluctance of implementation

- BIM maturity stage
- Hindrances

Objective 2: Explain current BIM guidelines and assess currently available BIM systems to establish maturity stages at which construction companies may be positioned

- BIM maturity stage
- Company sector

Objective 3: Determine the proportional representation of construction firm locations across the United Kingdom to provide a data set to validate research results against

- Location

Objective 4: Quantify advantages and pitfalls highlighted when reviewing the application of BIM software and hardware within construction firms

- Benefits
- Hindrances

Objective 5: Compare and evaluate this information to assess factors that may influence BIM adoption, cross-analysing both size and location of companies with their maturity stage

- Company size
- Location

Objective 6: Provide recommendations to enable more prosperous fulfilment of BIM operations at a national level

- Solutions

These headings were then split into objective company responses and subjective company beliefs on BIM implementation.

- Objective company information
    - Company size
    - Location
    - Company sector
    - BIM maturity stage
- Subjective company beliefs
    - Benefits
    - Hindrances
    - Solutions

Using these topics to derive firstly, objective, and secondly, subjective, information along with primary and secondary research data, the following questions were

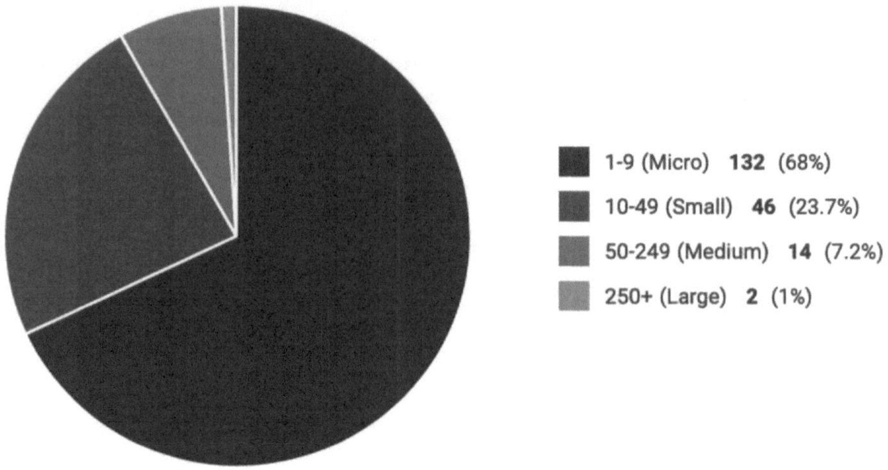

Figure 17.4 Company size of questionnaire survey respondents.

generated. Multiple-choice answer options can be seen in the questionnaire survey analysis.

Of the 195 responses, the majority were of those working for micro companies of fewer than nine employees, with 68% of results, followed by 24% from small companies, 7% from medium, and just 1% from large companies (Figure 17.4). Responses came relatively evenly across each of the regions (Figure 17.5), the majority coming from the South of England, aligning with the proportion of firms. As with all multiple answer questions, the overall percentage shown represents respondents who selected each answer, for example, their company may be based in more than one location.

Table 17.4 demonstrates the differential for responses based on expected results as in Figure 17.3 in comparison to responses received. Responses from Northern Ireland were excluded as the expected proportion of firms was unknown. Based on results anticipated, the main outlier was for responses from companies based in London being 7.7% lower than expected, while responses from Wales and the Southwest were 4.7% and 4.3% higher than expected. While higher response rates would give a greater overall impression of BIM implementation, the even spread across the regions allows for fair comparison, which can then be extrapolated for a broader picture of the United Kingdom.

Respondents worked across each of the proposed sectors (Figure 17.6), showing even distribution among grouped sectors, with architecture at 25%, engineering (structural and civil) at 27%, surveying (quantity and building) at 23%, and contractors and management services at 25%. This view across each of the sectors gives a comprehensive view of differing BIM implementation among the construction industry.

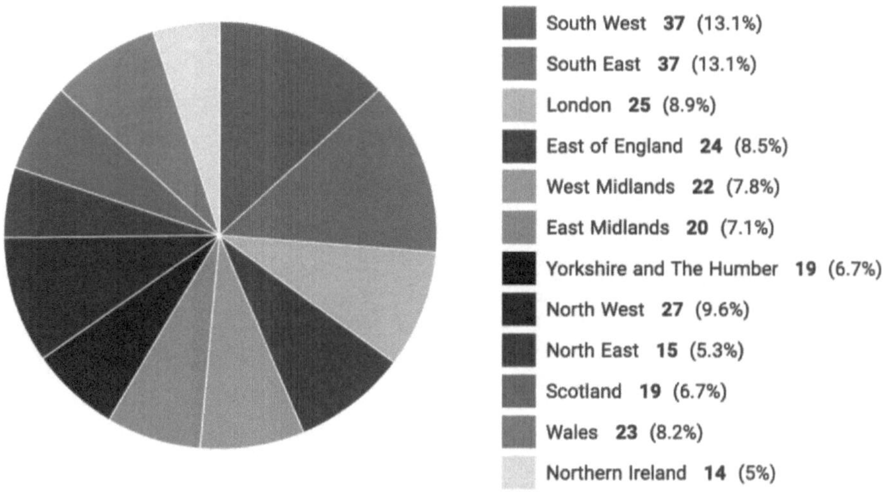

Figure 17.5 Location of questionnaire survey respondents.

Table 17.4 Survey responses comparative to expected proportion of firms

	Survey Respondents	*Expected Proportion of Firms*	*% Over/Under Expected*
London	9.3%	17.0%	−7.7%
Southeast	13.8%	17.0%	−3.2%
East of England	9.0%	12.7%	−3.7%
Southwest	13.8%	9.5%	4.3%
Northwest	10.1%	9.2%	0.9%
West Midlands	8.2%	7.7%	0.5%
East Midlands	7.5%	7.1%	0.4%
Yorkshire and The Humber	7.1%	7.0%	0.1%
Scotland	7.1%	6.1%	1.0%
Wales	8.6%	3.9%	4.7%
Northeast	5.6%	2.8%	2.8%

The BIM maturity stage at which companies currently sit correlated with secondary data collected and primary data from online interviews. Most companies were at earlier stages of BIM implementation, with 66% of respondents either not using BIM or sitting at Stage 1 of the BIM maturity stages (Figure 17.7). As outlined in the Construction 2025 Industry Strategy (HM Government, 2013), all public projects are expectant of companies being operable at BIM Stage 2. As only 30% of respondents currently sit at this level, and just 5% of companies can

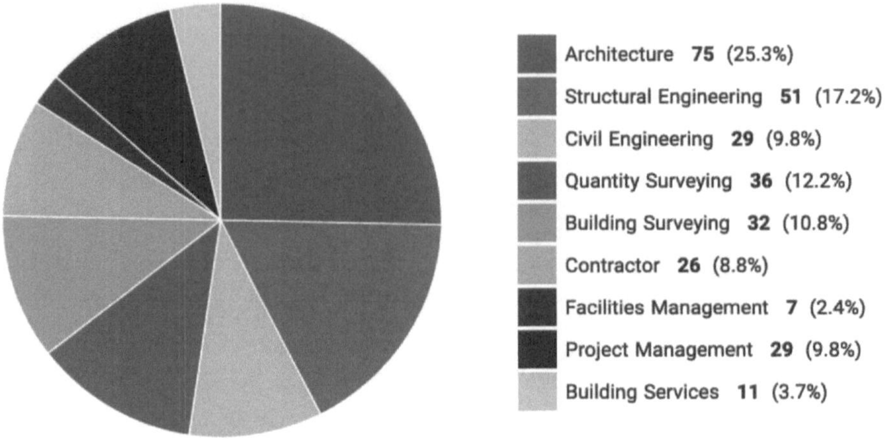

Figure 17.6 Company sector of questionnaire survey respondents.

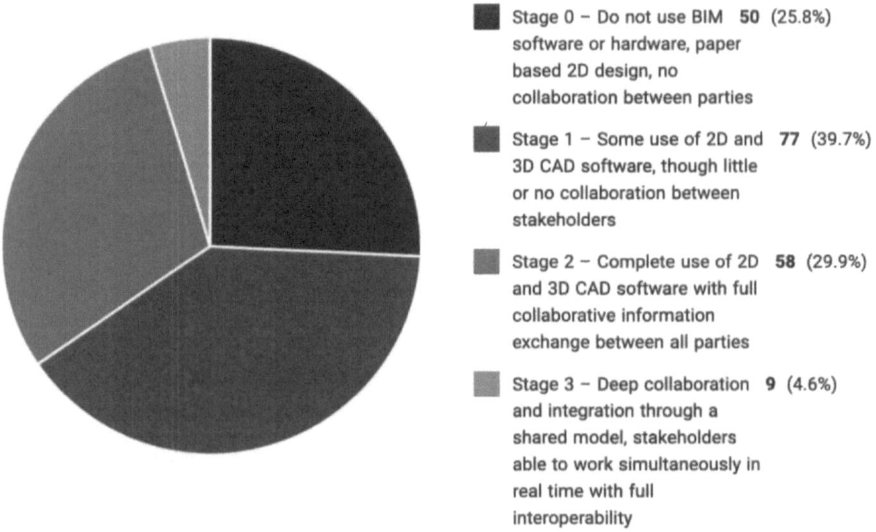

Figure 17.7 BIM maturity stage of questionnaire survey respondents.

collaborate at BIM Stage 3, this does not bode well for the recommended implementation of BIM.

BIM software and hardware used by respondents closely resembles the BIM maturity stage at which they operate, with 12% of companies not using any BIM software at all. The main software used, AutoCAD, largely used for 2D design,

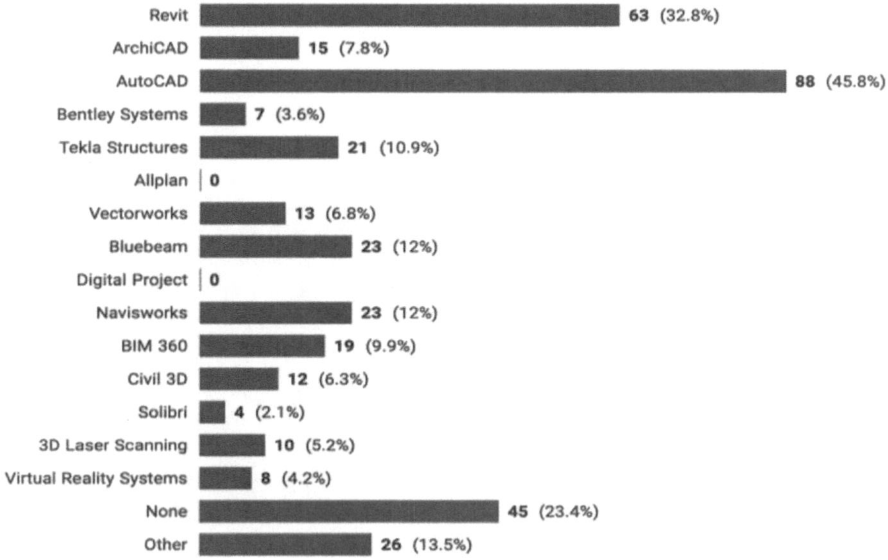

Figure 17.8 BIM software and hardware used by questionnaire survey respondents.

is commensurate with businesses largely operating at Stage 1. Of the options given, each software was used by at least four companies apart from Allplan and Digital Project (Figure 17.8). In addition, for respondents who answered Other, ten highlighted that SketchUp was being used, with numerous others using different CAD software.

By using bivariate data comparison for company size against BIM maturity stage (Figure 17.9), and company sector against BIM maturity (Figure 17.10), a greater picture of BIM implementation across the United Kingdom was achieved.

As only two responses were obtained from large companies, both of which sat at BIM Stage 2, these have been excluded from the graph shown in Figure 17.9 for greater visualisation purposes. For data analysis the percentage of each group was used, rather than the total number. Data shows that most micro and small companies operate at Stage 1, while most medium-size companies are operating at Stage 2.

The hypothesis from these data results is that as companies grow larger, their ability to work at higher maturity stages increases. As responses from medium-size companies only totalled 7% of responses, results could be enhanced further with a larger amount of data, particularly from medium-size and large businesses. However, to reinforce the study by Harty, Kouider, and Paterson (2015), the expectation was matched that micro and small companies either do not use BIM or operate at lower levels. As companies employ more staff, their ability to implement BIM at a greater level results in a peak of operability at Stage 2, and greater chance of integration at Stage 3.

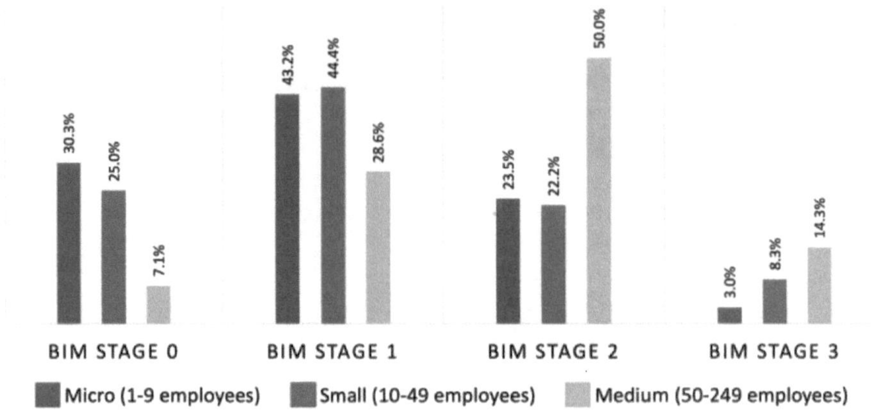

Figure 17.9 Comparison of company size and BIM maturity.

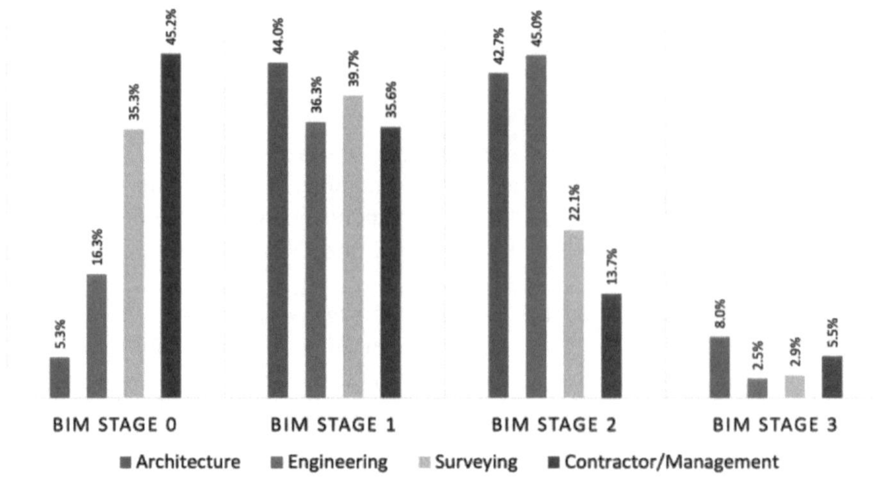

Figure 17.10 Comparison of company sector and BIM maturity of survey respondents.

By comparing location of companies against their BIM maturity, within 75% of the regions BIM Stage 1 is most prevalent. Responses from companies based in the Northwest have the highest proportion of companies sitting at Stage 1 (55%). Companies based in Scotland and West Midlands each have over 45% of companies working at Stage 2, whereas the East of England and Northern Ireland each have over a third of companies not using BIM at all. Yorkshire and The Humber, London, and the Northeast each have 20% or above of companies capable of working at BIM Stage 3.

Comparison of company sector against BIM maturity was made, grouping sectors as aforementioned into architecture, engineering, surveying and contractors, and management services (Figure 17.10). As anticipated, most of the surveying and contractor and management services sit at Stage 0 and Stage 1. This is largely due to the lack of BIM software and hardware dedicated to their services, with these companies having less need for design-based software, which encompasses a large portion of BIM systems.

Architecture and engineering companies have very low numbers not using BIM at all, with engineering firms being most prevalent at Stage 2. Due to a low number of companies operable at Stage 3, it was surprising to see 5.5% of contractors and management services operating at this level, so this is likely to be an outlier due to insufficient data. It is encouraging for the construction industry to see that a relatively high percentage of both architectural firms and engineering companies currently operate at BIM Stage 2. This assumes that most preconstruction phase processes currently meet the government mandate; however, it shows the need for greater drive of BIM implementation at both construction and post-construction phases if BIM is to recognise its full potential.

Subjective Company Beliefs

Of responses to the benefits that companies had seen from implementing BIM, there were a total of 92 respondents (48%) who did not use BIM. After excluding these, results show that, of respondents who use BIM, the greatest believed benefit is of having better collaboration and communication with all stakeholders, as shown in Figure 17.11. This feeds into the second most common answer, having a more enhanced, efficient work stream, as well as greater coordination between all disciplines.

Least prevalent answers included those benefits of being able to attract better candidates, gaining a tendering advantage, and the ability to win public sector projects requiring BIM Level 2. The latter could be expected as most companies, excluding those who do not use BIM, currently sit at Stage 1 and therefore are unable to benefit from bidding for public sector projects.

There are two essential takeaways from the results: firstly, that the main benefits respondents see are of logistical and cultural matters with the enhanced collaborative nature of BIM, and the ability to undertake greater quantities of work. Secondly, while still valuable, less perceived benefits come from a financial, educational, and political standpoint. The ability to attract more capable staff, gaining an edge over competitors at tender stage, and ability of companies to meet the government mandate are not seen as key benefits. Of respondents who answered Other, the majority stated that they had not seen any benefits with BIM.

Figure 17.12 displays the hindrances companies have found in their ability to implement BIM to its full potential within their company. The most frequent answer was the cost implication of software with 53% of responses, followed by the lack of work requiring BIM to be used (46%). Results show that of the

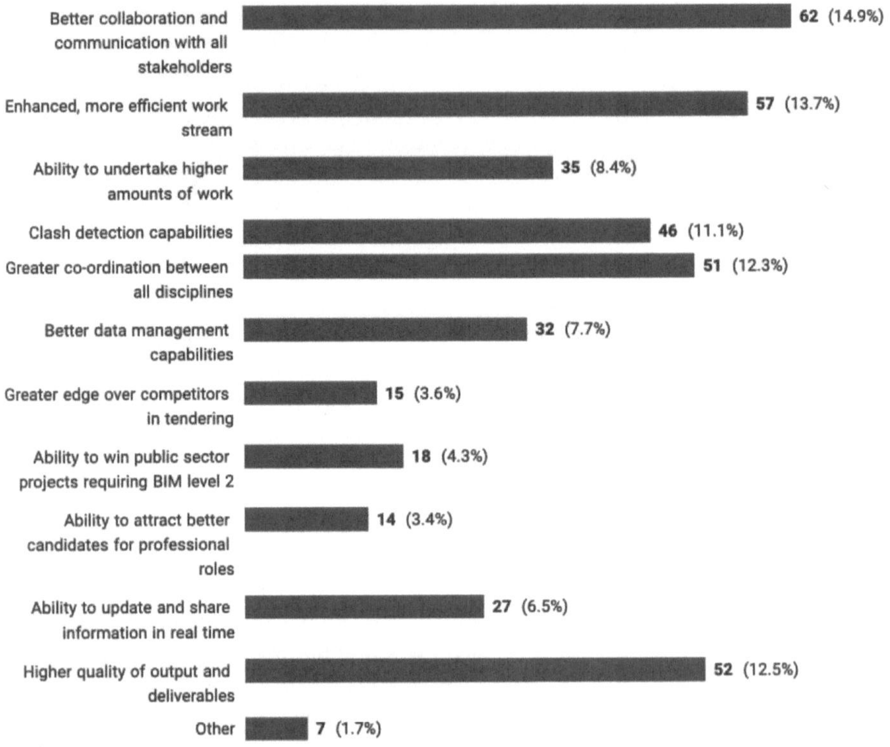

Figure 17.11 Benefits of BIM as believed by questionnaire survey respondents.

predetermined factors the largest issue companies found was of a financial position. This is closely followed by those of a temporal and educational manner, where 44% of respondents highlighted the time implications in training new staff and time taken to create designs.

Cultural issues were inherent, with 29% of companies indicating the issue of other companies less willing to embrace BIM causing collaborative issues, as well as the construction industry being slow to change. This collaborative issue is particularly poignant, as contrastingly, the main benefit companies had seen came from the interoperable nature of BIM. This reinforces that unless the cultural issue of all companies not embracing BIM is resolved, full integration within the industry will be impossible to achieve. Least frequent responses included those of a political level, and particularly those of a logistical nature. This indicates that, while there are still issues that can be found, logistically the implementation of BIM within companies should be completely feasible.

In relation to whether location of a company influenced their ability to implement BIM, many respondents (63%) indicated that location was not a factor (Figure 17.13). There was a lack of responses generated in an affirmative manner

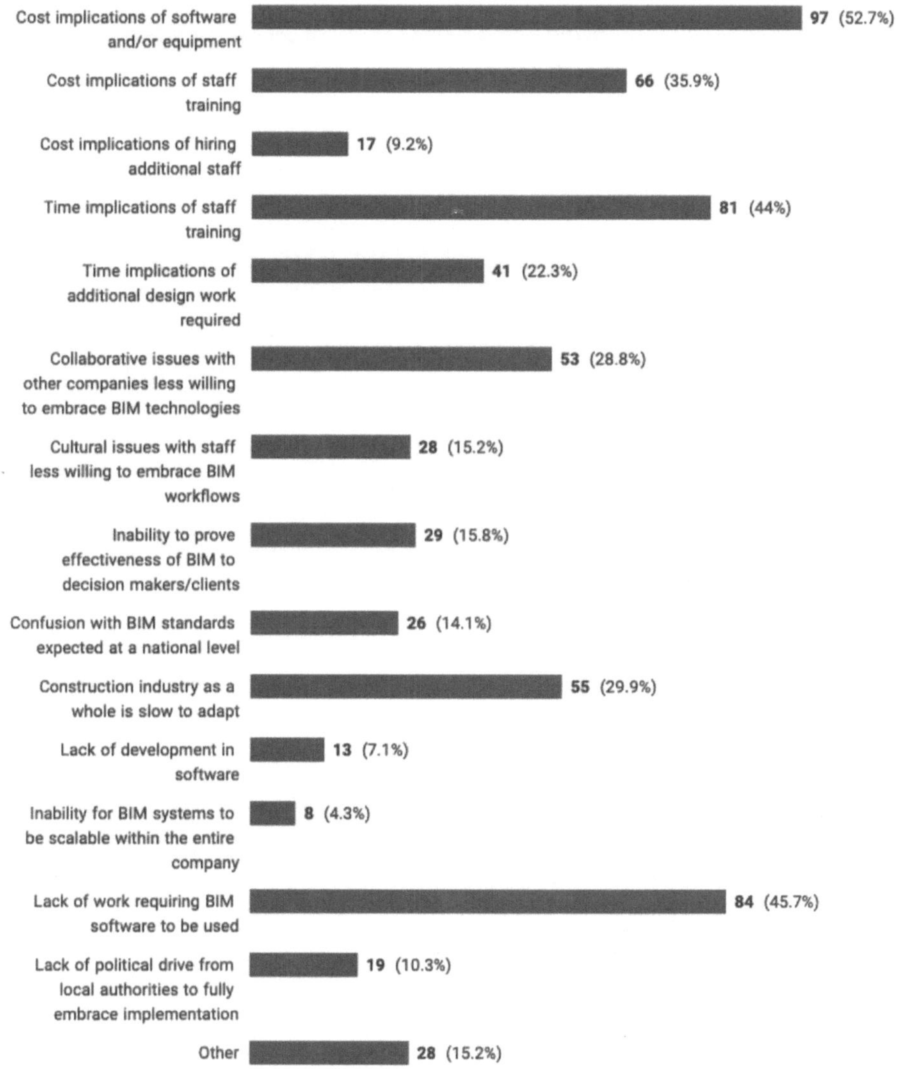

Figure 17.12 Hindrances found via questionnaire survey in ability to implement BIM.

of their location. Of the positive answers, across the various choices, responses came from East of England, Yorkshire and The Humber, and Scotland.

Collaborative issues were again reiterated, with the lack of implementation among other local stakeholders (22.5%) being the key issue companies had found with their location. This cultural issue was most prevalent, while the financial, educational, and political aspects each had over 12% of responses. Companies that responded in a negative manner regarding their ability to implement BIM due to their location had at least three responses from each region, with the majority (27%) of negative responses coming from the Southwest (Figure 17.14).

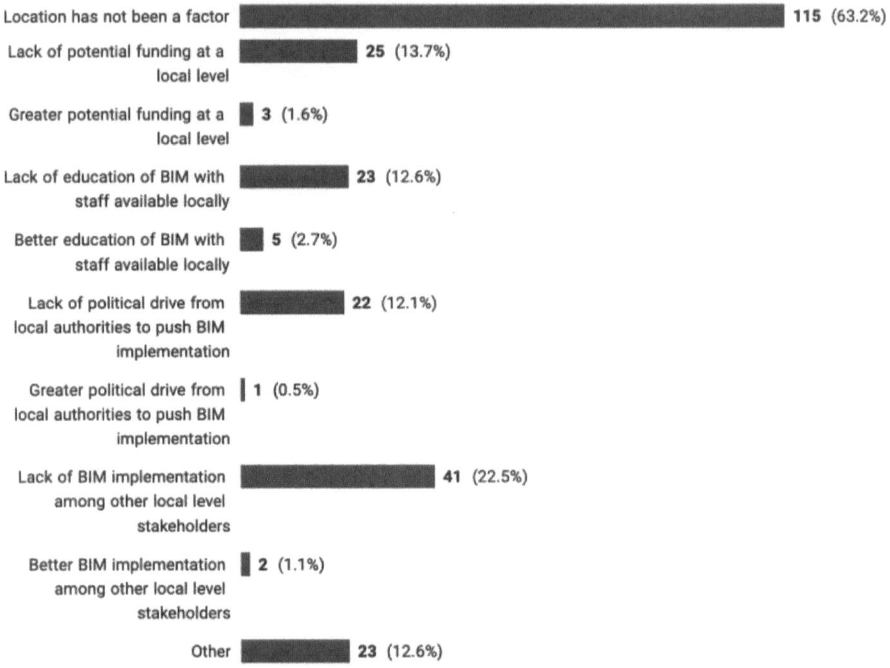

Figure 17.13 Issues found by the respondents in their ability to implement BIM.

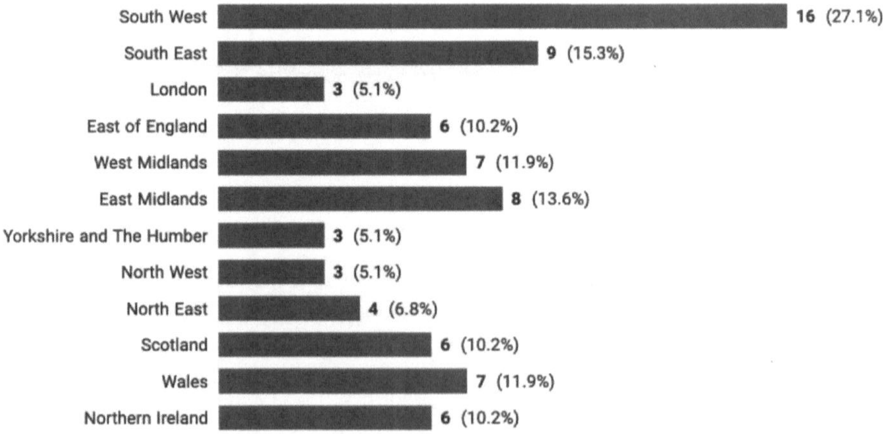

Figure 17.14 Responses with negative views on ability to implement BIM due to location.

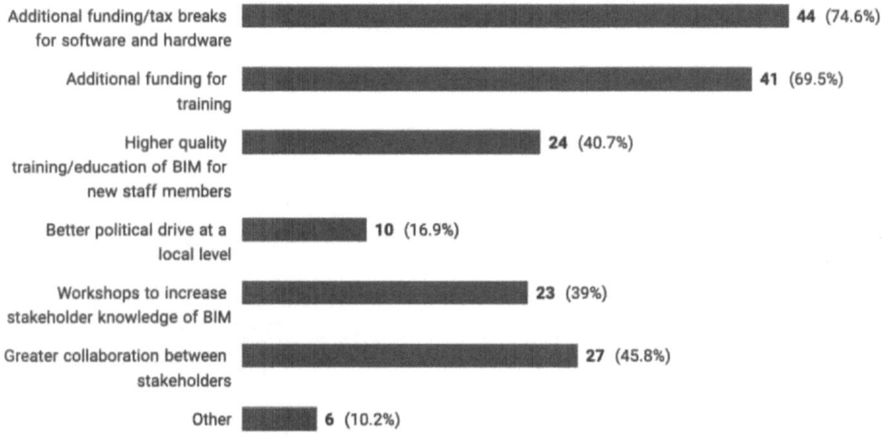

Figure 17.15 Potential solutions from respondents for improvements in BIM implementation.

In terms of potential solutions for additional support that companies felt would be most beneficial (Figure 17.15), most popular responses were regarding financial factors. Additional funding for software and hardware was the most common response at 75%, while 70% believe additional funding for training would be most beneficial. In a similar pattern to benefits and hindrances perceived, 46% of respondents believe that greater collaboration between stakeholders would be of assistance, again reinforcing the need for a cultural change within the industry. Educational factors were also considered to play a part; however, only 17% of responses felt that improved political implications would be supportive.

Results collected have allowed for the generation of a vast set of data analysis and further study can be conducted in greater detail for more specific analysis. Filters can be put in place on the data set to extract information in highly explicit detail, for example, analysis of where small engineering companies based in Scotland sit on the BIM maturity stage, and the benefits and hindrances they believe influence them in their area. This ability to interactively analyse the data in finite detail could be hugely beneficial, especially if the amount of data collected could be increased to grow the database, particularly regarding larger companies.

Discussion

For validity and reliability purposes it is essential to evaluate the comparison between secondary data collected in the literature review and primary data from both online interviews and the questionnaire survey. The study by Plan Radar (Construction Digital, 2021) identified that the United Kingdom has a high percentage of companies operating at Stage 2 and above, although it does not demonstrate the figure this represents. This correlated with qualitative data collected from interviews, with 59% of respondents operating at Stage 2 and 18% of companies stating the operability at Stage 3. However, this largely differed from bulk data returned from the survey, with 65% of respondents stating they could operate at Stage 1 or were not using BIM at all. As online interview questions were targeted to companies more likely to use BIM to gather their experiences of its use, this explains differing results between interviews and surveys. The disparity between survey responses and the Plan Radar study could be attributed to larger companies more likely to have been contacted, and to having responded to the latter. In contrast, the survey targeted companies of all sizes, with responses more likely to be returned by smaller companies.

Regarding company size in relation to the BIM maturity stage, Gledson et al. (2012) noted that SMEs were less likely to implement BIM. While results from online interviews and surveys both had a lack of participation from large companies, data corroborated that, as companies employed more staff, BIM capabilities within the business improved. The survey highlighted that most micro and small companies either did not implement BIM or sat at Stage 1, falling short of the political mandate for application of public sector contracts. This furthers the notion for differing data collected from the survey and the Plan Radar study.

In relation to company location when compared to BIM maturity, no prior study can be found, so cross-evaluation is not possible between studies. As qualitative data from interviews was limited, it is not feasible to realistically determine the location for the respondents related to BIM maturity; therefore, the only reliable source is the survey data. This revealed many regions had companies operable at Stage 1, with companies least likely to implement BIM based in Northern Ireland, and lower numbers than expected within the East of England using BIM. The largest proportion of companies currently operating at Stage 2 were based in Scotland and the West Midlands, while the highest percentage of respondents operable at BIM Stage 3 fell within London, Yorkshire and The Humber, and the Northeast.

Architectural and engineering companies had a higher percentage of firms operating at BIM Stage 2, while contractors and management services were most likely not to implement BIM at all. The surveying industry was most prevalent within BIM Stage 1, while all sectors had low numbers able to operate at BIM Stage 3, with architectural firms the most common at just 8%. These results correlated to research within the literature review, in which software for both architectural and engineering firms was much more available due to their inherent design needs. This unfortunately perpetuates the misconception of BIM being a design-based product rather than a system that can be applied across all sectors.

Subjective beliefs of companies were analysed to determine the most common benefits and hindrances found, as well as revealing potential solutions for enhanced BIM implementation. Using the predetermined factors, key benefits found by businesses were of a logistical and cultural manner within their own company. It uncovered that while still advantageous, financial, educational, and political benefits were not as prevalent. These beliefs correlated between interviews conducted and the survey. Construction 2025 (HM Government, 2013) highlighted key benefits with BIM implementation as being those of a financial and temporal manner. However, this comes from a macro perspective of the industry operating, in terms of projects costing less overall and being completed at a quicker rate, rather than from a company micro perspective. Businesses have to suffer costs required and time taken at a personal level, thus making them less likely to view these as genuine benefits.

This is amplified by the belief within both interview and survey responses that the biggest hindrances were those of a financial nature, additionally temporal and educational, notably the cost of purchasing software and time taken to train staff. While BIM was beneficial for the culture within a company, it was contradictorily a substantial hindrance within the industry culturally as a whole, with companies limited by not all stakeholders embracing BIM to the same level. This corroborates research by Harty, Kouider, and Paterson (2015) as well as Khosrowshahi and Arayici (2012), who found the culture within the industry, from stakeholder to stakeholder, was seen as one of the key factors behind BIM implementation reluctance.

Regarding solutions, it was strongly determined within both survey responses and interview answers that financial factors would be most beneficial, allowing companies to enhance BIM capabilities through both funding for software and staff training. It was again reinforced that a cultural shift would be required for all parties to be working from the same position, while highlighting that better educational systems would be beneficial, in turn advancing the cultural shift required.

Overall, both interview and survey responses corroborated one another at almost every stage, as well as adhering closely to prior studies researched within the literature review. It is therefore assumed that the reliability and validity of the data acquired is successful.

Conclusion

Based on primary data responses to potential solutions that could be offered to companies to enhance their ability to implement BIM, the key factor is that of a financial nature. If the UK government is hoping to achieve its mandate of ensuring construction firms operate at BIM Stage 2, then additional funding will need to be made available. This could largely be achieved through operating tax breaks for companies, showing that they are embracing the ethos behind BIM for greater collaborative means. The UK construction industry contributes £90bn to the UK economy, accounting for 6.7% overall (HM Government, 2013). For the industry to evolve culturally to fully embrace the digital environment, funding needs to be made available for both BIM software and the efficient, effective training required.

The issue most highlighted amongst responses, for both beneficial and detrimental matters, was of the culture within the construction industry towards BIM. By addressing financial restrictions placed on companies for necessary software, as well as being responsive with workshops and funding for training, this can start to revolutionise the culture of the industry. While internal company culture within firms is seeing a net benefit from BIM, particular attention needs to be focused on highlighting the perks of collaboration from business to business. Promoting effectiveness of BIM from the base level, ensuring staff within each sector recognise its true potential, is the only way of ensuring a cultural shift moving up the production chain, and from stakeholder to stakeholder.

The study has been limited by the result sample size returned. Despite a large sample set of potential respondents, and an acceptable return rate, the data collected would enormously benefit from a greater returned data population. There is a need for more medium and large companies to respond, to derive a more comprehensive view of the industry. While every effort was made to involve large companies, there were difficulties in obtaining personal email addresses, and of those who did reply when contacted for initial online interviews, there was a reluctance for larger companies being willing to participate. It was therefore anticipated that the response rate for a survey coming from an unknown sender would create response hesitancy.

Secondary data collected within the literature review answered the first objective by exploring the process through which BIM has become integrated within the construction industry. The common misinterpretation is that BIM is simply CAD; rather, it is the evolution of construction practice allowing for full-life design, data interpretation and analysis for buildings when fully integrated (Kensek and Noble, 2014). Despite inherent benefits that BIM implementation brings, there has been a well-recognised reluctance to embrace BIM within the construction industry sectors.

The literature review fulfilled the second objective by examining current BIM guidelines and exploring BIM software and hardware currently available to different construction sectors. This analysis helped to gain a greater understanding and establishment of maturity stages against which construction companies may be positioned, which was then fully evaluated through data analysis of online interviews and the questionnaire survey. Additionally, the third objective was achieved within the literature review by uncovering the proportion of construction firm locations across the United Kingdom and examination of potential political and economic external factors. The percentage of companies within each region was then cross-analysed with survey responses within the survey analysis to determine their validity (Table 17.4) and then represented visually with BIM maturity stages for each region (Figure 17.16).

Secondary data collected, along with data analysis of both online interviews and the survey conducted, assessed benefits and hindrances found by companies with the application of BIM software and technologies, answering the fourth objective. This allowed for the fifth objective to be met by determining external factors that may influence BIM implementation and assessing the size of companies and their geolocation with their BIM maturity stage. Through qualitative primary data collection from online interviews conducted, both objective information and subjective beliefs were analysed. By determining company size, location, sector, and BIM maturity stage, as well as understanding first hand benefits, hindrances, and potential solutions, qualitative data could be interpreted into a concept map. By scrutinising the concept map, derived information could be grouped into set issues: financial, cultural, temporal, educational, logistical, and political. Survey data collected allowed for analysis of objective company information, enabling bivariate comparison to determine how size, location, and business sector may each influence the BIM maturity stage of a company.

Within the Discussion section, the entire primary and secondary data collected has been cross-examined, giving a holistic view. It can be affirmed that the aim of the study has been met, through identifying opportunities and hindrances companies have encountered in the transition to BIM and exploring factors that have influenced BIM integration across the UK.

BIM implementation is still falling behind the expectations set by the UK government nation-wide. Cost and time implications to fully integrate BIM, seen within empirical data collected from companies across the United Kingdom, outweigh clear benefits that BIM could bring to the industry. The required cultural shift, which has been recognised for decades, for all stakeholders to embrace BIM to enable it to achieve its full potential, has still not been met. Through this study, key factors were established, each of which needs to be comprehensively addressed, which can work in turn to benefit and advance one another. Through a greater political drive, committing to financial incentives and compelling educational options, this can bring an enhanced cultural position for the AEC, enabling temporal, logistical, and financial benefits generated industry wide.

References

Aghimien, D. et al. (2021) *Construction Digitalisation: A Capability Maturity Model for Construction Organisations*. Florida: CRC Press.

Artech House Eynon, J. (2016) *Construction Manager's BIM Handbook*. New Jersey: John Wiley & Sons.

Atkin, B. and Brooks, A. (2015) *Total Facility Management*. 4th ed. New Jersey: John Wiley & Sons.

Autodesk (2023a) Civil 3D. Available at: www.autodesk.co.uk/products/civil-3d/overview (Accessed: 28 March 2023)

Autodesk (2023b) Features of Navisworks 2023. Available at: www.autodesk.com/products/navisworks/features (Accessed: 28 March 2023)

Autodesk (2023c) Revit for Architecture. Available at: www.autodesk.co.uk/products/revit/architecture (Accessed: 28 March 2023)

Bentley Systems (2023a) CAD Modelling and Visualisation. Available at: www.bentley.com/software/cad-modeling-and-visualization/ (Accessed: 28 March 2023)

Bentley Systems (2023b) Open Cities Map. Available at: www.bentley.com/software/opencities-map/ (Accessed: 28 March 2023)

Bluebeam (2023) Construction Takeoff Software. Available at: www.bluebeam.com/uk/workflows/takeoffs-and-estimation/ (Accessed: 28 March 2023)

BSI Group (2022) Transition Guide: Moving from PAS 1192-3 to BS EN ISO 19650-3. Available at: www.bsigroup.com/en-GB/iso-19650-BIM/bs-en-iso-19650-3-transition-guide/ (Accessed: 28 March 2023)

Cabinet Office (2011) Government Construction Strategy. Available at: https://assets.publishing.service.gov.uk/government/uploads/system/uploads/attachment_data/file /61152/Government-Construction-Strategy_0.pdf (Accessed: 28 March 2023)

Cohen, L., Manion, L. and Morrison, K. (2007) *Research Methods in Education*. 6th ed. London: Taylor & Francis.
Commonwealth and Development Office (2023) Small to Medium Sized Enterprise (SME) Action Plan. Available at: www.gov.uk/government/publications/fcdo-small-to-medium-sizedenterprise-sme-action-plan/small-to-medium-sized-enterprise-sme-action-plan (Accessed: 13 May 2023)
Construction Digital (2021) Available at: https://constructiondigital.com/digital-construction/which countries-are-leading-adoption-bim (Accessed: 28 March 2023)
Davies, M.B. and Hughes, N. (2014) *Doing a Successful Research Project: Using Qualitative or Quantitative Methods*. 2nd ed. London: Bloomsbury.
Denscombe, M. (2014) *Good Research Guide: For Small-Scale Social Research Projects*. 5th ed. New York: McGraw-Hill Education.
Dubois, A. and Gadde, L. (2002) 'Systematic combining: An abductive approach to case research', *Journal of Business Research*, 55, pp. 553–560. https://doi.org/10.1016/S0148-2963(00)00195-8
Eastman, C. et al. (2018) *BIM Handbook: A Guide to Building Information Modelling for Owners, Designers, Engineers, Contractors, and Facility Managers*. 3rd Ed. New Jersey: John Wiley & Sons.
Emmitt, S. (2018) *Barry's Advanced Construction of Buildings*. 4th ed. New Jersey: John Wiley & Sons.
Epstein, E. (2011) *Implementing Successful Building Information Modelling*. Massachusetts: Artech House.
Fellows, R.F. and Liu, A.M.M. (2015) *Research Methods for Construction*. 4th ed. New Jersey: John Wiley & Sons, Incorporated Foreign.
Gledson, B.J., Bleanch, P., and Henry, D. (2012) 'Does size matter? Experiences and perspectives of BIM implementation from large and SME construction contractors' 1st UK Academic Conference on Building Information Management (BIM) 2012, Northumbria University, Newcastle upon Tyne, UK, September 5–7, 2012, pp. 97–108. https://doi.org/10.13140/RG.2.1.5168.7280
Graphisoft (2023) *Why Graphisoft – Our Story*. Available at: https://graphisoft.com/whygraphisoft/our-story (Accessed: 28 March 2023)
Grenfell, M.J. (2014) *Pierre Bourdieu: Key Concepts*. 2nd rd. London: Routledge.
Gu, N. and London, K. (2010) 'Understanding and facilitating BIM adoption in the AEC industry', *Automation in Construction*, 19(8), pp. 988–999. https://doi.org/10.1016/j.autcon.2010.09.002
Hardin, B. and McCool, D. (2015) *BIM and Construction Management: Proven Tools, Methods, and Workflows*. 2nd ed. New Jersey: John Wiley & Sons, Incorporated.
Harty, J., Kouider, T., and Paterson, G. (2015) *Getting to Grips with BIM: A Guide for Small and Medium-Sized Architecture, Engineering and Construction Firms*. Florida: CRC Press.
HM Government (2013) *Construction 2025 – Industrial Strategy: Government and Industry in Partnership*. Available at: https://assets.publishing.service.gov.uk/government/uploads/system/uploads/attachment_data/file /210099/bis-13-955-construction-2025-industrial-strategy.pdf (Accessed: 28 March 2023)
HM Government (2018) *Industrial Strategy: Construction Sector Deal*. Available at: https://assets.publishing.service.gov.uk/government/uploads/system/uploads/attachment_data/file /731871/construction-sector-deal-print-single.pdf (Accessed: 28 March 2023)
Holzer, D. (2015) *The BIM Manager's Handbook, Part 1: Best Practice BIM*. New Jersey: John Wiley & Sons.

IFMA (2013) *BIM for Facility Managers*. New Jersey: John Wiley & Sons.
Kensek, K.M. (2014) *Building Information Modelling*. London: Taylor & Francis.
Kensek, K.M. and Noble, D. (2014) *Building Information Modelling: BIM in Current and Future Practice*. New Jersey: John Wiley & Sons.
Khosrowshahi, F. and Arayici, Y. (2012) 'Roadmap for implementation of BIM in the UK construction industry', *Engineering, Construction and Architectural Management*, 19(6), pp. 610–635. https://doi.org/10.1108/09699981211277531
Mason, J. (2017) *Qualitative Researching*. 3rd ed. London: SAGE Publications.
Matthews, B. and Ross, L. (2010) *Research Methods: A Practical Guide for the Social Sciences*. London: Pearson Education.
Migilinskas, D. et al. (2013) 'The Benefits, Obstacles and Problems of Practical BIM Implementation', *Procedia Engineering*, 57, pp. 767–774. https://doi.org/10.1016/j.proeng.2013.04.097
Office for National Statistics (2020) *Construction Statistics. Great Britain*. Available at: www.ons.gov.uk/businessindustryandtrade/constructionindustry/articles/constructionstatisti cs/2020 (Accessed: 28 March 2023)
Reddy, K.P. (2012) *BIM for Building Owners and Developers: Making a Business Case for Using BIM on Projects*. New Jersey: John Wiley & Sons.
Reed, R. and Sims, S. (2014) *Property Development*. 6th ed. Florida: CRC Press.
RIBA (2020), RIBA Plan of Work 2020 Overview, Published by RIBA, 66 Portland Place, London, W1B 1AD. Available at: https://www.architecture.com/knowledge-and-resources/resources-landing-page/riba-plan-of-work?srsltid=AfmBOoqTLPiDCDSCwJ8MneHGYgX8zwWwa3PDSzEE2yv1-pHvLzQEVILt (Accessed: 28 March 2023)
RICS (2020) *The Future of BIM: Digital Transformation in the UK Construction and Infrastructure Sector*. Available at: www.rics.org/globalassets/rics-website/media/upholding-professionalstandards/sector-standards/construction/future-of-bim_1st-edition.pdf (Accessed: 28 March 2023)
Rowley, J. (2014) 'Designing and using research questionnaires', *Management Research Review*, 37(3), pp. 308–330. https://doi.org/10.1108/MRR-02-2013-0027
Sapsford, R. and Jupp, V. (2006), *Data Collection and Analysis*. 2nd ed. London: SAGE Publications.
Saunders, M., Lewis, P., and Thornhill, A. (2019) *Research Methods for Business Students*. 8th Ed. London: Pearson Education.
Siebelink, S. et al. (2020) 'Understanding barriers to BIM implementation: Their impact across organizational levels in relation to BIM maturity', *Frontiers of Engineering Management*, 8. https://doi.org/10.1007/s42524-019-0088-2
Smith, D.K. and Tardif, M. (2009) *Building Information Modelling: A Strategic Implementation Guide for Architects, Engineers, Constructors, and Real Estate Asset Managers*. New Jersey: John Wiley & Sons.
Stoop, I. et al. (2010) *Improving Survey Response: Lessons Learned from the European Social Survey*. 2nd ed. New Jersey: John Wiley & Sons.
Tal, D. and Altschuld, J. (2021) *Drone Technology in Architecture, Engineering and Construction: A Strategic Guide to Unmanned Aerial Vehicle Operation and Implementation*. New Jersey: John Wiley & Sons.
Torres, P.L. and de Cássia Veiga Marriott, R. (2009) *Handbook of Research on Collaborative Learning Using Concept Mapping*. Pennsylvania: IGI Global.
Trimble Solutions Corporation (2023) *The Home of Truly Constructible BIM*. Available at: www.tekla.com/uk (Accessed: 28 March 2023)

UK BIM Framework (2021) *About UK BIM Framework*. Available at: www.ukbimframework.org/about/ (Accessed: 28 March 2023)
UK Construction Media (2018) *Issues Affecting the Adoption of 3D BIM Modelling*. Available at: www.ukconstructionmedia.co.uk/press-releases/issues-affecting-the-adoption-of-3d-bim modelling-xs-cad/ (Accessed: 28 March 2023)
Whyte, J. (2002) *Virtual Reality and the Built Environment*. Florida: CRC Press.
Yori, R., Kim, M., and Kirkby, L. (2019) *Mastering Autodesk Revit 2020*. New Jersey: John Wiley & Sons.

18 Understanding Fossil Fuel Consumption in Commercial Buildings

An Experimental Analysis of Energy Performance Labelling in England

Kevin Muldoon-Smith, Jonthan Pearson, Paul Greenhalgh, Mark Kirk, and Steve Philips

Introduction

The experimental research in this chapter adapts information embedded in existing Energy Performance Certificates (EPCs) to compare hypothetical energy performance rating data with an original fossil fuel consumption metric. This is to understand for the first time whether those commercial buildings considered most environmentally efficient are also relatively less reliant on fossil fuels. In this chapter, commercial buildings are taken to mean office, retail, leisure, and industrial properties. A conservative estimate is that global real estate consumes 40% of global energy annually and accounts for more than 20% of international carbon emissions (WEF, 2016). The UK government aims to reduce UK CO2 emissions to close to zero by 2050 to attain its energy efficiency targets. This may lead to mass devaluation of property markets without intervention – so-called stranded assets (Pitman, 2022). In this research, climate-related stranded assets in the property sector have the potential to suffer from unanticipated or premature write-downs, devaluation, or conversion to liabilities. The geographical impact is abandonment, environmental degradation, and/or socioeconomic decline in urban areas, while the global impact is systemic risk to asset valuations in global capital markets. More positively, greater understanding of climate-related stranded assets also has the potential to influence and provoke sustainable retrofit.

The aim of the chapter is to question the assumption that those buildings considered most efficient are also consuming relatively fewer fossil fuels. The authors argue that it is regularly taken for granted in built environment practice, government policy, valuation of assets, and bank lending that a positive EPC rating is correlated with reduced fossil fuel consumption – investment and financing decisions are then made on this basis. The authors find that the assumed link between energy performance and fossil fuel exposure is not correlated, with some of the best-performing properties also consuming the most fossil fuel for their heating, cooling, and business processes. This indicates that there is a pronounced

DOI: 10.1201/9781003380559-18

disparity in awareness of the risks posed by energy policy transformation and the ability to respond to this situation. It is important to note that the chapter does not take issue with the principle of EPCs; they are a powerful tool in benchmarking performance and highlighting energy efficiency – and, in England, supporting building regulations. Rather, the chapter aims to tackle the issue of EPCs being used as a proxy for fuel consumption. This original research is significant for several interlinked reasons:

1) Valuers and banks now factor in energy performance when valuing and securing debt against properties, with those properties performing poorly receiving adverse lending conditions. This lending is important for the overall financing of commercial real estate but also the financing and targeting of retrofit and adaptation measures designed to achieve 2050 climate change targets. Currently, the energy labelling system functions as a means of understanding societies' energy use, and how it can be monitored and improved (Muldoon-Smith and Greenhalgh, 2019). However, this same system has the potential to hardwire and connect valuation risk into global capital markets if fossil fuel exposure is not fully understood. In this sense, EPCs and associated minimum energy rules prime already financialised real estate assets (for example through international mortgage markets, real estate investment trusts, unit trusts, and property companies) for stranding. The approach in this chapter seeks to uncover this situation by moving beyond passive econometric assumptions of fossil fuel stranding to actively understand the climate-related performance of property assets. It is a central contention in this chapter that the increasingly passive techniques of real estate investment and finance, which are typically based on index-based performance monitoring, do not proactively recognise or manage the potential for fossil fuel-based stranded assets. Rather, this issue only becomes apparent after valuation and performance begin to shrink in comparison with market baselines.
2) The current geo-political situation has coalesced with environmental imperatives to reduce fossil fuel exposure. At the moment, 78% of UK energy comes from fossil fuels – oil, gas, and a small amount of coal (EAC, 2023). Dependence on oil and gas has recently left the United Kingdom exposed to the biggest fossil fuel price increase since the 1970s oil crisis, as Russia has restricted the supply of gas to continental Europe since 2022 and OPEC+ nations reduced the production of oil in 2023. This has increased the impetus to transition away from fossil fuels due to fuel poverty, consistency of supply, and energy security.
3) In light of this, the chapter's findings question the inherent validity of energy performance assessments and whether it is meaningful for quality-based consideration of efficiency to have little to no correlation with fossil fuel exposure. If society is going to achieve its climate-based targets ahead of 2050, then the primary means of understanding the energy situation in the built environment should capture and understand fossil fuel consumption.
4) Perhaps most importantly government retrofit and adaptation policies are primarily based upon energy performance ratings – the best current estimate of

> the relationship between society's complex energy-based technical procedures (building design) with human behaviour (building use). The motivation for this research is to provide a sound basis for policymakers when governments and practice evaluate ideas for climate change transition and adaptation in the real estate sector. For those property professionals involved in the day-to-day management of real estate assets in the developed world, the chapter provides an approach to understanding the wider significance of climate-related threats, which we hope will contribute to more knowledgeable and effective practice in relation to real estate–based stranded assets. Expanding knowledge in this area will help city leaders as well as investment portfolio and asset managers in mature urban areas deal with the challenges of adapting an ageing property stock. However, it is also hoped that this approach will help city leaders and property professionals dealing with the demands of accelerating urbanisation in the less developed world, which requires an understanding of urban development processes and the potential impact of stranded assets.

The remainder of this chapter sets out the significance of stranded assets for investors, financiers, government, built environment practitioners, and society. It then considers energy labelling in the international built environment before concentrating on energy labelling in England. It then introduces the experimental method in this chapter – based upon an adaptation of existing energy performance assessments situated within an exhaustive sample of EPC-related commercial property stock in Newcastle-upon-Tyne. It then introduces the central findings and final conclusions alongside limitations and opportunities for further research. In setting out the conclusions, the chapter considers how the existing EPC system, supported by other baseline datasets, could be used to better reflect cost (to help tackle fuel poverty), carbon emissions (to understand what is going on in each building), and consumption (what fuel is being used) – the so-called three Cs (Elmhurst, 2022).

Stranded Assets and Energy Labelling in the Built Environment

Caldecott et al. (2016) indicate that stranded assets are assets that have 'suffered from premature or unanticipated write-downs, devaluations, or conversions to liabilities'. Stranded assets are not new; buildings have been rendered redundant and/or obsolescent since their original development. However, what is new is

> the influence, systemic reach and disruption of climate change and associated environmental policy on some property assets, related capital markets (at the macro scale) and individual communities (at the micro scale) that are reliant on homes to live, and commercial property to work. (Muldoon-Smith and Greenhalgh 2019:62)

While fossil fuel–related stranded assets received attention from the likes of the UN (McGrath, 2014), the Friedman (2014), the Gurría (2013), and the Carney (2015), the real estate sector has been relatively slow to react to the specific issue of fossil fuel–related stranded assets. The current authors previously argued that real assets make up a large part of total global investment worth and are a significant store of national, corporate, and individual wealth; the omission of real estate from the stranded asset discourse poses significant threat for underlying property valuation (Muldoon-Smith and Greenhalgh, 2019).

In the United Kingdom, initial environmental-related intervention was based on minimum energy standards – targeted policy action and enforcement was relatively low key, with the *Estates Gazette* arguing in 2019 that no properties was fined for failing to improve their environmental ratings in the early years of minimum energy standards (Tomusk, 2019). In addition, the UK government only recently (April 2023) started to extend minimum energy rules to all commercial properties; previously the regulations only related to those properties available to let or purchase. While environmental improvement was limited, there is growing momentum around sustainable retrofit, circular economy-based interventions, and conversion to renewable energies, with considerable investment expected in the proceeding decades.

This is why it is crucial that the energy performance scaffolding that helps connect and signpost society, government, and funding with regard to environmental performance is meaningful and accurate. However, in England climate change targets and fossil fuel–stranded assets have been conflated with EPCs that do not actually measure fossil fuel consumption; instead, they are a hypothecated measurement of energy emissions per square metre of floor area. Generally, those assets with lower hypothetical emission rates attain a better EPC rating, thereby measuring the asset's environmental performance based on the efficiency of the building components as opposed to the environmental impact of the fuel it consumes. The ensuing sections of this chapter presents further information on the international evolution of energy labelling before grounding this situation in England.

Energy Labelling in the Built Environment

In recent decades, sectors associated with the built environment began to develop and introduce legislation to reduce its impact on climate change and global warming. Environmental labelling was regularly the primary tool in initial efforts to improve the environmental credentials of building stock (Reed et al., 2009). In this sense, energy labelling is an example of a socio-technical system that connects building design with how humans use buildings. The primary contention is that an adverse rating will make relative stock less attractive to tenants and thus incentivise landlords, investors, and policymakers to take remedial action to improve the situation. They are the foremost driver of building improvement as they influence decision-making in transactions and provide a guide for cost-optimal recommendations for energy performance improvement (BPIE, 2014). According to Sayce et al. (2010), these initiatives have either adopted a multi-criteria sustainability approach or narrowed their attention on energy. During the middle of the 1990s, the BREAAM1 (multi-criteria) tool set the standard in the United Kingdom. France quickly created its own HQE2 multi-criteria model. Others later followed suit in the 1990s, including the Swiss Minergie and the North American Energy Star (both connected to energies). In the 2000s, additional innovation led to the creation of the multi-criteria schemes LEED (North America), CASBEE (Japan), Green Globe (Canada), and Green Star (Australia).

One of the most ambitious approaches in Europe – in 2010 the European Union adopted the Energy Performance of Building Directive (European Union, 2010). The directive sets out the framework for, and central ambition of, ensuring all building stock is zero-emission and fully decarbonised by 2050. In doing so, energy efficiency would be boosted, energy poverty would be tackled, people's vulnerability to energy prices would be reduced, and green transitions would support economic resilience and job creation. The directive was further updated in 2021 to accelerate this transition, facilitating more targeted financing for the worst buildings – the revised directive (alongside the Renovation Wave strategy presented in October 2020) sets out strategies to at least double the energy renovation rate by 2030.

The evolving directive made it mandatory for all European properties to hold an EPC. The England and Wales Government used EPCs as the foundation for Minimum Energy Efficiency Standards (MEES), which were legislated for in 2015 via the Energy Efficiency (Private Rented Property) (England and Wales) Regulation Act. This Regulation adopted a minimum energy standard for both domestic and non-domestic privately rented properties. From April 2018 it was illegal for any commercial property to be let with an EPC rating below E – this meant any properties with F or G ratings were illegal to let (in 2020 the same rule began to apply to domestic properties). In England and Wales, it is estimated that 10% of residential property stock and 18% of commercial stock are under this threshold (Muldoon-Smith and Greenhalgh, 2019). The Government of England and Wales has also committed to a forward plan for MEES. The minimum energy performance standard gradually increased in the lead up to 2050 – from 1 April 2023, these regulations applied to all non-domestic properties, not only those agreeing a new let, but lease-renewal as well if an EPC is already in place, or tenants wishing to sublet.

While ambitious, there are several exemptions to MEES that have the potential to reduce the strategies potency. For example, there is heritage-related protection against any building improvement that would alter the character or appearance of an historical building or associated conservation area, where any improvement would reduce the market value by more than 5%, where the interventions do not pay for themselves through energy cost savings within a seven-year period, or if the landlord cannot get consent from the relevant planning authority and/or sitting tenant. While exemptions obviously give landlords and investors leeway to avoid environmental improvements, there is an Exemption Register that details those properties still available on the market with an EPC of F or G – the strategy is that inefficient buildings will be publicly named and may suffer value and reputational depreciation. Nevertheless, MEES has not been a panacea for environmental improvement. Hosgood (2017) argued that up to 70% of EPC ratings in England and Wales could be inaccurate (either too high or too low) due to inconsistent application of EPC assessment and the evolving criteria of the underlying method of EPC calculation – the Simplified Energy Model (SBEM).

Methodology: Creating an Energy Consumption Metric

The first challenge for those with an interest in the built environment is recognising that climate-related stranded assets exist, particularly those associated with fossil fuels. The next challenge is creating the information baselines that reflect this existence, to then in turn help society counteract these issues through sustainable retrofit strategies, divestment, and appropriate new development. To this end, this section sets out an experimental methodology that questions the relationship between fossil fuel consumption and EPC labelling.

The underlying hypothesis that is tested is:

> *Are those commercial properties with the best energy performance ratings consuming least fossil fuels?*

The methodology and data analysis make use of EPC information freely available on the Energy Performance of Building Register – the Register holds all EPCs in England. It then adapts existing metrics in the underlying EPC data set and creates a new fossil fuel consumption metric, to aid comparison between hypothetical energy emissions and energy, particularly fossil fuel–related, consumption. In theory, there should be a clear correlation between energy rating, hypothetical emissions, and consumption. However, our method and new consumption metric indicates that fossil fuel consumption is a) widespread across property stock and EPC grade, and b) regularly associated with the best properties and EPCs.

While individual property information is available in the data set, we have chosen to anonymise building names and addresses, instead aggregating information into property type (office, industrial, leisure and retail) at the Newcastle-upon-Tyne administrative area scale. While the data set is available at regional and national scale, the authors wanted to first concentrate on method refinement at the finer grain single administrative area. This is an original contribution to knowledge with city-wide EPC modelling via fossil fuel consumption not having been previously conducted. This city-wide scale also offers the original opportunity to compare asset types (office, retail, leisure, and industrial), revealing key differences between the respective asset types.

The EPC open data set for Newcastle-upon-Tyne was filtered to retain relevant variables, most notably EPC rating, emissions rate ($kgCO2/m^2$), and floor area. The first step in the creation of a new consumption metric was to calculate the total emissions for each EPC unit/building. This was calculated by multiplying the emissions rate by floor area. This new 'total emission' data field was then broken down into its constituent elements (space and water heating and cooling). Then, using ECUK 2022 End Use Tables Split[1] as a guide, servicing arrangements were broken down into their estimated fuel use – EPCs detail what fuel source is used but not in what proportion. Although only an estimation, this is an important step in innovation as it is not currently possible to understand actual consumption of energy in the commercial real estate sector. In the public sector, there is the Display Energy Certificate (DEC), which measures actual consumption of energy. However, there was never a DEC process put in place for either the commercial or residential real estate markets. Hence, the ECUK data allowed us to create sector-wide estimates of the apportionment of energy used between space and water heating and cooling. The emissions for each service were converted to fossil fuel consumption using the appropriate conversion factor based on the service fuel source.

The research then examined the consumption processes that the EPC does not account for. EPCs provide estimated data on the heating and cooling of a building but not on its 'day-to-day' activities – the embodied use of the building. Layering in this additional information to the model allowed a more holistic understanding of overall energy consumption in the building – beginning to form a basis for energy cost (currently the primary function of EPCs), consumption, and type of energy used.

In the incorporation of additional data to estimate the consumption characteristics of 'day-to-day' activities, it is important to note that each asset sector uses different fuels for their 'day-to-day' processes, that is, retail and leisure uses oil for cooking and industrial uses solid fuel for some high temperature processes. The conversion factors and splits for each element are largely based on a set of assumptions found in the 2022 Government Greenhouse Gas Conversion Factors for Company Reporting process. The report, which is published by the Department for Business, Energy & Industrial Strategy, provides the current official set of UK government conversion factors. These factors enable organisations and individuals to calculate GHG emissions from a range of activities, including energy use. For instance, a conversion factor can be used to calculate the amount of GHG emitted because of burning a particular quantity of oil in a heating boiler. By reverse, the same conversion factor can be used to calculate the quantity of oil burnt in a heating boiler by the amount of GHG it emits (e.g., carbon emissions). Again, although these are estimates, this is an important subsequent step in innovation as it enables for the first time an estimation of in-building energy consumption. The description and assumptions inherent in these conversion factors can be seen in Table 18.1.

The importance of the initial service breakdown is that it allowed the authors to estimate the fossil fuel consumption of the various servicing arrangements of a building that consumes fuel. Each service may (and often does) utilise different fuel sources, and therefore the services need to be disaggregated across the data set to enable the appropriate fuel conversion factor to be applied – for example, Some buildings use gas-fired boilers for heating and mains-supplied electricity for cooling. If taken together, the carbon emitted from both the gas and mains-electricity cannot accurately be converted to consumption as each needs a different conversion factor. When split, the gas and grid-supplied electricity conversion factors can be applied to the emissions appropriately. The final stage in the adaptation of EPC information and creation of a new consumption metric process was to calculate the fossil fuel consumption for each service (heating, cooling, and other). This was achieved via the following calculation:

Table 18.1 Conversion factors

Fuel Source	*Multiplier*	*Description and Rationale*
Gas	2.53925	Natural Gas. 2.53925 kgCO2 per kg
Grid-Supplied Electricity	1.01167	Electricity supplied by the National Grid. Gas (38.5% x 2.53925) + Coal (1.5% x 2.27045) = 1.01167
Other	1	Nominal figure as the fuel source is not identified in the EPC dataset.
Oil	3.22920	Fuel Oil 3.22920 kgCO2 per kg
LPG	2.93929	Liquified Petroleum Gas 2.93929 kgCO2 per kg
Biomass	0	Wood pellets 0.05055 kgCO2 per kg *Biomass is a bioenergy, therefore fossil fuel conversion factor 0.*
District Heating	2.53925	Taken as natural gas, assuming the district heating system is supplied by conventional boilers. 2.53925 kgCO2 per kg
Coal	2.27045	Coal (electricity generation) 2.27045 kgCO2 per kg
Dual Fuel (Mineral and Wood)	0.04303	Assumed to be wood burning stoves and fuel source is 'wood logs'. 0.04303 kgCO2 per kg *Dual fuel is a bioenergy, therefore fossil fuel conversion factor = 0.*
Solid Fuel	1	Not defined in the baseline data so given score of 1

*Total Emissions * (Heating/Cooling/Other Split) * conversion factor*

Figure 18.1 describes this process via three layers of interrogation. Layer 1 (fabric) describes the building and associated EPC. Layer 2 (services) describes the division of energy emissions into space and water heating, cooling, and 'other'. Layer 3 (exposure) describes the actual fuel type in relation to space and water heating, cooling, and other processes.

There is scope to further improve the methodology by making use of class information held in the EPC Register and Valuation Office Agency (VOA) valuation data set. The use class information would potentially allow further sub-division of asset type, for example into storage, light industrial, and heavy industrial – all of which would have different energy consumption dynamics. Each building will have a different split, especially in the case of industrial properties due to their heterogeneous nature, in which manufacturing and industrial processes can differ considerably. In this study, the industrial class is dealt with as one homogenous entity.

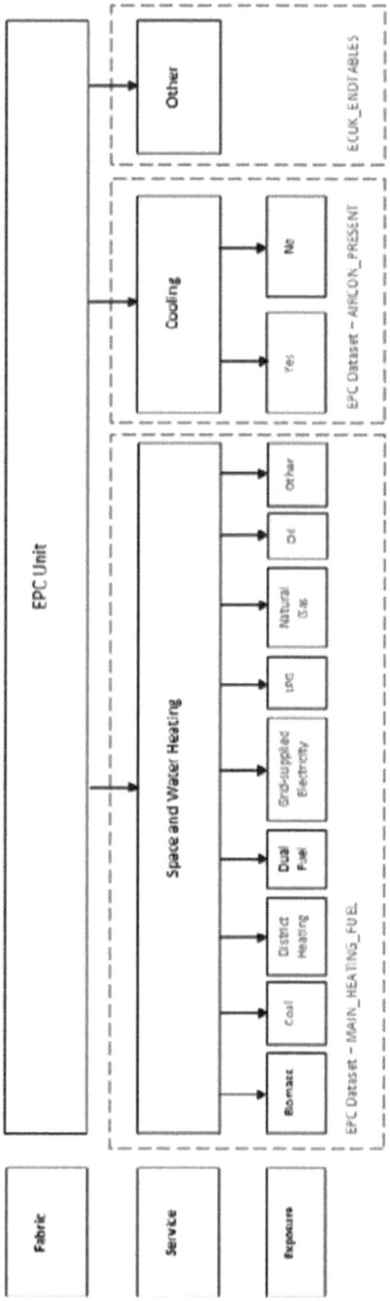

Figure 18.1 Consumption metric method process development.

In addition, the asset-related detail held in the experimental data set is restricted by the nature of minimum energy rules and ownership activity. For example, office, leisure, and retail properties tend to transact regularly on the open market, thus triggering the need for an EPC (until 2023, a landlord only needed an EPC if the property was being let or sold), while industrial properties tend to transact less and are more typically owner occupied. This is reflected in the data set for Newcastle-upon-Tyne, where there are only 507 EPC entries related to industrial properties, compared with the VOA valuation data set, where there are 1974 hereditaments (taxable properties) in Newcastle-upon-Tyne. This indicates that our database represents 25.5% of the industrial market in Newcastle-upon-Tyne.

Findings and Discussion

The following section presents the main findings in the chapter. First, as a foreground to the analysis, the research presents the current EPC situation in Newcastle-upon-Tyne, detailing the grade of EPC for each asset class on a sliding scale of A–G (A being best and G worst). The research then presents the original calculations of total emission and total consumption for each asset type. The chapter then goes on to describe fossil fuel consumption for each asset type, before finally contrasting the respective EPC and fossil fuel consumption data sets to examine and reveal the relationship between energy performance rating and actual fossil fuel consumption.

Energy Performance and Fossil Fuel Use

Table 18.2 and Figure 18.2 describe the energy performance rating in Newcastle-upon-Tyne for the respective asset types.

In each asset category, A-grade EPCS only account for 1% of total properties, while B-grade EPCS account for between 8% and 10% in each asset category. For all asset categories, the vast majority of EPCs were in the C–E range, with retail and leisure performing better than industrial and then office categories. F- and G-rated properties accounted for between 8% and 16% of properties. This indicates that there was still a minority of properties operating below the minimum energy performance threshold, likely taking advantage of exemption criteria.

Table 18.3 describes total emissions (and emission rate per m^2) and total consumption rate (and consumption rate per m^2). The 'emissions' represent the total amount of hypothetical carbon dioxide that is emitted by the building's services, whereas 'consumption' represents an estimation of the total amount of fossil fuel the building's services and day-to-day activities consume. Leisure properties consume significantly more fossil fuels than the other asset categories, with an average consumption rate of 357.48 $kgCO2/m^2$. More than double the next highest consumer, industrial property is at 140.33 kg/m^2, followed by retail with 135.02kg/m^2 and offices with 104.12 kg/m^2.

Taking forward the methodology set out in Figure 18.1, Figure 18.3 describes the contribution of 'space and water heating', 'cooling', and 'other' to overall fossil fuel consumption. The 'other' category accounts for the actual business process

Table 18.2 EPC rating in commercial property sectors in Newcastle-upon-Tyne

Energy Performance Rating	*Offices*	*Retail*	*Leisure*	*Industrial*
A	12 (1%)	13 (1%)	9 (1%)	5 (1%)
B	116 (8%)	182 (9%)	80 (10%)	38 (8%)
C	415 (27%)	707 (34%)	252 (31%)	141 (28%)
D	444 (29%)	664 (32%)	295 (36%)	181 (36%)
E	293 (19%)	283 (14%)	117 (14%)	81 (16%)
F	123 (8%)	99 (5%)	40 (5%)	24 (5%)
G	124 (8%)	130 (6%)	24 (3%)	35 (7%)

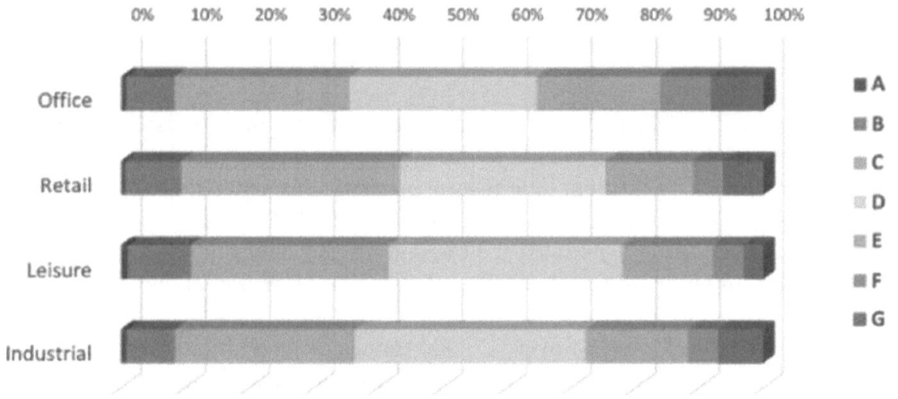

Figure 18.2 EPC rating in commercial property sectors in Newcastle upon-Tyne.

Table 18.3 Emission rate compared with consumption rate

	Average Total Emission kg	*Average Emission Rate kgCO2/m2*	*Average Total Consumption kg*	*Average Consumption Rate kg/m2*
Office	45.931	62.95	80.890	104.12
Retail	32.906	108.42	47.343	135.02
Leisure	86.550	173.49	190.364	357.48
Industrial	71.580	72.22	140.221	140.33

taking place in the building, in addition to its heating and cooling. This is an original element of analysis, in that it displays the actual consumption of fossil fuels in commercial buildings, not just the building components. It demonstrates that heating and cooling account for varying degrees of the consumption footprint. In office, heating accounts for most consumption, followed by other and then a small

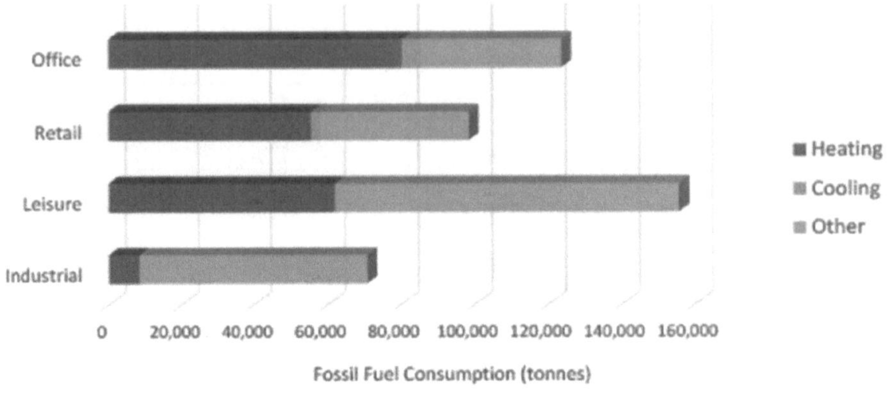

Figure 18.3 City-wide fossil fuel consumption by element.

contribution from cooling. A very similar pattern of consumption is seen in retail property. In the leisure asset category, 'other' presents the greatest quantity of consumption, followed by heating. In industrial property, the 'other' category accounts for the greatest quantity of fossil fuel consumption (60,000 tonnes of fossil fuel pa) with a small proportion accounted for by heating and only minimal contribution from cooling. In terms of actual quantity of fossil fuel consumed, the leisure asset category consumes the most fossil fuel (over 150,000 tonnes of fossil fuels per year), followed by office (just over 120,000 tonnes of fossil fuels per year), retail (just under 100,000 tonnes of fossil fuels per year) and then industrial (just under 70,000 tonnes of fossil fuels per year). Industrial likely consumes a far greater cumulative quantity of fossil fuels in Newcastle-upon-Tyne because, as already noted, the sample size for industrial is not a representation of all industrial stock. Across all asset categories, less than 0.2% of stock consumed no fossil fuels – all leisure properties consumed some form of fossil fuel.

Figures 18.4 to 18.7 describe what fuel source is used for 'space and water heating', 'cooling', and 'other'. In the office asset category, gas accounts for the greatest quantity of fossil fuel consumption (just over 70,000 tonnes), followed by grid-supplied electricity (48,403 tonnes) and oil (4,603 tonnes). Other,[2] liquified petroleum gas (LPG), and district heating account for a tiny proportion of fuel consumption (328 tonnes). In the retail asset category, grid-supplied electricity accounts for the most energy consumption (57,815 tonnes), followed by gas (39,680 tonnes) and oil (798 tonnes). LPG, district heating, and dual fuel account for a small fraction of underlying consumption (111 tonnes). In the leisure asset category, gas accounts for the most consumption (70,049 tonnes), followed by grid-supplied electricity (47,055 tonnes), oil (31,006 tonnes), and other (7,441 tonnes). LPG and district heating account for only a small fraction (27 tonnes). In the industrial asset category, gas accounts for the most consumption (37,247 tonnes), followed by grid-supplied electricity (26,444 tonnes) and smaller contributions

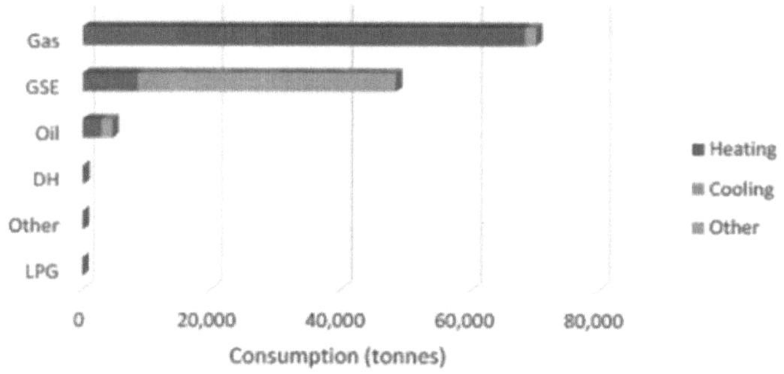

Figure 18.4 Office fossil fuel consumption.

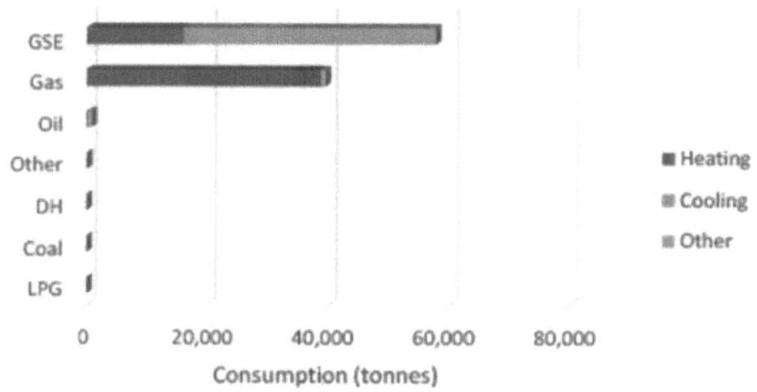

Figure 18.5 Retail fossil fuel consumption.

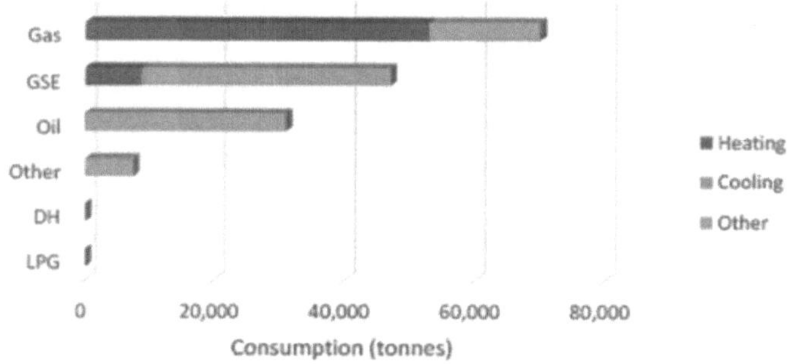

Figure 18.6 Leisure fossil fuel consumption.

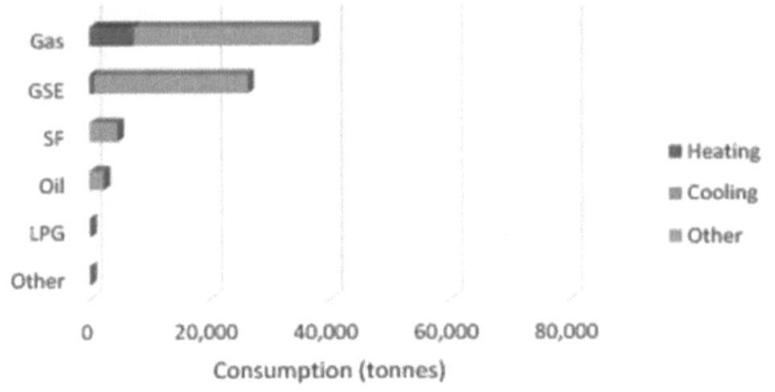

Figure 18.7 Industrial fossil fuel consumption.

from solid fuel (4,662 tonnes) and oil (2,395 tonnes). Other and LPG accounted for a tiny fraction of consumption (63 tonnes). While there are variations between the respective asset categories, the broad proportions of consumption are similar, with gas and grid-supplied electricity accounting for most of the overall consumption, followed in distant third and fourth place by oil and other. More progressive types of fuel consumption, such as district heating, account for only a small fraction of consumption across all asset categories.

Understanding the Relationship Between EPC and Fossil Fuel Consumption

Figures 18.8 to 18.11 represent a comparison of EPC rating and total consumption rating for each asset category. In each figure, there is no correlation between a given EPC asset rating and total fossil fuel consumption. Most of the energy consumption falls between 0 and 100 tonnes of fossil fuel per year. However, this is spread relatively evenly between EPC ratings, rather than grouped in relation to relative EPC rating, while those properties consuming the most energy between 100 and 950 tonnes of fossil fuel per year are also grouped, broadly, evenly between EPC rating, rather than in relation to relative EPC rating.

An EPC asset rating therefore provides little indication as to the total amount of fossil fuels consumed by a building or hereditament. The EPC assessment methodology is intended to calculate the energy efficiency rating of a building in terms of its running cost, rather than consumption of energy. The higher the rating, the more energy efficient the building is and the lower the fuel bills are likely to be. While there is consideration for the fuel type of each service, this is in the context of fuel bills from an economic perspective rather than fuel consumption from an environmental impact standpoint. Therefore, larger buildings that consume far greater amounts of fossil fuels may achieve better EPC asset ratings than those buildings that consume little to no fossil fuels.

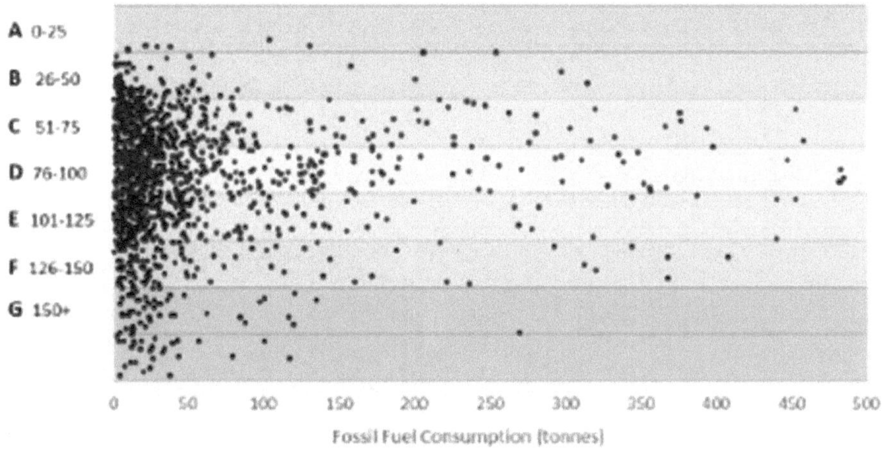

Figure 18.8 Cross tabulation of EPC and fossil fuel in the office sector.

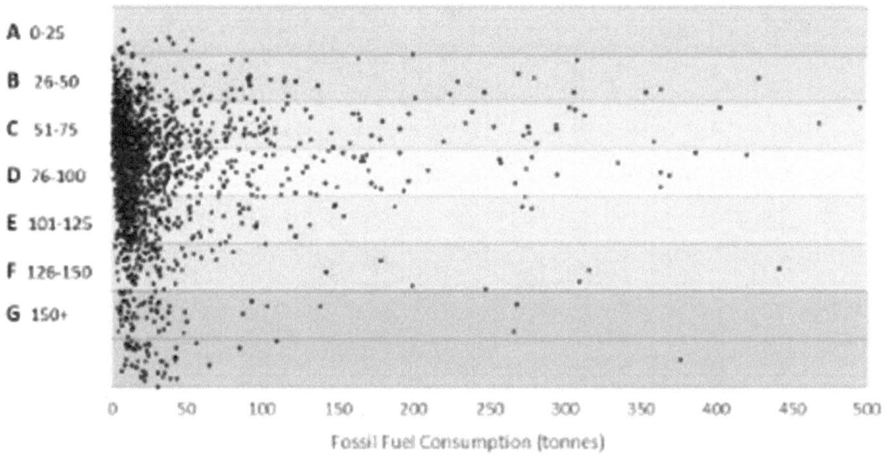

Figure 18.9 Cross tabulation of EPC and fossil fuel in the retail sector.

Conclusion

In response to the underlying hypothesis in this chapter – that hypothetical energy performance would be positively correlated with fossil fuel use – the original findings suggest that there is little positive correlation between those commercial properties with the best energy performance ratings and those properties consuming most fossil fuels. Instead, it can be the case that

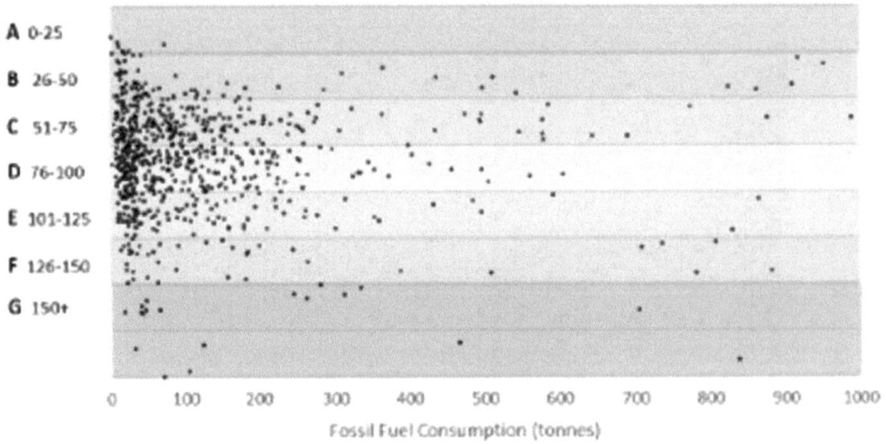

Figure 18.10 Cross tabulation of EPC and fossil fuel in the leisure sector.

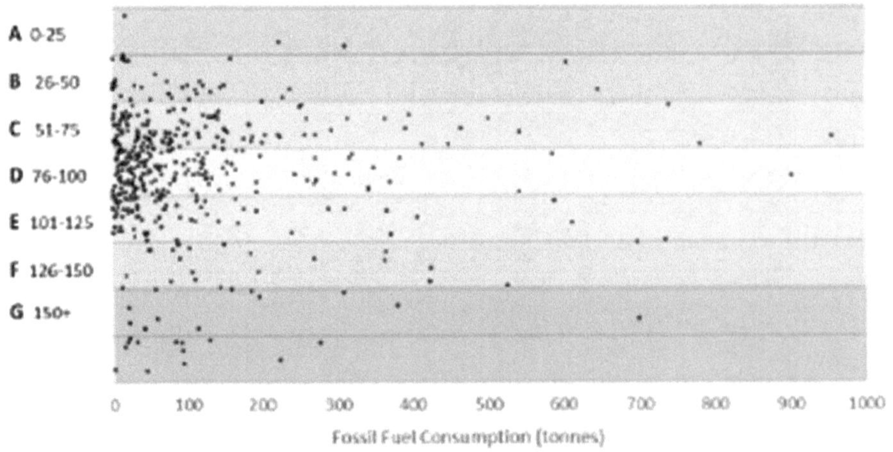

Figure 18.11 Cross tabulation of EPC and fossil fuel in the industrial.

those properties with the very best energy performance assessments also consume the most fossil fuel. This is counterintuitive, and both complicates and confuses strategies to divest from fossil fuels and improve environmental performance.

This research developed a model that converts the emissions rate of a building (produced by the EPC assessment method) to fuel consumption rate, which enables us to hypothesise and estimate the amount and type of fuel each building consumes on an annual basis. To our knowledge, no other labelling model exists that estimates commercial fossil fuel consumption based on EPC data and allied data sets in the public domain. By extrapolation, we can estimate: a) the sectors that consume the most fossil fuels, b) the building services that are the biggest contributors to fossil fuel consumption, and c) the fossil fuels that each sector and service is most reliant upon. This has the potential to inform strategies and policies to tackle the reduction of fossil fuel reliance in the commercial real estate sector and chart a trajectory towards a holistic environmental policy that focuses on the three Cs of cost, carbon emission, and consumption. There is obvious potential to complement this with DECs in all commercial buildings (measuring actual performance and consumption) and with live data from increasingly smart monitoring technology.

However, if the government wants to understand a building's impact on the environment (in terms of the amount of fossil fuels consumed and carbon emitted) and make an immediate improvement, they would be better off measuring the building than producing an EPC asset rating per m^2. This is because an EPC asset rating provides little indication of the total amount of fossil fuels consumed by a building. Large buildings with good EPCs consume a lot more fossil fuels (due to their size) than smaller, inefficient buildings with low EPC ratings. If we want less fossil fuel dependence, converting these larger buildings to renewables may be a good strategy.

In addition, the new findings in this chapter help to understand the fossil fuel–related stranded asset situation in more detail. It is hoped that revealing the fossil fuel consumption situation underneath energy labelling will play a positive role in influencing and provoking better targeted built environment retrofit measures – connecting the 'what is needed with the how it can be implemented' at the global level (Dixon et al., 2018). This is because, certainly in Europe, energy labelling is the key adopted socio-technical signifier in the built environment (and indeed appliances within the home and working environment) for sustainability, environmental improvement, and fossil fuel exposure (rather than a Display Energy Certificate). Institutions of financial lending and valuation utilise the EPC as one of the key barometers of fossil fuel use.

While the research in this chapter only regards England, financialisation means that the ownership of these properties and indeed their financing is connected to practices, techniques, and capital flows in the international property market and financial system. Mispricing and taking for granted assumptions in relation to fossil fuel consumption has the potential to undermine underlying valuations and lead to rapid devaluation, while previous research (Muldoon-Smith and Greenhalgh, 2019) revealed the presence of this fossil fuel risk in the international property market. This research argues that this risk is currently misunderstood and largely hidden from the international energy labelling conventions – and particularly those that follow the underlying models adopted in England and wider Europe. If this risk is suddenly apparent, there is potential for revaluation to be sudden, which could destabilise international property markets and related financial institutions, particularly if they have similar property attributes in terms of design, construction, heating, and ventilation.

Notes

1 ECUK stands for Energy Consumption in the UK 2022 – National Statistics, which is a data set released by the Department for Energy Security and Net Zero and the Department for Business, Energy & Industrial Strategy.

2 In this instance, the EPC dataset classes 'other' as a heating fuel source, which is not encompassed by the rest. This should not be confused with the authors' 'other' designation for building use in Figure 18.3.

References

BPIE (2014) *Energy Performance Certificates Across the EU: A Mapping of National Approaches*, BPIE, Brussels.

Caldecott, B., Harnett, E., Cojoianu, T., Kok, I. & Pfeiffer, A. (2016) *Stranded Assets: A Climate Risk Challenge*, Inter-American Development Banks, Washington, DC.

Carney, M. (2015) *Breaking the Tragedy of the Horizon – Climate Change and Financial Stability*, Speech given at Lloyd's of London by the Governor of Bank of England.

Dixon, T, Lannon, S. & Eames, M. (2018) Reflections on disruptive energy innovation in urban retrofitting: methodology, practice and policy. *Energy Research and Social Science*, 37, 255–259.

Elmhurst Energy (2022) *The Elmhurst Almanac: Turning the Dail*. Elmhurst Energy, Lutterworth, UK.

Energy Audit Committee (2023) Accelerating the Transition from Fossil Fuels and Securing Energy Supplies. Fourth Report of Session 2022–23.

European Union (2010) Energy Performance of Buildings Directive. 2010/31/EC, the 'EPBD. European Union. Brussels.

Friedman, T.L. (2014) Obama on Obama on Climate, *The New York Times*, available at: https://www.nytimes.com/2014/06/08/opinion/sunday/friedman-obama-on-obama-on-climate.html. Accessed 7 June 2014.

Gurría, A. (2013) The Climate Challenge: Achieving Zero Emissions, Available at: Lecture by OECD Secretary-General, www.oecd.org/about/secretary-general/theclimate-challenge-.

Hosgood, G. (2017) *Understanding the New Minimum Energy Efficiency Standards*, Allsop, London, UK.

McGrath, P. (2014) *Ban Ki-Moon Urges Pension Funds to Dump Fossil Fuel Investments*, ABC, New York.

Muldoon-Smith, K. & Greenhalgh, P. (2019) Suspect foundations: Developing an understanding of climate-related stranded assets in the global real estate sector. *Energy Research and Social Science*, 54, 60–67.

Pitman, K. (2022) *Climate-Related Stranded Assets in Real Estate Part 1: Driving Change and De-Risking Real Estate*. World Built Environment Forum. London, UK.

Reed, R., Bilos, A., Wilkinson, W. & Schulte, K.W. (2009) International comparison of sustainable rating tools. *Journal of Sustainable Real Estate*, 1(1), 1–22.

Sayce, S., Sundberg, A. &. Clements, W. (2010) *Is Sustainability Reflected in Commercial Property Prices: A Review of the Evidence Base, RICS Research*, RICS, London.

The Non-Domestic Minimum Building Energy Performance Standards Working Group (2014) Report to UK Government.

Tomusk, K. (2019) MEES One Year On: No Enforcement Proceedings in First 12 Months. *Estates Gazette*, accessed 4 May 2023.

World Economic Forum (2016) *Environmental Sustainability Principles for the Real Estate Industry: Industry Agenda*, World Economic Forum Industry Agenda Council on the Future of Real Estate and Urbanisation, January.

19 Artificial Intelligence as the Way Forward for Scan-to-BIM Practice in Cultural Heritage

Aleksander Gil, Yusuf Arayici, and Bimal Kumar

Introduction

The success of building information modelling (BIM) across the built environment inspires researchers and industry professionals to explore similar working methods applied in the cultural heritage sector. Initially, heritage BIM (HBIM) was interpreted as a 3D representation of a heritage asset generated by BIM software, but over a decade since this term was originally coined, the definition, obtainable benefits, and general outlook on HBIM evolved. Thus, the state-of-the-art HBIM publications are starting to shape the vision for heritage digital twin, and an efficient process of transforming point cloud data into semantic HBIM geometry and how the HBIM model can be used during its lifecycle.

The automation of segmentation and classification of heritage building elements is critical in research and practice as it allows cultural heritage (CH) stakeholders to understand, operate, and conserve their buildings more effectively. Thus, artificial intelligence workflows capable of classification such as machine learning and deep learning gained importance and popularity in research due to the increasing demand for accurate HBIM modelling generated from point cloud data. Furthermore, point cloud data is a common file type also used in autonomous driving, computer-aided manufacturing, and general surveying sectors, and the automation of its classification is an unsolved common goal.

The chapter is structured in six sections. The introduction is followed by a theoretical background, which discusses the process of Scan-to-BIM in cultural heritage. Then, research aim and methodology are presented; this is then followed by the state-of-the-art Scan-to-BIM case studies. Next, the machine learning and deep learning methods are discussed in the context of cultural heritage point cloud classification. Lastly, the chapter concludes with the findings and insights for the future works.

Scan-to-BIM in Cultural Heritage Documentation and Challenges

The process of documenting a heritage building follows a Scan-to-BIM workflow streamlined by advanced surveying technologies. Thus, the data capture sensor

DOI: 10.1201/9781003380559-19

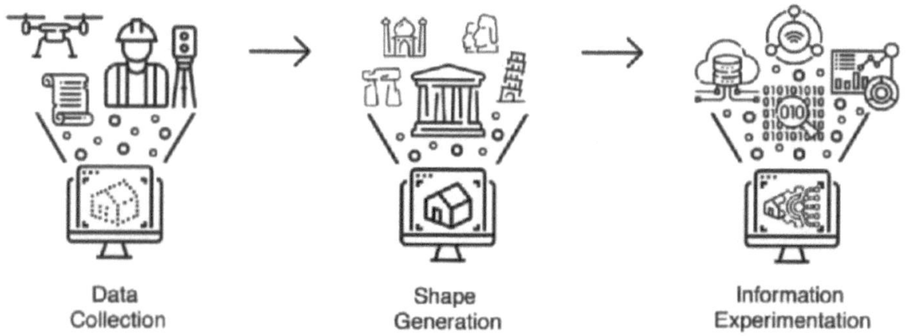

Figure 19.1 Scan-to-BIM process workflow.

may vary depending on site suitability. The overall process follows three steps, as illustrated in Figure 19.1.

- Data collection – a process where building data is acquired by terrestrial laser scanners, drone-mounted LiDAR, RGB-D sensors, or photogrammetry and processed into a holistic point cloud representation of the built asset
- Shape (HBIM) generation – a process where the point cloud file is used as reference for generating a HBIM model using system tools and libraries of components
- Information experimentation – a process where valuable information is mapped onto 3D geometry, 2D outputs are generated, analysis is undertaken, and outcomes are visualised in a tailored way to suit the project outcomes

This process allows researchers and practitioners to utilise HBIM as a project management tool, improve stakeholder engagement, make better conservation and refurbishment decisions, undertake various advanced analysis studies, and conduct 4D historical analysis. Nevertheless, this process is hindered by regulatory, process-based, and technical challenges.

The lack of regulatory compliance requirements or standards for HBIM or Scan-to-BIM processes limits the adoption. Although a government BIM mandate affects conservation projects over a certain cost threshold, there is no single entity or organisation to define the process of building conservation in the United Kingdom, which leads to many misinterpretations.

The annual BIM report by NBS states that there are approximately six million heritage assets erected before 1919 and an application of BIM for heritage should be considered. Yet, only best practice case studies and working guides have been released by organisations such as Historic England, and there are no signs of large-scale implementation of HBIM.

Following the lack of clear standards and the communication of clients, HBIM need and definition of level of information need also proves to be challenging. Our research compared five surveying businesses offering Scan-to-BIM and found that the "LOD levels" vary significantly from each other. In a study by Banfi (2017), a novel approach to geometry and modelling detail was developed to accommodate for HBIM project requirements. The proposed Grades of Geometry 9 and 10 (GOGs) can be interpreted as the current state-of-the-art but their application across global HBIM case studies is nearly non-existent. As a result, HBIM project implementation requires significant research or consultancy.

One of the most common criticisms of Scan-to-BIM in the cultural heritage context found in our review is the complex and laborious process of HBIM modelling using BIM software. This challenge has many factors that contribute to the inefficiency of this process, such as:

- Damaged, decayed, or weathered building elements
- The geometrical complexity of handcrafted building elements
- Interdisciplinary systems found following adaptive building reuse
- BIM software limitations in shape generation
- BIM software limitation in large-point cloud file processing
- Lack of heritage building ontology in BIM software for semantic data mapping

It is underlined in the literature that the need for automation is crucial as digital upskilling or further interdisciplinary recruitment for project delivery does not seem appropriate for a reasonably low-funded sector.

Research Aim and Methodology

The semi-systematic literature review enabled a foundation to be provided for UK-specific case study experimentation via the Scopus database, providing a collection of 1,541 articles. Results were filtered for duplicates and the possibility for download, and the final collection only included studies published after 2016. The research process workflow is shown in Figure 19.2.

The following part of this research project will also involve design science research whereby the knowledge obtained from experimentation will continue to refine our hypothesis for an automated method of HBIM generation from point cloud data. Furthermore, the currently open-source method and our prototype will be tested against a UK-specific case study dataset and evaluated based on the results.

This methodology depicted in Figure 19.2 allowed us to achieve the aim of this study, which is to develop an automation method for HBIM development from point cloud data via the use of artificial intelligence.

State-of-the Art in Scan-to-BIM

Centennial Hall in Wroclaw, Poland, is a great example of semantic HBIM, as the point cloud data was manually replicated using BIM software to a high level of geometrical and structural detail. Hidden structural elements were reverse-engineered and simulations were undertaken. Furthermore, this UNESCO site has

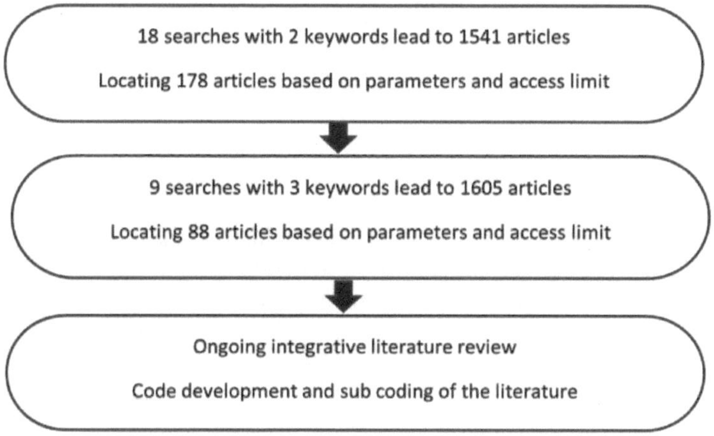

Figure 19.2 Research process workflow.

HBIM preservation and construction material information and catalogues all elements into a specific library. This is a great example of an early HBIM adoption for building management and future planning (Janisio-Pawłowska, 2021).

The Zee Ain Village is a heritage site in Saudi Arabia, which has 49 dwellings that date back over 400 years. The study focuses on developing an HBIM model from photogrammetry and terrestrial laser scanning data at a large scale. The novelty of this study lies in the aspect of projecting textures from the captured data directly onto the HBIM model. This can be considered as one of the states of the art as textures are commonly missing in HBIM and this workflow proves to be efficient and very informative (Alshawabkeh et al., 2021).

An excellent example of how HBIM can benefit the restoration of a heritage building is shown in a study by Biagini et al. (2016), where earthquake damage on Nome di Maria church is repaired using HBIM methodology. The authors export 4D and 5D solutions, which lead to much clearer documentation and improved project management.

Baik (2017) proposes a more efficient workflow of delivering HBIM models by producing Jeddah HBIM component library, a set of parametric components, that could be reused on various other similar sites. Thus, this study demonstrated improvements in the overall efficiency of assembling a HBIM model; however, the initial development efforts and likelihood of similar buildings finding the right assets in the library are critiqued.

Literature also shows some state-of-the-art visualisation of HBIM models such as the Isabel II Dam in Spain, surveyed using advanced UAV photogrammetry methods and cross-referenced with legacy data. The large-scale dam model was taken into a game engine and artistically textured, providing a photo-realistic appearance (Martínez-Carricondo et al., 2021).

When looking at the value of heritage information, Parisi et al. (2019) showcase their work on the Escuela Tecnica Superior de Arquitectura in Grenada. This evolutionary work pioneers semantic information linking and building ontology implementation while working with a BIM platform. The origins of this study showcase how HBIM can contain all previous construction works that occurred to an asset even dating back five centuries.

One of the most impressive practical approaches to defining LOD and LOA for heritage was published by Chow et al. (2019) using the West block, central block, and Library of Parliament as a case study. It is considered that this is one of the most practical articles published on HBIM implementation and the most reasonable to justify the accuracy and the detail needed for heritage regeneration and record purposes.

Artificial Intelligence as the Way Forward

Machine Learning Techniques

The application of random forest (RF) to a large complex dataset of the Milan Cathedral and Pomposa Abbey in a study by Teruggi et al. (2020) exposed an accuracy drawback, which led to a proposal of a novel multi-level, multi-perception method (MLMP). The authors decided to develop a bespoke ontology for building elements that correlated to the three different subsamples of the input dataset, as shown in Figure 19.3.

At each dataset scale, they applied RF, nearest neighbour algorithm, and octree to isolate each building ontology class. Although this supervised method achieved 94–97% accuracy for each subsample following a five-minute training procedure, the study concluded that the RF classifier struggled to correctly segment building features that combine similar segmentation classes.

Zhang et al. (2022) replicated this method on two buildings from the Wutai Monti UNESCO Heritage Site in China. They achieved an F1 result of 93–96% depending on the subject case study building and reflected on the small building features that

Figure 19.3 Three classification steps with random forest.
Source: Teruggi et al., 2020.

were missed in the classification results. The final recommendations from their study advised that extra attention should be devoted to the dataset quality and colour, but generally, they believe this method is suitable for heritage building application.

Grilli et al. (2019) compared four different ML and DL classifiers on cultural heritage datasets including RF, one vs. one classifier, 1D and 2D CNN, and Bi-LSTM RNN. This study provided a valuable insight into how different combinations of point features such as coordinates, radiometric values, and geometric features affect the output accuracy and concluded that ML methods outperformed DL. Using the same RF method, Croce et al. (2021) conducted further tests with various combinations of point features as the first part of their study. Their results proved that the best combination of features were nine geometric features (linearity, planarity, omnivariance etc.) and z coordinate, which achieved an F1 score of 98.81%. The second part of their study focused on HBIM generation automation, which was approached on a general and trivial basis depending on the building element. General elements were replicated algorithmically using parametric adaptive components and trivial elements leveraged a BIM library of components and primitive fitting using RANSAC.

In a study by Grilli et al. (2019), RF and K-means clustering were tested to develop a heritage point cloud framework proven on four case studies varying in size, classification purpose, and overall complexity. This framework showcases the use of RF on orthoimage to identify material types, the use of RF on point cloud to identify façade components, and k-means clustering to analyse eroded and deteriorated surfaces. Although the results vary significantly across the different case studies, this framework paves the way for the future studies and proves that 2D image classification and result projection onto 3D datasets is an efficient method, especially if the image quality and uniform regions are optimised during the processing stages.

Deep Learning Techniques

The application of CNNs using 2.5 aerial datasets and additional features such as heat map data can be also found in a study by Pan et al. (2018), whose novel CNN architecture allows for 2D classification and fusion onto the point cloud. Addressing the limitation of 2.5D input data, Boulch et al. (2018) proposed a method called SplatNet, which has a 2D–3D and a direct 3D variation that leverages CNN models and feature fusion for classification. Using a heritage façade as the input dataset, the 2D–3D variation that applies two layers of CNN proved to be most effective with an instance average IoU of 85.4%. Even more impressive results can be found in a study by Xu et al. (2018), where a four-layer CNN process applied directly to 3D point cloud input data achieved 92.4% accuracy. In the context of 2.5D CNN methods applied on CH datasets, Pellis et al. (2022b) proposed a pioneering method that achieves depth-based image classification from the ArCH benchmark dataset and applies a feature fusion process to classify the complementary point clouds. The results of this method achieved a global accuracy of 87–90% based on three buildings tested.

The application of a multi-view classification method in the CH context can be found in a study by Bacharidis et al. (2018), where a cost-efficient processing method is applied to georeferenced, stereoscopic images to reconstruct heritage building façades. This method extracts façade features and plots them into a point cloud format that allows for meshing and texture projection. The texture and colour-driven feature bias found in this method was also addressed shortly after, whereby improving the feature detection was undertaken via depth maps. This helped to significantly reduce the deviation of final mesh accuracy (Bacharidis et al., 2020).

In contrast to the meshing workflow, Pellis et al. (2022a) addressed point cloud classification using a multi-view method on the ArCH benchmark dataset. As the benchmark contains 2D RGB-D images and point clouds, the authors compared results using the available 2D data and 2D shots generated directly from the 3D datasets. The 93.4% accuracy results were initially obtained from the DeepLabv3+ CNN on the full image classification, which gave a promising start to the study, but the following test conducted on other buildings lowered the quantity of data used for training and prediction, which proved that the accuracy decreased significantly due to lower image definition. Furthermore, the back-projection of prediction image classification introduced significant data loss and mapping inaccuracies, which led to the final point cloud only achieving an accuracy of 33.4% and 57.3%.

Qi et al. (2017) developed PointNet++ as an improvement to their original pioneering framework, which introduces a sampling layer and a grouping layer before classifying the dataset using PointNet. The idea of the additional layers is to construct local regions using several points as centroids that are paired with a local region growing module. All regions have a certain overlap, which then allows for feature detection at various scales. Malinverni et al. (2019) applied PointNet++ to four different CH case study datasets, which initially had to be translated, scaled, and subsampled as the typical point cloud dataset used for Scan-to-HBIM is too heavy. In this study, only four classes (arc, column, wall, and window) were able to classify due to the limited amount of object commonality in the training data and their results proved that this network was only able to achieve an average F1 score of 30.8%.

Application of edge convolutions were explored beyond the foundation laid by Y. Wang et al. (2019). This led to DGCNN-mod (Dynamic Graph CNN), which considers additional features such as point normal and HSV values (Pierdicca et al., 2020). 3DLEBNet, a novel method proposed by Cao & Scaioni (2021), combines a DGCNN encoder that computes a "codeword" feature with a folding decoder inspired by FoldingNet. Through a process of linking the codewords with the edge convolution outputs, a semantic segmentation was achieved with an accuracy of 77% using the ArCH dataset that can be interpreted as one of the DL state-of-the-art methods applied in the CH context.

One of the most impressive aspects of this method is that it only requires 10% of the training data compared to other methods such as PCNN, PointNet, or DGCNN. A great extent of comparison of their method with others was also carried out in their research. 3DLEBNets ablation study explores the effectiveness of each part of the network architecture and highlights that generally an increased number of features improves the classification results. However, the increase in features should also be paired with an increase in the training data used.

Matrone et al. (2020) undertook a study that compares classification approaches using both ML and DL methods in the CH context. This study applies various ML and DL methods to the ArCH dataset and proposes DGCNN-mod-3Dfeat, which considers additional handcrafted features extracted using machine learning. Overall, the proposed method achieves 82.2% F1 score, which is only beaten by the ML RF classifier. This proves that including ML computed covariance features improves the performance of DGCNN methods.

Conclusion

This chapter presents a thorough review of existing point cloud segmentation methods applicable to cultural heritage to automate the generation of heritage building information modelling. Firstly, the quantitative insight found across the available literature is documented and the fundamental principles of machine learning methods are defined. ML and HBIM studies are discussed in terms of their context and classification success. Multiple DL types and techniques applied to point cloud files are described. Their approach, results, and potential for CH application are reviewed.

The classification of CH point clouds is an important topic growing in popularity, and thanks to the development of many improvements to ML and DL methods, some early success trends are emerging across the literature. Nonetheless, the current state-of-the-arts still face significant challenges and the diversity of CH built assets still has not been tested on an international level. With this chapter, we present cutting-edge research published in the literature. This will form the baseline to the development of the ML and DL method for automated HBIM modelling from point cloud data for the UK-specific case study dataset, as a novel method to amplify HBIM use in cultural heritage.

References

Alshawabkeh, Y., Baik, A., & Fallatah, A. (2021). As-textured as-built BIM using sensor fusion, Zee Ain historical village as a case study. *Remote Sensing*, *13*(24). https://doi.org/10.3390/rs13245135

Bacharidis, K., Sarri, F., Paravolidakis, V., Ragia, L., & Zervakis, M. (2018). Fusing georeferenced and stereoscopic image data for 3D building facade reconstruction. *ISPRS International Journal of Geo-Information*, *7*(4). https://doi.org/10.3390/ijgi7040151

Bacharidis, K., Sarri, F., & Ragia, L. (2020). 3D building façade reconstruction using deep learning. *ISPRS International Journal of Geo-Information*, *9*(5). https://doi.org/10.3390/ijgi9050322

Baik, A. (2017). From point cloud to Jeddah Heritage BIM Nasif Historical House – case study. *Digital Applications in Archaeology and Cultural Heritage*, *4*, 1–18. https://doi.org/10.1016/j.daach.2017.02.001

Banfi, F. (2017). BIM orientation: Grades of generation and information for different type of analysis and management processes. *International Archives of the Photogrammetry,*

Remote Sensing and Spatial Information Sciences – ISPRS Archives, *42*(2W5), 57–64. https://doi.org/10.5194/isprs-archives-XLII-2-W5-57-2017

Biagini, C., Capone, P., Donato, V., & Facchini, N. (2016). Towards the BIM implementation for historical building restoration sites. *Automation in Construction*, *71*, 74–86. https://doi.org/10.1016/j.autcon.2016.03.003

Boulch, A., Guerry, J., le Saux, B., & Audebert, N. (2018). SnapNet: 3D point cloud semantic labelling with 2D deep segmentation networks, *Computers & Graphics*, 71, 189–198. https://doi.org/10.1016/j.cag.2017.11.010

Cao, Y., & Scaioni, M. (2021). 3DLEB-net: Label-efficient deep learning-based semantic segmentation of building point clouds at LoD3 level. *Applied Sciences (Switzerland)*, *11*(19). https://doi.org/10.3390/app11198996

Chow, L., Graham, K., Grunt, T., Gallant, M., Rafeiro, J., & Fai, S. (2019). The evolution of modelling practices on Canada's parliament hill: An analysis of three significant heritage building information models (HBIM). *ISPRS Annals of the Photogrammetry, Remote Sensing and Spatial Information Sciences*, *42*(2/W11), 419–426. https://doi.org/10.5194/isprs-Archives-XLII-2-W11-419-2019

Croce, V., Caroti, G., de Luca, L., Jacquot, K., Piemonte, A., & Véron, P. (2021). *From the Semantic Point Cloud to Heritage-Building Information Modeling: A Semiautomatic Approach Exploiting Machine Learning*. https://doi.org/10.3390/rs

Grilli, E., Özdemir, E., & Remondino, F. (2019). Application of machine and deep learning strategies for the classification of heritage point clouds. In H. Arefi & S. S. M (Eds.), *ISPRS International GeoSpatial Conference 2019, Joint Conferences of 5th Sensors and Models in Photogrammetry and Remote Sensing, SMPR 2019 and 3rd Geospatial Information Research, GI Research 2019* (Vol. 42, Issue 4/W18, pp. 447–454). International Society for Photogrammetry and Remote Sensing. https://doi.org/10.5194/isprs-archives-XLII-4-W18-447-2019

Grilli, E., & Remondino, F. (2019). Classification of 3D digital heritage. *Remote Sensing*, *11*(7). https://doi.org/10.3390/RS11070847

Janisio-Pawłowska, D. (2021). Analysis of the possibilities of using HBIM technology in the protection of cultural heritage, based on a review of the latest research carried out in Poland. *ISPRS International Journal of Geo-Information*, *10*(10). https://doi.org/10.3390/ijgi10100633

Malinverni, E. S., Pierdicca, R., Paolanti, M., Martini, M., Morbidoni, C., Matrone, F., & Lingua, A. (2019). Deep learning for semantic segmentation of 3D point cloud. *International Archives of the Photogrammetry, Remote Sensing and Spatial Information Sciences – ISPRS Archives*, *42*(2/W15), 735–742. https://doi.org/10.5194/isprs-archives-XLII-2-W15-735-2019

Martínez-Carricondo, P., Carvajal-Ramírez, F., Yero-Paneque, L., & Agüera-Vega, F. (2021). Combination of HBIM and UAV photogrammetry for modelling and documentation of forgotten heritage. Case study: Isabel II dam in Níjar (Almería, Spain). *Heritage Science*, *9*(1). https://doi.org/10.1186/s40494-021-00571-8

Matrone, F., Grilli, E., Martini, M., Paolanti, M., Pierdicca, R., & Remondino, F. (2020). Comparing machine and deep learning methods for large 3D heritage semantic segmentation. *ISPRS International Journal of Geo-Information*, *9*(9). https://doi.org/10.3390/ijgi9090535

Pan, X., Gao, L., Marinoni, A., Zhang, B., Yang, F., & Gamba, P. (2018). Semantic labeling of high resolution aerial imagery and LiDAR data with fine segmentation network. *Remote Sensing*, *10*(5). https://doi.org/10.3390/rs10050743

Parisi, P., lo Turco, M., & Giovannini, E. C. (2019). The value of knowledge through H-BIM models: Historic documentation with a semantic approach. *ISPRS Annals of the Photogrammetry, Remote Sensing and Spatial Information Sciences*, *42*(2/W9), 581–588. https://doi.org/10.5194/isprs-archives-XLII-2-W9-581-2019

Pellis, E., Murtiyoso, A., Masiero, A., Tucci, G., Betti, M., & Grussenmeyer, P. (2022a). An image-based deep learning workflow for 3D heritage point cloud semantic segmentation. *International Archives of the Photogrammetry, Remote Sensing and Spatial Information Sciences – ISPRS Archives*, *46*(2/W1-2022), 429–434. https://doi.org/10.5194/isprs-archives-XLVI-2-W1-2022-429-2022

Pellis, E., Murtiyoso, A., Masiero, A., Tucci, G., Betti, M., & Grussenmeyer, P. (2022b). 2D to 3D label propagation for the semantic segmentation of heritage building point clouds. *International Archives of the Photogrammetry, Remote Sensing and Spatial Information Sciences – ISPRS Archives*, *43*(B2-2022), 861–867. https://doi.org/10.5194/isprs-archives-XLIII-B2-2022-861-2022

Pierdicca, R., Paolanti, M., Matrone, F., Martini, M., Morbidoni, C., Malinverni, E. S., Frontoni, E., & Lingua, A. M. (2020). Point cloud semantic segmentation using a deep learning framework for cultural heritage. *Remote Sensing*, *12*(6). https://doi.org/10.3390/rs12061005

Qi, C. R., Yi, L., Su, H., & Guibas, L. J. (2017). *PointNet++: Deep Hierarchical Feature Learning on Point Sets in a Metric Space*. http://arxiv.org/abs/1706.02413

Teruggi, S., Grilli, E., Russo, M., Fassi, F., & Remondino, F. (2020). A hierarchical machine learning approach for multi-level and multi-resolution 3d point cloud classification. *Remote Sensing*, *12*(16). https://doi.org/10.3390/RS12162598

Wang, Y., Sun, Y., Liu, Z., Sarma, S. E., Bronstein, M. M., & Solomon, J. M. (2019). Dynamic graph CNN for learning on point clouds. *ACM Transactions on Graphics*, *38*(5). https://doi.org/10.1145/3326362

Xu, Y., Fan, T., Xu, M., Zeng, L., & Qiao, Y. (2018). *SpiderCNN: Deep Learning on Point Sets with Parameterized Convolutional Filters*. http://arxiv.org/abs/1803.11527

Zhang, K., Teruggi, S., & Fassi, F. (2022). Machine learning methods for UNESCO Chinese heritage: Complexity and comparisons. In L. Fregonese, F. Fassi, & F. Remondino (Eds.), *9th International Workshop on 3D Virtual Reconstruction and Visualization of Complex Architectures, 3D-ARCH 2022* (Vol. 46, Issues 2/W1-2022, pp. 543–550). International Society for Photogrammetry and Remote Sensing. https://doi.org/10.5194/isprs-archives-XLVI-2-W1-2022-543-2022

20 Summary and Conclusion

Bimal Kumar

In the 18 chapters presented in this seminal volume, the book explores diverse topics, each having implications for sustainability through the digital transformation of the built environment. While these chapters do not directly focused on sustainability through digital transformation in the built environment, they do allude to such implications. Each chapter offers insights that contribute to the broader understanding of how digital transformation intersects with the built environment. The indirect connections between these chapters and sustainability underscore the multifaceted nature of the relationship between technology and the built environment, providing a nuanced perspective on the potential impact of digital advancements on sustainability initiatives.

Approximately half of the 18 chapters in the book, specifically those authored by Bahriye, Valentah, Gokhan, Bertolino, Pavithra, Oluwadamilare Olamide Ilesanmi, Muldoon-Smith, and Pearson, focus on the direct intersection of digital solutions with sustainability in the built environment. These contributions delve into comprehensive details about their respective works, emphasising the impact of digital applications on the development and promotion of sustainable communities and the built environment. This subset of chapters provides a focused exploration of the tangible ways in which digital advancements contribute to the advancement of sustainability goals within the context of urban development and community planning.

Bahriye (Circular Economy for Sustainable Built Environment through Building Information Modelling) focuses on the role of BIM in circularity and proposes a holistic circular building life cycle approach that includes the main steps and principles for design, construction, use, and end-of-life phases along with the indicators. She suggests that offering an integrated database to exchange information among different parties through a common model can easily incorporate circular economy principles into the practices of different phases of the building life cycle. In addition, if circularity data is embedded into the BIM software, then measuring the performance (achieving the indicators) as well as strategic decisions to foster circular business models can be easily achieved. To conclude, this chapter summarises key BIM inputs for each phase for achieving the main circularity indicators, underlining the importance of IFC to extract the data for the analysis.

DOI: 10.1201/9781003380559-20

Manase and Manase (Digital Transformation as a Business Differentiation Strategy for Sustainable Built Environment) contend that while digital transformation may open new opportunities and benefits, they also give rise to new challenges like higher costs for infrastructure and training needs and change management.

Demirdöğen et al. (Sustainable and Lean Healthcare Facility Management) bring out the differences between healthcare and other facilities largely because of the nature of 24/7 usage and the main users (the patients). The chapter is focused on facilities management and the impact of facility environmental conditions on patients. They propose 50 FM success indicators grouped under environmental, user-oriented, emergency and safety management, maintenance and renovation, physical, and life cycle related factor groups. In addition to these success indicators, the types of knowledge required for the FM were identified under managerial, financial, and technical factor groups. In total, 64 knowledge types were defined. Subsequently, a framework to integrate building energy performance simulation, YBM and BDA, while considering both the types of FM information types and FM KPIs, was developed and evaluated from the perspective of the LM philosophy. In light of the findings, a prototype was developed to analyse the most important FM information types and KPIs. According to the findings, 100% of the respondents confirmed that the proposed system would be efficient in solving the FM information queries and KPI analysis, while 97% stated that it is practical in carrying out FM activities. In terms of usage, 60% of the participants stated that the proposed system would be practical to use during the FM process.

Bertolino (Digital Technologies and Alternative Social Imaginaries: Producing Architectural Counternarratives in Eco-communities) concludes that based on data collected in the three eco-communities discussed in the chapter, the potential for introducing digital networks into eco-communities lies in their ability to amplify collective efforts, foster collaboration, and leverage technology for positive environmental and social impact, sharing practices of what was formerly defined as "radical ecology." Moreover, as an interpretive analytical category, radical ecology unfurls as a powerful lens for architectural researchers to critically engage with and unravel established frameworks in the spatial discourse.

Pavithra et al. (Promoting Built Asset Flood Resilience for Sustainable Development Through Data Advantage) directly addresses issues for improving the sustainability of communities and how data-driven approaches are rapidly growing in the context of flood resilience. They claim that despite the wide spectrum of data and its potential to improve flood resilience, current studies often focus on limited data usage for built asset resilience. They present a synthesis of existing literature on data-driven built asset flood resilience to discover the prospects of data utilisation, challenges, and available strategies to enhance the deployment of data potentials for built asset flood resilience towards sustainability. They claim that their findings are significant for decision-makers to create data-inspired plans for resilient built assets to enhance sustainable outcomes in communities.

Oluwadamilare et al. (Post-Occupancy Evaluation for Sustainability Performance of Green Buildings) propose that the characteristics that contribute to the sustainability of a building encompass various aspects such as the efficient utilisation of energy, water, and other resources, the incorporation of renewable energy sources like solar energy, the implementation of measures to reduce pollution and waste, and the facilitation of reuse and recycling. The chapter assessed the performance of a green building Heritage Place located in Ikoyi, Lagos, to determine the level of building performance, the level of user satisfaction, and the relationship that exists between the level of user satisfaction and level of building performance. The users express a level of dissatisfaction with the materials and resources parameter. The occupants of the building express high levels of satisfaction with the indoor environmental quality, the building facilities, and the various factors contributing to water efficiency. Based on their research, they suggest that significant consideration should be given to the criteria about materials and resources in the context of green building, as it exhibited the lowest average score in the building they assessed. Furthermore, the implementation of waste management strategies, including the establishment of efficient waste collection systems and the promotion of recycling practices, should also be prioritised.

Muldoon-Smith et al. (Understanding Fossil Fuel Consumption in Commercial Buildings: An Experimental Analysis of Energy Performance Labelling in England) propose a hypothesis that hypothetical energy performance would be positively correlated with fossil fuel use. Interestingly, they conclude that there is little positive correlation between those commercial properties with the best energy performance ratings and those properties consuming the most fossil fuels. Instead, it can be the case that those properties with the very best energy performance assessments also consume the most fossil fuel. This is counterintuitive and both complicates and confuses strategies to divest from fossil fuels and improve environmental performance. They arrive at this conclusion by developing a model that converts the emissions rate of a building (produced by the EPC assessment method) to fuel consumption rate, which enables one to hypothecate and estimate the amount and type of fuel each building consumes on an annual basis. This finding has major implications for informing strategies and policies to tackle the reduction of fossil fuel reliance in the commercial real estate sector and chart a trajectory towards a holistic environmental policy that focuses on the three Cs of cost, carbon emissions, and consumption.

In another chapter by the same team, Muldoon-Smith et al. (Commercial Property Vacancy for Building Adaptation and Re-use) propose a conceptual framework for commercial vacancy that moves beyond the positive facets of vacancy, such as initial, frictional, and cyclical vacancy types and the general approximation of structural vacancy. The conceptual framework introduced in this chapter has been tested in six major commercial locations. They conclude that each location in the world contains a variety of comparable but highly specific real estate markets that are contingent and socially produced in each context. It is therefore likely that the operation of vacancy will be different in alternative market contexts. Moreover, grounding the conceptual framework in the six locations made use of vacancy duration as the primary means of separating vacancy. Duration of vacancy is only an approximation of the various types of vacancy introduced in this chapter.

The following chapters address sustainability issues directly, although all the research presented has implications for sustainable communities through digital transformation of the built environment.

Doan et al. (Building Information Modelling Adoption in the New Zealand Construction Industry: Insights from BIM Experts) present an in-depth analysis of implementing building information modelling in New Zealand's construction industry. They point out that BIM-driven project delivery results in fewer reworks, increased productivity, time-saving, and better collaboration among stakeholders. Several challenges have been listed and some pointers have been proposed to address them. Some of the key issues highlighted are high cost of implementation, resistance to change, lack of collaboration, awareness and training, weak leadership, and legal contractual issues. To address these challenges, the study recommends prioritising training for construction practitioners and aligning building design with the requirements of the New Zealand construction industry. Additionally, it is also recommended that it is essential to understand the legal contracts prevailing in New Zealand and determine minor adjustments that will not have legal ramifications for the construction company.

Olamide Ilesanmi et al. (Robotics / Automation Awareness in Foreign and Local Firms in Nigeria) provide insights into the difference between Nigerian and other firms operating there regarding awareness and applications of robotics and automation technologies. The study finds that the understanding of automation and robotics is about average, with drones being the most often used robotics technology among construction firms. The need to improve performance is the primary driver of automation and robotics, which are driven by the numerous issues the construction sector faces. Interestingly, it discovers that indigenous firms demonstrated a greater level of awareness of automation and robotics, contrary to the general expectation of foreign firms having an upper hand in robotics and automation awareness levels.

Adekunle (Human Augmentation Technology Adoption for the Construction Industry) reports on a study on emerging technologies focusing on human augmentation technology and its adoption in the construction industry to improve productivity. The study presents human augmentation as a viable solution to solve problems of lower productivity, loss of workforce, and the unattractive image of the construction industry due to its high accident rate. The study identified human augmentation in the following dimensions for the construction industry: cognitive and genetic enhancement, brain enhancement and prosthetics, ubiquitous computing, virtual reality and wearable computers, and robotics and exoskeleton. It proposes that prosthetics present the solution to workers losing their jobs in the event of life-threatening accidents during site operations. Instead of construction workers becoming unable to work in the industry due to physical disability, they are offered a second chance and reintegrated into the industry workforce through replacing missing or defective body parts. Despite the array of implementation challenges, the study focused on two critical issues, viz. ethics and cost, as these are deemed to be more important due to the human dimension of the proposed solutions.

Jin and Zhu (Rendering the Digital Future: Understanding the Social-Spatial Perception of a Dynamic Data Intensive Environment as an Architect) suggest that the proliferation of digital data has enabled the creation of an immense amount of information, and the development of sophisticated algorithms has made it possible to process and analyse this data to identify patterns and trends. As a result, digital data has become a powerful tool for providing customised and liberated services and policies. They claim that digitalisation can be used to detect even the most nuanced unintentional patterns, allowing for the provision of services that are tailored to an individual's specific needs and preferences. This has led to several important implications for people's personal and professional lives. As a result of this, the concept of adaptive environments has been gaining traction in the field of architecture in recent years and architects have an important role to play in shaping the experiences and interactions of occupants in a dynamic data-intensive environment. The chapter proposes that beyond the functional design of human–computer interfaces, interactivity can be used as a tool for storytelling and engagement, to create new social experiences and atmospheres that stimulate the senses and encourage users to interact with their surroundings in new ways, with important implications for creating sustainable communities.

Wiles et al. (Measuring Walkability Through Data-Led Analysis and Spatial Modelling) present their findings to corroborate the argument that well-structured and connected street networks promote AST behaviour, particularly in primary schools in England, though they may be extended to other contexts. These findings suggest that planning policy should have an emphasis on siting new schools in densely populated locations where distances to school are walkable, and where street layouts have been designed to create a dense, well-connected, and walkable network. They argue that the same principles should be applied when building new neighbourhood developments. Their study's findings are best understood in the context of the larger body of research on the syntactic analysis of urban environments, since they may have an impact on street classification systems, particularly as they inform the design of street sections.

Shashwat et al. (The Effect of Building Façade Materials in Mitigating UHI Effect: An Urban City Centre Study) present an investigation into the effect of building façade materials on the mitigation of the urban heat island (UHI) effect, which causes increased energy costs and summertime outdoor discomfort. A UHI of 1.3 °C is observed between the city centre and the weather station collection point (the average air temperature for a representative summer day). The evaluation of building façade material is conducted through the numerical simulation tool ENVI-met. Outdoor thermal comfort indices such as the urban thermal climate index, physiologically equivalent temperature, and standardised effective temperature were evaluated using the ENVI-met microclimate tool. Among the four simulation cases investigated post-validation process, they conclude that the application of high reflectance material on the building façade impacts the microclimate by reducing the outdoor temperature the most. They suggest that adopting a high solar reflective façade has a positive impact on mitigating the effect of the UHI effect in the summer, which will help reduce the heat gain through the façade.

Anaç et al. (Evaluation of Historical Buildings With Heritage Building Information Modelling) present a study on the importance of documentation systems for the protection and transfer of original features to the next generations. They claim that although there are various studies to document the current conditions of buildings in historical building studies, no platform has been identified to enable different participants to access common data and share the process. Their position is that this situation makes historical building research, survey, restitution, and restoration processes difficult, resulting in data losses because the survey and restoration projects, which are documents, are prepared in 2D, but sections and elevations are produced only as much as the restorer or architect draws, and not all details are drawn. Therefore, in their study, which has a limited scope of only covering Gaziantep traditional civil buildings, a digital archive model integrated with 3D HBIM was developed to be used for digital storage of data on historical buildings.

Vicente and Arayici (3D Printing Applicability in Housing Projects in Australia) present a study conducted to provide additional information about implementing 3D printing technologies in Australian housing construction projects. In this research, the researcher explored the use of 3D printing technology in the worldwide construction industry, providing examples of good practices and progressing to the advantages and disadvantages of using this disruptive technology, focusing on current concerns such as quality and standards. Their findings suggest that, although some 3D printing companies in Australia are trying to execute pilot tests to showcase the technology to the construction industry and population, a considerable amount of work still needs to be done. Furthermore, they compared three housing projects in Australia, considering traditional construction methods versus the implementation of 3D printing technology, using data collected from three case studies to carry out an analysis of these projects' cost, time, quality, and environmental performance. They concluded that the 3D printing projects have performed substantially better than the traditional construction housing projects. They also conclude that implementing this technology in housing projects could represent a cost reduction of 16% when compared against benchmarking data of square meterage average cost for a three-bedroom house in Australia. If economies of scale were considered, they estimate that could result in even more savings.

Laurance and Arayici (BIM Implementation in Construction Companies in the United Kingdom) conducted a primary as well as secondary data-based study to analyse BIM implementation in the United Kingdom as a result of the government's mandate in 2011. Based on primary data responses to potential solutions that could be offered to companies to enhance their ability to implement BIM, they conclude that the key factor is of a financial nature. They claim that if the UK government is hoping to achieve its mandate of ensuring construction firms operate at BIM Level 2, then additional funding will need to be made available. They suggest this could largely be achieved through operating tax breaks for companies, showing that they are embracing the ethos behind BIM for greater collaborative means. They believe that by addressing financial restrictions placed on companies for necessary software, as well as being responsive with workshops and funding for training, this can start to revolutionise the culture of the industry. In their final analysis, they claim that BIM implementation is still falling behind the expectations set by the

UK government nationwide. They suggest that the required cultural shift, which has been recognised for decades, for all stakeholders to embrace BIM to enable it to achieve its full potential, has still not been met.

Gil et al. (Artificial Intelligence as the Way Forward for Scan-to-BIM Practice in Cultural Heritage) present a thorough review of existing point cloud segmentation methods applicable to cultural heritage to automate the generation of heritage building information modelling (HBIM). Firstly, they document quantitative insights found across available literature and provide an overview of the fundamental principles of machine learning (ML). They then discuss ML and HBIM studies in terms of their context and classification results. Multiple deep learning types and techniques applied to point cloud files are also described. Their approach, results, and potential for CH application are then further reviewed. The classification of CH point clouds is an important topic growing in popularity, and thanks to the development of many improvements to ML and DL methods, some early success trends are emerging across the literature. Nonetheless, they contend that the current state-of-the-art still faces significant challenges, and the diversity of CH-built assets still has not been tested on an international level. With this chapter, they present cutting-edge research published in the literature. They believe this will form the baseline for the development of the ML and DL method for automated HBIM modelling from point cloud data for UK-specific case study datasets and will prove to be a novel method to amplify HBIM use in cultural heritage.

Index

Note: Page locators in **bold** and *italics* represents tables and figures, respectively. Endnotes are indicated by the page number followed by "n" and the note number e.g., 111n5 refers to note 5 on page 111.